AF352951

The Mode of Actions of the Current Point-of-Care Treatments for Osteoarthritis of the Knee

The Mode of Actions of the Current Point-of-Care Treatments for Osteoarthritis of the Knee

Editor

Masato Sato

Basel • Beijing • Wuhan • Barcelona • Belgrade • Novi Sad • Cluj • Manchester

Editor
Masato Sato
Department of Orthopaedic Surgery
Tokai University School of Medicine
Isehara
Japan

Editorial Office
MDPI
St. Alban-Anlage 66
4052 Basel, Switzerland

This is a reprint of articles from the Special Issue published online in the open access journal *International Journal of Molecular Sciences* (ISSN 1422-0067) (available at: www.mdpi.com/journal/ijms/special_issues/POCT_OCK).

For citation purposes, cite each article independently as indicated on the article page online and as indicated below:

Lastname, A.A.; Lastname, B.B. Article Title. *Journal Name* **Year**, *Volume Number*, Page Range.

ISBN 978-3-7258-1108-3 (Hbk)
ISBN 978-3-7258-1107-6 (PDF)
doi.org/10.3390/books978-3-7258-1107-6

Contents

Preface

Osteoarthritis of the knee (OAK) is the most common cause of disability in the worldwide population and is an intractable degenerative disease of the joint that progresses slowly and is globally prevalent. It has heterogeneous phenotypes, including cartilage breakdown, synovitis, osteophyte formation, and bone marrow lesions. Histological studies show that OAK patients have variable degrees of synovitis, with higher levels of pro-inflammatory cytokines and infiltration of immune cells, predominantly macrophages. A fundamental treatment for OAK does not exist, and its development is imperative. Currently, a conservative approach using intra-articular injections, such as stem cell therapy and platelet-rich plasma, has become widely available. Mesenchymal stem cells express a variety of growth factors and cytokines that aid in the repair of degraded tissue, restoration of normal tissue metabolism, and, most importantly, counteracting inflammation. However, the mode of action of these point-of-care treatments remains unclear. This Special Issue contains original research papers, full reviews, and perspectives that address the progress and current knowledge of the molecular mechanisms of these point-of-care treatments for OAK.

I would like to thank Professor Ichiro Sekiya of Tokyo Medical and Dental University, a leading stem cell researcher and orthopedic surgeon, and Dr. Jennifer Woodell-May of Zimmer Biomet, a leading developer of autologous protein solutions, for taking time out of their very busy schedules to write review articles. I would also like to thank my colleagues who have always participated in our research activities in an upbeat and positive manner.

Masato Sato
Editor

International Journal of
Molecular Sciences

Review

Candidates for Intra-Articular Administration Therapeutics and Therapies of Osteoarthritis

Eriko Toyoda [1,2], **Miki Maehara** [1,2], **Masahiko Watanabe** [1,2] and **Masato Sato** [1,2,*]

[1]   Department of Orthopaedic Surgery, Surgical Science, Tokai University School of Medicine, 143 Shimokasuya, Isehara 259-1193, Japan; etoyoda@tokai-u.jp (E.T.); m-maehara@tsc.u-tokai.ac.jp (M.M.); masahiko@is.icc.u-tokai.ac.jp (M.W.)
[2]   Center for Musculoskeletal Innovative Research and Advancement (C-MiRA), Tokai University Graduate School of Medicine, 143 Shimokasuya, Isehara 259-1193, Japan
*   Correspondence: sato-m@is.icc.u-tokai.ac.jp; Tel.: +81-463-93-1121; Fax: +81-463-96-4404

Abstract: Osteoarthritis (OA) of the knee is a disease that significantly decreases the quality of life due to joint deformation and pain caused by degeneration of articular cartilage. Since the degeneration of cartilage is irreversible, intervention from an early stage and control throughout life is important for OA treatment. For the treatment of early OA, the development of a disease-modifying osteoarthritis drug (DMOAD) for intra-articular (IA) injection, which is attracting attention as a point-of-care therapy, is desired. In recent years, the molecular mechanisms involved in OA progression have been clarified while new types of drug development methods based on gene sequences have been established. In addition to conventional chemical compounds and protein therapeutics, the development of DMOAD from the new modalities such as gene therapy and oligonucleotide therapeutics is accelerating. In this review, we have summarized the current status and challenges of DMOAD for IA injection, especially for protein therapeutics, gene therapy, and oligonucleotide therapeutics.

Keywords: disease-modifying osteoarthritis drug; gene therapy; oligonucleotide therapeutics

Citation: Toyoda, E.; Maehara, M.; Watanabe, M.; Sato, M. Candidates for Intra-Articular Administration Therapeutics and Therapies of Osteoarthritis. *Int. J. Mol. Sci.* **2021**, 22, 3594. https://doi.org/10.3390/ijms22073594

Academic Editor: Maurizio Battino

Received: 1 March 2021
Accepted: 26 March 2021
Published: 30 March 2021

Publisher's Note: MDPI stays neutral with regard to jurisdictional claims in published maps and institutional affiliations.

1. Introduction

Osteoarthritis (OA) of the knee is a disease in which degeneration of articular cartilage gradually progresses, and the cartilage is eventually worn to expose the subchondral bone. OA significantly decreases the quality of life due to joint deformation and pain. While most patients with OA are elderly, it has been reported that nearly 40% of people over the age of 60 years have signs of early OA [1,2]. In addition to aging, OA is a disease involving multiple etiologies such as a history of joint injury, gender, genetic background, and obesity [3,4]. Since degenerated cartilage does not repair spontaneously, it is important to prevent the degeneration of cartilage from an early stage. However, current drug treatment for early OA is limited to the administration of analgesics for pain, viscosupplement such as hyaluronic acid that enhances joint lubricity, and intra-articular (IA) injection of corticosteroids [5]. None of these therapies are considered to have the effect of suppressing the progression of cartilage degeneration and destruction and ultimately, replacement arthroplasty is the only radical treatment for late-stage OA. Since OA is not a life-threatening disease, minimally invasive treatment at an outpatient level is preferred; IA injection is most commonly employed for OA. Since OA progresses over a long period of time, early intervention is important so that symptoms can be controlled throughout life. The development of a disease-modifying osteoarthritis drug (DMOAD) that can be administered by IA injection, which is attracting attention as a point-of-care therapeutic, is desired.

Currently, corticosteroids and hyaluronic acid are the drugs most commonly administered by IA injection, with the main action being pain control. So far, no DMOADs have been developed [6]. Novel treatment options are being developed with the expectation of controlling inflammation and promoting cartilage regeneration in OA. Options include recombinant protein, gene therapy, platelet-rich plasma, and cell therapy [7]. This review outlines the current status and challenges that the new biologics bring to treating OA, especially protein therapeutics, gene therapy, and nucleic-acid therapeutics.

2. Target Biological Pathways for DMOADs

OA is a disease mainly characterized by cartilage degeneration, pathological changes in entire joint, such as joint capsule thickening, osteophyte formation, subchondral sclerosis, and synovitis. Although the molecular mechanism underlying these pathological conditions remains not fully understood, many excellent reviews that discuss the molecular pathology of OA have been published [6,8–12]. As a consensus, prolonged low-grade inflammation is thought to play a pivotal role in OA initiation and progression. In summary, activation of innate immunity by the decomposition products of the joint matrix generated by trauma or mechanical overload causes synovitis. Synovitis induces production of proinflammatory mediators from synovial cells, immune cells, chondrocytes, or cells in subchondral bone. Risk factors for OA, such as aging, injury history, obesity, and some genetic backgrounds are thought to trigger or prolong inflammation.

Increased proinflammatory cytokines such as interleukin (IL)1β, tumor necrosis factor α (TNF-α), IL-6, IL-15, IL-17, and IL-18, and the imbalance of anti-inflammatory cytokines such as transforming growth factor-β (TGFβ), IL-4, IL-10, and IL-13 suggested contribute to OA pathogenesis [13–15].

Further, inflammatory joint condition induces alteration of the chondrocyte phenotype, such as cell proliferation, cluster formation, production of both matrix proteins and matrix-degrading enzymes, and chondrocyte hypertrophy and apoptosis [16]. Matrix-degrading enzymes such as matrix metalloproteinases (MMPs) and a disintegrin and metalloproteinase with thrombospondin motifs (ADAMTS) family degrade cartilage component type II collagen and proteoglycan. Alteration of chondrocyte phenotype considered as regenerative response caused by inflammation. The chondrocytes follow the process of chondrocyte maturation and resulting matrix remodeling, improper hypertrophy-like maturation and cartilage calcification. Cartilage homeostasis disruption is considered to be involved in cartilage degeneration and osteophyte formation.

The importance of each factor in OA progression remains unclear. A blockage of the inflammatory signal induced by proinflammatory mediators, a supplementation of anti-inflammatory factors, or an alteration of chondrocyte phenotype is considered a promising target for DMOADs.

3. The Advantage of IA Delivery in OA Treatment

For the control of inflammation, protein therapeutics blocking TNFα, IL-1, and IL-6 have been already developed for treatment of rheumatoid arthritis (RA). RA is a systemic inflammatory disease and also causes joint destruction. Inflammation of RA is generally more severe than that of OA. Thus, the control of inflammation by these protein therapeutics is of benefit to the RA patients and increases in the occurrence of infections is tolerated [17,18].

In contrast to RA, the daily symptoms in OA such as morning stiffness, heat, pain, and joint effusions are initially comparatively mild [8]. The systemic administration of immunosuppressive protein therapeutics poses an unacceptable risk for OA patients.

In OA, usually only one or two large joints are affected, and inflammation is limited to the affected joints. Furthermore, articular cartilage, which is the target tissue for OA treatment, is an avascular tissue, which makes it difficult for a drug to be distributed by systemic administration. IA drug delivery is relatively easy for the diarthrodial joint, thereby ensuring that the drug is delivered to the affected area. The merit of IA injection is that sufficient drug concentration can be achieved at the target site while avoiding the disadvantages of potential side effects. Thus, IA delivery of DMOADs, including already developed immunosuppressive protein therapeutics, has been considered as an option to address the issues, adverse effect, and drug distribution.

4. OA Treatment by IA Injection of Protein Therapeutics

As mentioned above, various inflammatory cytokines and cartilage matrix-degrading enzymes in joint fluid increase. The expression of genes encoding these proteins in synovial cells and chondrocytes also increases to the extent that suppression of TNFα and IL-1β actions is considered to be a promising target for the treatment of OA. In addition, IA injection of humoral factors that have cartilage repair and cartilage protection properties is also of great interest because they may be a DMOAD for OA of the knee (Table 1).

IL-1Ra, which is a natural antagonist of IL-1β, and anti-IL-1β antibody, which neutralizes IL-1β, have already been shown to be effective in RA. Clinical trials targeting OA have also been conducted. Considering the potent induction of matrix-degrading enzymes by IL-1β, IL-1Ra may suppress the degradation of articular cartilage. However, in a clinical trial (NCT00110916) of IA injection of IL-1Ra (anakinra) [19], the drug failed to achieve the primary endpoint, a change in the Western Ontario and McMaster Universities osteoarthritis index (WOMAC) score from baseline to week 4 [20]. It has been pointed out that there is the possibility the cartilage protective effect had not been exhibited because of the short half-life of IL-1Ra (4 h) and because patient-relevant pain assessment data were used as the primary endpoint [21]. In future studies, it is considered necessary to maintain the IA concentration of IL-1Ra at an effective concentration by improving the dwell time, such as with a controlled release of the recombinant protein. Currently, a clinical trial of anakinra aimed at preventing OA by administration within 7 days after anterior cruciate ligament (ACL) injury is scheduled (NCT03968913 [22]). Since it is estimated that 50% of patients with ACL injury develop OA, the trial plan is to track the progression to OA by measuring marker proteins in joint fluid for early inflammation and cartilage damage. The effect of IL-1Ra could be observed even with a temporary effect.

Table 1. Clinical trials of intra-articular (IA) injection of recombinant protein for osteoarthritis (OA) patients.

Study Name (ClinicalTrials.gov [1])	Mode of Action Biologicals	Study Phase Outcome Measures	Study Identifier Current Status [1] (Completion Year)
Biologic Therapy to Prevent Osteoarthritis After ACL Injury	Inhibit IL-1β IL-1 Receptor antagonist r-metHuIL-1ra (Anakinra)	Early Phase 1 cytokine level/Knee pain and function/marker level	NCT03968913 [22] Not yet recruiting
Study of Safety, Tolerability, Preliminary Efficacy of Intra-articular LNA043 Injections in Patients with Articular Cartilage Lesions and Knee Osteoarthritis	Assist cartilage repair. A modified human angiopoietin-like 3 protein	Phase 2 MRI/AEs/protein level/antibodies/Others	NCT03275064 [23] 2017~Recruiting
A Study to Investigate the Safety and Effectiveness of Different Doses of Sprifermin in Participants with Osteoarthritis of the Knee (FORWARD)	Assist cartilage repair Fibroblast growth factor 18 (Sprifermin)	Phase 2 MRI/WOMAC/PGA/mJSW/ Protein level	NCT01919164 [24] Completed (2019) has results

Table 1. *Cont.*

Study Name (ClinicalTrials.gov [1])	Mode of Action Biologicals	Study Phase Outcome Measures	Study Identifier Current Status [1] (Completion Year)
Dose Finding Study of Bone Morphogenetic Protein 7 (BMP-7) in Subjects with Osteoarthritis (OA) of the Knee	Assist cartilage repair Bone Morphogenetic Protein 7 (BMP-7/OP-1)	Phase 2 WOMAC	NCT01111045 [25] Completed (2011)
To Determine the Safety, Tolerability, Pharmacokinetics and Effect on Pain of a Single Intra-articular Administration of Canakinumab in Patients with Osteoarthritis in the Knee	Inhibit IL-1β humanized monoclonal antibody to interleukin-1β (Canakinumab)	Phase 2 AEs/VAS/WOMAC/Others	NCT01160822 [26] Completed (2010) has results
Treatment of Knee Osteoarthritis with Intra-Articular Infliximab	Inhibit TNFα chimeric monoclonal antibody to TNF-α (Infliximab)	Phase 4 Cellular infiltrates/Effusion/WOMAC/Others	NCT01144143 [27] Completed (2011) has results
Study of Intra-articular DLX105 Applied to Patients with Severely Painful Osteoarthritis of the Knee	Inhibit TNFα a single-chain (scFv) antibody fragment against TNF-α (DLX105)	Phase 1/2 AEs/VAS/WOMAC	NCT00819572 [28] Completed (2010)

[1] https://clinicaltrials.gov (accessed on January 2021), the most recent phase and status study of the same drug candidate. ACL, anterior cruciate ligament; WOMAC, Western Ontario and McMaster osteoarthritis index (WOMAC) pain subscale; MRI, quantification or assessment of cartilage by MRI; AEs, incidence of adverse events; PGA, Patient's Global Assessment; mJSW, Minimal Joint Space Width in the Medial and Lateral Compartments as Evaluated by X-ray; VAS, Visual Analogue Scale to assess the intensity of knee pain.

The IL-1β neutralizing antibody, canakinumab, is effective against juvenile idiopathic arthritis [29]. A clinical trial of IA injection of canakinumab in OA patients was conducted in 2012 (NCT01160822 [26]). Further study of this drug for the treatment of OA has not been conducted. Interestingly, in the Canakinumab Anti-inflammatory Thrombosis Outcomes Study, a clinical trial in atherosclerosis patients involving 10,061 participants, patients received drug subcutaneously once every 3 months. Study results report that inhibition of IL-1β with canakinumab may substantially reduce rates of total hip arthroplasty/total knee replacement and ameliorate osteoarthritis-related symptoms [30]. Although verification is necessary, these results suggested that persistent inhibition of IL-1β might suppress OA progression. If subcutaneous administration of canakinumab is safe and tolerated, it may be an option for DMOAD.

Clinical trials have been conducted to determine the efficacy of a TNFα neutralizing antibody (infliximab) and a single-chain (scFv) antibody fragment against TNF-α, both of which inhibit the action of TNFα. For infliximab, the suppression of synovial inflammation had not been observed. For antibody fragment, no results have been posted and subsequent studies have not been performed.

An alternative to suppression of inflammation as an approach to DMOADs, IA injection of protein factors that promote chondrocyte differentiation and inhibit cartilage hypertrophy, has been carried out in anticipation of cartilage protection and promotion of regeneration. Human recombinant bone morphogenetic protein/osteogenic protein-1 (BMP-7/OP-1) is a factor that has been reported to suppress chondrocyte hypertrophy and protect cartilage. It had been used to promote bone regeneration [31]. Clinical trials of BMP-7 in OA showed that in participants who received BMP-7, there was a trend toward greater symptomatic improvement than with a placebo [25,32]; however, rhBMP-7 have been withdrawn from the market because of lingering safety concerns; vertebral osteolysis, ectopic bone formation, radiculitis, or cervical soft tissue swelling [33], and further study for OA treatment have not been conducted.

FGF18 has been confirmed as both inducing cartilage differentiation and having a protective action [34,35]. A clinical trial of IA injection of rhFGF18 (Sprifermin) was conducted, but no clear improvement was observed in the pain evaluation [24,36,37]. However, a dose-dependent increase in cartilage thickness has been observed, and there are high expectations for its potential as a DMOAD [38–40].

Angiopoietin-like 3 (ANGPTL3) has been identified as an inducer of chondrogenesis and cartilage repair. Clinical trials of IA injection of a modified form of ANGPTL3 are under way [41].

Provided that articular cartilage repair is expected, chondrocytes that respond to these factors or progenitor cells capable of cartilage differentiation need to be present or supplied to the degenerated cartilage. Chondroprogenitor cells have been reported to be present in synovial cells, cartilage surface cells, and periosteal cells, and are expected to contribute to joint protection [42,43]. Encouraging results with rhFGF18 [38–40] imply that the ability to respond to regenerative stimuli may remain in the joints of OA patients.

The development of protein therapeutics has been accelerated and many other proteins have been shown to be effective in animal OA models (Table 2). Protein engineering is increasingly used aiming at enhancing the efficacy and improving biodistribution and bioavailability.

Table 2. Protein therapeutics candidates in OA animal models with reported efficacy.

Protein		OA Animal Models	References
GH	Growth hormone	Rat TMJ-MIA model	[44]
		Rabbit collagenase injection model	[45]
HB-IGF-1	Humanized insulin like growth factor-1 fusion protein with a heparin-binding domain for targeting to cartilage	Rat MMx model	[46]
FzD7 CRD	Recombinant-Frizzled 7-cysteine-rich domain designed to inhibit Wnt3a/β-catenin signaling	Mouse DMM model	[47]
rhMidkine	rhMidkine	Mouse DMM model	[48]
IL4-10 FP	A fusion protein, the biological activity of IL-4 and IL-10 are preserved	Canine groove model	[49,50]
Sclerostin		Mouse tibial compression OA injury model	[51]
Atsttrin	An engineered protein composed of three tumor necrosis factor receptor (TNFR)-binding fragments of progranulin (PGRN)	Rat noninvasive ACL rupture model mouse ACLT model	[52]
rhGDF5	rh growth differentiation factor-5	Rat MMx model	[53]
rhPRG4	rh lubricin	Yucatan minipigs DMM model	[54]
		Rat ACLT model	[55]
CRB0017	recombinant monoclonal antibodies directed against ADAMTS5	STR/ort	[56]

Efficacy reported in an experimental animal model by IA injection in last 10 years in PubMed. TMJ-MIA, monosodium iodoacetate injection to temporomandibular joint; rh, recombinant human; MMx, the medial meniscus was resected; DMM, destabilizing the medial meniscus surgery; ACLT, anterior cruciate ligament transection; STR/ort, spontaneous osteoarthritis mouse.

5. Gene Therapy for OA Treatment by IA

In recent years, gene therapy with viral vectors containing plasmid deoxy ribonucleic acid (DNA) or messenger ribonucleic acid (mRNA) that incorporates genes with therapeutic effects has been put into practical use. Gene therapy exerts its effects through intracellular nucleic-acid transfection and translation into protein. In the case of gene therapy with mRNA, it does not need to be transcribed intracellularly, protein can be produced directly from the mRNA. When aiming for continuous expression, mRNA transfection is considered to be inferior to gene transfer by vectors. Since mRNA is of very low stability, drug delivery system (DDS) techniques that incorporate mRNA into cells while avoiding degradation are necessary. Lipid nanoparticles (LNPs) are currently the mainstream DDS for mRNA transfection but they are difficult to target to particular cell types, other than hepatocytes, because most LNPs are trapped by the liver when administered systemically. IA injection of the LNP-encapsulated mRNA may maintain protein bioavailability and action on synovial cells or chondrocytes in the target joint before the LNPs circulate to the whole body. Gene therapy with mRNA administration could be suitable for transient gene expression aimed at controlling inflammation after trauma and suppressing chondrocyte apoptosis.

Currently, research is being conducted on therapeutic methods in which vectors and plasmids are administered to cells in joints to produce factors with anti-inflammatory and cartilage repair effects instead of IA administration of recombinant proteins. Since OA is not a life-threatening disease, a vector with a higher safety profile than those used for cancer treatment is required. Table 3 shows the clinical studies of IA gene therapy for OA conducted to date. The viral vectors currently in clinical trials were adeno-associated virus (AAV) vectors, helper-dependent adenovirus (HDAd) vectors, and plasmids.

AAV is a nonpathogenic virus that can be transfected into cells in a tissue-specific manner, depending on their serotype, and can transfer genes into nondividing cells. Their use is expanding due to those characteristics. The introduction of AAV vectors into joints has been investigated in rat [57], rabbit [58], and equine [59] species, and is introduced into chondrocytes. Since adult chondrocytes usually do not proliferate and are rarely replaced, it is expected that long-term transgene expression will be possible, if genes can be introduced into chondrocytes [60]. If anti-inflammatory proteins can be continuously supplied through gene therapy, it may be possible to modify OA progression. In this context, gene therapy utilizing an IL-1Ra gene-containing vector has major implications for OA treatment based on the results obtained with protein administration. Viral transfection of the potent anti-inflammatory cytokines, IL-10, and IFNβ is also being investigated.

Table 3. Clinical trials of intra-articular delivery of gene therapy for OA patients.

Study Name (ClinicalTrials.gov [1])	Mode of Action Biologicals	Study Phase Outcome Measures	Study Identifier Current Status [1] (Completion Year)
Safety of Intra-Articular Sc-rAAV2.5IL-1Ra in Subjects with Moderate Knee OA (AAVIL-1Ra)	Inhibit IL-1β sc-rAAV2.5IL-1Ra	Phase 1 AEs	NCT02790723 [61] 2019~Recruiting
Study to Evaluate the Safety and Tolerability of FX201 in Patients with Osteoarthritis of the Knee	Inhibit IL-1β humantakinogene hadenovec IL-1Ra (FX201)	Phase 1 AEs/biodistribution	NCT04119687 [62] 2020~Recruiting
Efficacy and Safety of XT-150 in Osteoarthritis of the Knee	Supply IL-10 plasmid DNA with a variant of human IL-10 transgene (XT-150)	Phase 2 KOOS/WOMAC /Others	NCT04124042 [63] 2020~Recruiting
A Single Dose Clinical Trial to Study the Safety of ART-I02 in Patients with Arthritis	Supply IFN-β Recombinant AAV type 2/5 containing a hIFN-b gene (ART-I02)	Phase 1 AEs/clinical scores distribution/immune response /Others	NCT02727764 [64] Active (2022)

[1] https://clinicaltrials.gov (accessed on January 2021), the most recent phase and status study of the investigational drug. AEs, incidence of adverse events; KOOS, Osteoarthritis Outcome Score; WOMAC, Western Ontario and McMaster Osteoarthritis Index pain subscale.

A clinical trial of protein therapeutics of IL-1Ra showed that IL-1Ra rapidly disappeared from the joint. However, clinical trials of FX201 (humantakinogene hadenovec) aiming to express IL-1Ra in the joint are under way (NCT04119687 [62]). FX201 is an HDAd vector based on human serotype 5 (Ad5) that is designed to express IL-1Ra under the control of an inflammation-sensitive promoter. HDAd is a vector that lacks all viral genes except the regions required for viral genome replication and packaging to increase the safety of adenovirus vectors. It is a nonintegrating, nonreplicating vector that can incorporate relatively long genes. IA expression of IL-1Ra using HDAd in mice and rats has been reported [65]. It has also been reported that cotransfection with the HDAd vector containing the gene coding for lubricin exhibited improved protection against OA than single IL-1Ra coding HDAd vector transfection [66].

Similarly, aiming for IA expression of IL-1Ra, sc-rAAV2.5rIL-1Ra (a self-complementing, recombinant adeno-associated viral vector carrying IL-1Ra complementary DNA) has been investigated in a clinical trial (NCT02790723 [61]). The conventional AAV vector is single-stranded, and gene expression occurs after the complementary strand is synthesized in the cell to become double-stranded. In the process of self-complementing, the positive and negative strands are connected and packaged into a viral capsid to quickly become double-stranded, and hence gene expression occurs rapidly [67]. It is inferior to conventional AAV in that large-sized genes cannot be introduced. Preclinical studies have confirmed that more than 90% of the post-IA vector in rats remains in the joint and that IL-1Ra levels are also elevated [68]. Furthermore, improvement in total joint pathology following gene therapy has been reported in an equine OA model [69].

IL-10 is a potent anti-inflammatory cytokine and is thought to affect OA. IL-10 suppresses the production of IL-1β, IL-6, and TNF-α and induces the expression of IL-1Ra [70]. In addition, it suppresses matrix metallopeptidases (MMPs) expression. IL-10 is expected to be able to broadly suppress the inflammatory response and cartilage destruction seen in OA [71]. A plasmid DNA-based therapy (XT-50) that expresses a long-acting human IL-10 variant has been developed. The variant (hIL-10var) is inserted under the control of a C-X-C motif chemokine ligand 10 promoter and the promoter is expected to regulate IL-10 expression according to the onset of inflammation in the joint [72]. Administration of XT-150 to canine joints has reportedly increased IA IL-10 levels and reduced pain [73]. Clinical studies of a plasmid vector (XT-150) on OA have been conducted (NCT04124042 [59]).

IFNβ also regulates and suppresses the production of TNFα, IL-1β, and IL-6, and an AAV vector (ART-I02) that expresses IFNβ is being studied in rheumatoid arthritis and OA. ART-I02 is designed to express IFNβ under the control of a nuclear factor κB (NF-κB) promoter, which acts as a transcription factor during the induction of many inflammatory cytokines. More than 90% of the virus stayed in the joint after IA administration of ART-I02 and protein expression was confirmed even after 7 weeks [74]. In the administration of ART-I02 for collagen-induced arthritis in rhesus monkeys, a beneficial effect on joint swelling, histological inflammation, and bone erosion scores has been reported [75]. ART-I02 is also undergoing clinical research targeting OA (NCT02727764 [60]).

Introduction of PRG4/lubricin, which plays an important role in cartilage integrity, was investigated. Using 10mabHDV-PRG4, HDAd vector conjugated to an α-10 integrin monoclonal antibody, the increase of transduction into chondrocytes and production of lubricin have been reported [76]. Introduction of *IGF-I* by AAV vector has been also investigated [77]. Since the protein can be continuously expressed, gene therapy considered one of the options for improving the biodistribution and bioavailability. Using gene therapy, it is possible in principle to express a range of receptors and transcription factors, and thus, it may be possible to use gene therapy to change cellular properties such as the differentiation potential of chondrocytes. The safety and usefulness of gene therapy in OA treatment is yet unknown. It is expected that the knowledge of the production of neutralizing antibodies and the efficiency of introduction into target cells by IA in human bodies will be clarified from the ongoing clinical trials in the future.

6. Oligonucleotide Therapeutics as a Candidate for OA Treatment by IA Injection

In recent years, the development of oligonucleotide therapeutics that introduce oligo nucleic acids such as antisense and small interfering RNA (siRNA) into cells has progressed. The major oligonucleotide therapeutics are double-stranded RNA, siRNA [78], micro RNA (miRNA) [79], and single-strand RNA antisense [80]. Antisense is classified into splicing-controlled antisense that interferes with pre-mRNA splicing [81], RNA-degrading antisense that collaborates with RNaseH to degrade mRNA, and miRNA-inhibiting antisense that inhibits miRNA function [80]. In the past, in vivo administration of nucleic acid has had problems of intracellular delivery and in vivo stability. Over the past decade, modified nucleic acids have been developed that have improved nuclease resistance and increased stability in the body. By chemically modifying the nucleic acid, the binding property to the target sequence and the efficiency of uptake into cells have also been improved [82–85]. So far, antisense and siRNA have been put into clinical use [86–95].

For OA treatment, attempts are being made to suppress cartilage destruction by controlling the production of protein molecules involved in pathological conditions such as cytokines and signal transduction molecules using oligonucleotide therapeutics. In recent years, many comprehensive analyses of miRNAs and mRNAs in OA joints have been carried out, and the networks of mRNAs and miRNAs involved in the progression of OA have been clarified [96–98]. Many miRNAs involved in OA and chondrogenesis have been reported [99–101]. For genes involved in promoting OA, there is the possibility that OA progression could be suppressed by degrading specific mRNAs with siRNA or antisense RNA and suppressing protein production. In the case where miRNAs that target proteins associated with cartilage destruction in OA are reduced, administration of oligonucleotide therapeutics may be able to supplement the anabolic miRNAs and suppress OA progression. Conversely, using an antisense RNA to inhibit miRNAs that are increased in OA, it may be possible to restore cartilage protective protein production, which is suppressed by these miRNAs and control OA progression.

Table 4 shows oligonucleotides that exhibited in vivo efficacy in experimental OA models, those are limited in rodent models. Application of oligonucleotide therapeutics for IA injection is still underway. So far, no oligonucleotide therapeutics for DMOADs is in clinical trials yet.

Table 4. Oligonucleotide therapeutics candidates [1] with in vivo efficacy via IA injection.

Mode of Action	Oligonucleotide	Target Gene(s)	Outcomes	References
miRNA inhibitor	miRNA inhibitors antisense oligonucleotide	*miR-141/200c*	Recover SIRT1/modify IL-6/STAT3 pathway/prevent OA in mouse DMM model	[102]
		miR-203	Recover Erα/decrease cartilage degradation in postmenopausal OA rats	[103]
		miR-21-5p	Recover FGF18/attenuate the severity of OA in the mouse DMM model	[104]
		miR-34-5a	Protect cartilage in the DMM and high-fat diet/DMM mice	[105]
		miR-146b	Recover α2-macroglobulin/prevent OA in mouse DMM model	[106]
		miR-181a-5p	Attenuate cartilage destruction, hypertrophic, apoptotic/cell death, and type II collagen breakdown markers in mouse DMM model	[107]
		miR-128a	Recover ATG12/slow articular tissue destruction in rat ACLT model	[108]
		miR-449a	Recover SIRT 1/Prevent cartilage degradation in rat DMM model	[109]
mRNA inhibition	siRNA for target genes	*Hif-2α*	Prevent cartilage degeneration in ACLT/DMM mice	[110]
		Mmp13	Delay cartilage degradation in mouse DMM model	[111,112]
		Yap	Ameliorate OA development and reduce subchondral bone formation in ACLT mice	[113]
		Thr	Reduce angiogenic activities in subchondral bone ameliorated cartilage degradation in mouse DMM model	[114]
		FoxC1	Decrease β-catenin, ADAMTS-5, fibronectin, MMP3, and MMP13/decrease cartilage destruction in mouse DMM model	[115]
miRNA supplement	miR-210 mimic	Not mentioned	Upregulate *Col2a1* expression in the meniscus cells and VEGF and FGF2 expression in the synovial cell/enhance repair of the meniscus and prevent cartilage degeneration in rat DMM model	[116]
	miR-26a/26b mimic	*Fut4*	Promote chondrocytes proliferation and inhibit apoptosis/attenuate OA progression in rat ACLT-MMx model	[117]
	miR-145 mimic	*Mkk4*	Suppress the expression of MMP-3 and MMP-13, as well as p-MKK4, p-c-Jun, and p-ATF2/reduce cartilage destruction in rat MCLT-DMM model	[118]
	miR-140 mimic	Not mentioned	Reduce pathological scores and MMP-13 and ADAMTS-5 expression in rat ACLT-MMx model	[119]

[1] Efficacy reported in an experimental animal model by IA injection. DMM, destabilizing the medial meniscus surgery; ACLT, anterior cruciate ligament transection; ACLT-MMx, ACL was transected, and the medial meniscus was completely resected; MCLT-DMM, medial collateral ligament transection and DMM.

7. Discussion

The molecular mechanism of cartilage degeneration in OA and an understanding of target molecules that are the key to OA control are rapidly increasing. Once a target molecule is identified, it is possible to obtain DMOAD candidates by various drug discovery methods such as protein therapeutics, gene therapy, and nucleic-acid therapeutics (Figure 1). Each approach has pros and cons and appropriate modality for DMOADs differs depending on the expected mechanism of action and the target cells (Table 5).

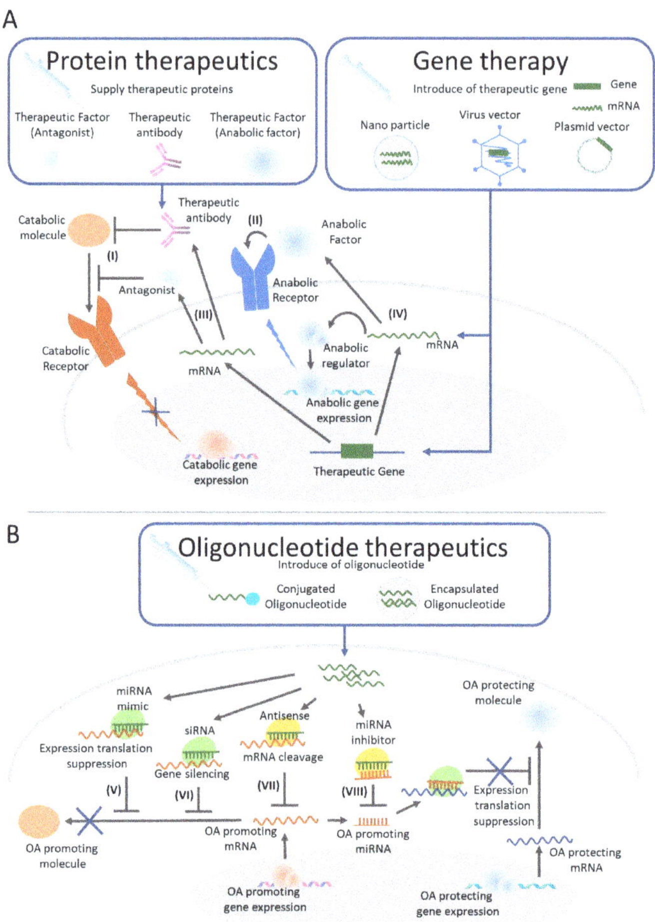

Figure 1. Current and potential targets of DMOADs for IA injection. (**A**) Current drug targets of protein therapeutics candidates for OA aim to block the interaction between proinflammatory cytokines and their receptors (I) or to supply cartilage protective proteins to the joint (II). The attempts of gene therapy candidates aim to supply proteins, which have anti-inflammatory (III) or cartilage repair effects (IV) instead of IA administration of recombinant protein. (**B**) OA progression may be controlled by reducing the protein production involved in pathological conditions through expression translation suppression by miRNA (V), gene silencing by siRNA (VI), or mRNA cleavage by antisense (VII). MicroRNA inhibitors inhibit OA promoting miRNAs (VIII) and may restore the expression of OA protecting proteins and may contribute to suppress OA progression.

Table 5. Summary of new approaches for disease-modifying osteoarthritis drugs (DMOADs).

	Protein Therapeutics	Gene Therapy	Oligonucleotide Therapeutics
Mechanism of action	Supply the required protein	Transduce target gene in cells Gene expression and translation is needed	Transfer to target cells Modulate the function or the fate of target mRNA
Application range of targets	For proteins that act extracellularly	Limitation in size of gene	Oligonucleotide sequence can be designed without off target effect
Delivery and distribution	Distribute to whole joint by IA	AAV vector provide cell type specific gene transfer	Drug delivery system for target need to be established
Retention time	Short Rapidly excreted from the joint	Vector: Continuous expression can be expected when transfected cells and vector retained mRNA: transient protein production	Effect will continue as long as oligonucleotide remain in cytosol
Control dosage and time	Possible	Amount of protein depend on transfection efficiency, host cell activity and etc. Promotor design provide regulated induction of protein	Possible
Relative manufacturing cost	High Biological manufacturing	High Biological manufacturing	Low Chemical synthesis"
Technical establishment and Safety concerns	Established Predictable	Approved mainly on life threatening disease Remain unknown risks?	Approved mainly in specific genetical disorder Remain unknown risks?

Since protein therapeutics do not penetrate the cell membrane, this approach is applicable to target molecules such as cell surface receptors and secretory factors. Protein therapeutics can be expected to have high specificity for target molecules. Many drugs have been put into practical use and drug discovery procedures have been established. Although IA injection can deliver the protein therapeutics directly, even macromolecules such as proteins and hyaluronic acid are excreted from the joints in few hours or a day [120]. Technological advances such as modification of amino acid sequence [121], chemical modification [122], and development of sustained-release carrier [7,120,123] will contribute to the development of DMOADs. For oligonucleotide therapeutics, once the gene sequence of the target molecule is known, a method for quickly designing drug candidates has almost been established. Once the method of transfection into cells is established, the existing method can be used for oligonucleotide therapeutics for other targets [124]. Since oligonucleotide therapeutics are based on gene sequences, they are expected to act specifically on the target. They can be manufactured by chemical synthesis making it easy to set standards and control quality. Oligonucleotide therapeutics act as siRNA and can hybridize only to the target mRNA to reduce specific proteins' synthesis. Oligonucleotide therapeutics aimed at supplementing or inhibiting miRNAs may act on multiple proteins. After being taken up into cells, their effect lasts for a certain period of time [110,125]. Oligonucleotide therapeutics can be inexpensively produced by chemical synthesis as distinct from biologically produced recombinant proteins and viral vectors. This is important for a candidate for the treatment of OA, which may be used for long periods in large numbers of patients. Using gene therapy, both intracellular and extracellular molecules can be expressed, either transiently or in response to a specific signal depending on the design of the vector [126]. There is also the possibility of the practical application of vectors that express siRNA [127–129], miRNA [130,131], and antisense RNA [132]. The ability to localize and supplement proteins through oligonucleotide therapeutics for long-term is considered to be one of the great advantages of gene therapy. However, so far, gene therapy has rarely been put to practical use in areas other than cancer and serious congenital diseases [133]. In those who have antibodies to viral vectors, there are concerns about reduced introduction efficiency and adverse events. In addition, because viral vectors or plasmid is biologically manufactured, it is difficult to control the quality standard, and the manufacturing cost is considered to be higher than that of oligonucleotide therapeutics. Gene therapy for OA may make significant progress if the use of gene therapy in high-severity diseases accumulates and safety concerns diminish.

There are several obstacles for development of DMOADs. Generally, surgically induced OA models have been used to evaluate in vivo efficacy. These models reflect some aspect of post traumatic OA in human, however heterogeneous nature of human OA should be considered. It has been reported that 12% of the overall prevalence of symptomatic OA is attributable to post-traumatic OA [134]. The pathophysiology of disease models used in preclinical research may be distinct from that of the majority of patients with OA. It should be considered to select appropriate experimental model, such as naturally occurring or genetically modified spontaneous models, surgically or chemically induced model or non-invasive models, which can replicate the targeted etiology in human OA [135]. For heterogeneity of human OA, the new classification for distinguishing different phenotype of OA is proposed [136]. It is expected that more precise OA classification will enable the selection of appropriate target patients for clinical trials according to the mechanism of action of the test drug [137]. Meanwhile establishment of reliable outcome measures, which can follow the course of disease and effect of intervention, are desired. The great potential of functional or compositional MRI for noninvasive assessment of tissue-structure changes in OA has been reported [138]. The selection of suitable experimental models, target patients, and appropriate outcome measure would be crucial for the successful development of new therapies.

8. Conclusions

OA progresses gradually over a long period of time and degeneration and damage to articular cartilage are irreversible. It is important to intervene in cartilage degeneration from an early stage and control its progression. Thus, IA injection of DMOAD at point-of-care may be one of the treatment options. In addition to conventional chemical compounds and protein therapeutics, DMOAD may be created from state-of-the-art gene therapy and oligonucleotide therapeutics.

Author Contributions: Conceptualization, E.T. and M.S.; writing—original draft preparation, E.T. and M.M.; writing—review and editing, E.T. and M.S.; supervision, M.W. and M.S.; funding acquisition, E.T., M.W. and M.S. All authors have read and agreed to the published version of the manuscript.

Funding: This research was supported by AMED under Grant Numbers 20bk0104101h0001, JSPS KAKENHI Grant Number JP20K07297 and TERUMO LIFE SCIENCE FOUNDATION.

Data Availability Statement: No new data were created or analyzed in this study.

Acknowledgments: The image of Antibody and plasmid is from TogoTV (©2016 DBCLS TogoTV /CC-BY-4.0)

Conflicts of Interest: The authors declare no conflict of interest.

References

1. Dillon, C.F.; Rasch, E.K.; Gu, Q.; Hirsch, R. Prevalence of Knee Osteoarthritis in the United States: Arthritis Data from the Third National Health and Nutrition Examination Survey 1991-94. *J. Rheumatol.* **2006**, *33*, 2271–2279.
2. Yoshimura, N.; Muraki, S.; Oka, H.; Mabuchi, A.; En-Yo, Y.; Yoshida, M.; Saika, A.; Yoshida, H.; Suzuki, T.; Yamamoto, S.; et al. Prevalence of Knee Osteoarthritis, Lumbar Spondylosis, and Osteoporosis in Japanese Men and Women: The Research on Osteoarthritis/Osteoporosis against Disability Study. *J. Bone Miner. Metab.* **2009**, *27*, 620–628. [CrossRef]
3. Silverwood, V.; Blagojevic-Bucknall, M.; Jinks, C.; Jordan, J.L.; Protheroe, J.; Jordan, K.P. Current Evidence on Risk Factors for Knee Osteoarthritis in Older Adults: A Systematic Review and Meta-Analysis. *Osteoarthr. Cartil.* **2015**, *23*, 507–515. [CrossRef] [PubMed]

4.   Vina, E.R.; Kwoh, C.K. Epidemiology of Osteoarthritis: Literature Update. *Curr. Opin. Rheumatol.* **2018**, *30*, 160–167. [CrossRef]
5.   Bannuru, R.R.; Osani, M.C.; Vaysbrot, E.E.; Arden, N.K.; Bennell, K.; Bierma-Zeinstra, S.M.A.; Kraus, V.B.; Lohmander, L.S.; Abbott, J.H.; Bhandari, M.; et al. OARSI Guidelines for the Non-Surgical Management of Knee, Hip, and Polyarticular Osteoarthritis. *Osteoarthr. Cartil.* **2019**, *27*, 1578–1589. [CrossRef]
6.   Mathiessen, A.; Conaghan, P.G. Synovitis in Osteoarthritis: Current Understanding with Therapeutic Implications. *Arthritis Res. Ther.* **2017**, *19*, 18. [CrossRef]
7.   Rai, M.F.; Pham, C.T. Intra-Articular Drug Delivery Systems for Joint Diseases. *Curr. Opin. Pharmacol.* **2018**, *40*, 67–73. [CrossRef]
8.   Sellam, J.; Berenbaum, F. The Role of Synovitis in Pathophysiology and Clinical Symptoms of Osteoarthritis. *Nat. Rev. Rheumatol.* **2010**, *6*, 625–635. [CrossRef]
9.   Loeser, R.F.; Goldring, S.R.; Scanzello, C.R.; Goldring, M.B. Osteoarthritis: A Disease of the Joint as an Organ. *Arthritis Rheumatol.* **2012**, *64*, 1697–1707. [CrossRef]
10. Sokolove, J.; Lepus, C.M. Role of Inflammation in the Pathogenesis of Osteoarthritis: Latest Findings and Interpretations. *Ther. Adv. Musculoskelet. Dis.* **2013**, *5*, 77–94. [CrossRef]
11. Robinson, W.H.; Lepus, C.M.; Wang, Q.; Raghu, H.; Mao, R.; Lindstrom, T.M.; Sokolove, J. Low-Grade Inflammation as a Key Mediator of the Pathogenesis of Osteoarthritis. *Nat. Rev. Rheumatol.* **2016**, *12*, 580–592. [CrossRef] [PubMed]
12. Chen, D.; Shen, J.; Zhao, W.; Wang, T.; Han, L.; Hamilton, J.L.; Im, H.-J. Osteoarthritis: Toward a Comprehensive Understanding of Pathological Mechanism. *Bone Res.* **2017**, *5*, 1–13. [CrossRef]
13. de Lange-Brokaar, B.J.E.; Ioan-Facsinay, A.; van Osch, G.J.V.M.; Zuurmond, A.-M.; Schoones, J.; Toes, R.E.M.; Huizinga, T.W.J.; Kloppenburg, M. Synovial Inflammation, Immune Cells and Their Cytokines in Osteoarthritis: A Review. *Osteoarthr. Cartil.* **2012**, *20*, 1484–1499. [CrossRef]
14. Sohn, D.H.; Sokolove, J.; Sharpe, O.; Erhart, J.C.; Chandra, P.E.; Lahey, L.J.; Lindstrom, T.M.; Hwang, I.; Boyer, K.A.; Andriacchi, T.P.; et al. Plasma Proteins Present in Osteoarthritic Synovial Fluid Can Stimulate Cytokine Production via Toll-like Receptor 4. *Arthritis Res. Ther.* **2012**, *14*, R7. [CrossRef]
15. Wojdasiewicz, P.; Poniatowski, Ł.A.; Szukiewicz, D. The Role of Inflammatory and Anti-Inflammatory Cytokines in the Pathogenesis of Osteoarthritis. *Mediat. Inflamm.* **2014**, *2014*, 561459. [CrossRef]
16. Goldring, M.B. Chondrogenesis, Chondrocyte Differentiation, and Articular Cartilage Metabolism in Health and Osteoarthritis. *Ther. Adv. Musculoskelet. Dis.* **2012**, *4*, 269–285. [CrossRef] [PubMed]
17. Singh, A.; Lester, C.; Drapp, R.; Hu, D.Z.; Glimcher, L.H.; Jones, D. Tetraspanin CD9 and Ectonucleotidase CD73 Identify an Osteochondroprogenitor Population with Elevated Osteogenic Properties. *Development* **2015**, *142*, 438–443. [CrossRef] [PubMed]
18. Minozzi, S.; Bonovas, S.; Lytras, T.; Pecoraro, V.; González-Lorenzo, M.; Bastiampillai, A.J.; Gabrielli, E.M.; Lonati, A.C.; Moja, L.; Cinquini, M.; et al. Risk of Infections Using Anti-TNF Agents in Rheumatoid Arthritis, Psoriatic Arthritis, and Ankylosing Spondylitis: A Systematic Review and Meta-Analysis. *Expert Opin. Drug Saf.* **2016**, *15*, 11–34. [CrossRef]
19. Clinicaltraials.gov. Available online: https://clinicaltrials.gov/ct2/show/NCT00110916 (accessed on 31 January 2021).
20. Chevalier, X.; Goupille, P.; Beaulieu, A.D.; Burch, F.X.; Bensen, W.G.; Conrozier, T.; Loeuille, D.; Kivitz, A.J.; Silver, D.; Appleton, B.E. Intraarticular Injection of Anakinra in Osteoarthritis of the Knee: A Multicenter, Randomized, Double-Blind, Placebo-Controlled Study. *Arthritis Rheumatol.* **2009**, *61*, 344–352. [CrossRef] [PubMed]
21. Martel-Pelletier, J.; Pelletier, J.-P. A Single Injection of Anakinra for Treating Knee OA? *Nat. Rev. Rheumatol.* **2009**, *5*, 363–364. [CrossRef]
22. Clinicaltrials.gov. Available online: https://clinicaltrials.gov/ct2/show/NCT03968913 (accessed on 31 January 2021).
23. Clinicaltraials.gov. Available online: https://clinicaltrials.gov/ct2/show/NCT03275064 (accessed on 25 January 2021).
24. Clinicaltraials.gov. Available online: https://clinicaltrials.gov/ct2/show/NCT01919164 (accessed on 31 January 2021).
25. Clinicaltraials.gov. Available online: https://clinicaltrials.gov/ct2/show/NCT01111045 (accessed on 31 January 2021).
26. Clinicaltraials.gov. Available online: https://clinicaltrials.gov/ct2/show/NCT01160822 (accessed on 31 January 2021).
27. Clinicaltraials.gov. Available online: https://clinicaltrials.gov/ct2/show/NCT01144143 (accessed on 31 January 2021).
28. Clinicaltraials.gov. Available online: https://clinicaltrials.gov/ct2/show/NCT00819572 (accessed on 31 January 2021).
29. Quartier, P.; Alexeeva, E.; Constantin, T.; Chasnyk, V.; Wulffraat, N.; Palmblad, K.; Wouters, C.; I Brunner, H.; Marzan, K.; Schneider, R.; et al. Tapering Canakinumab Monotherapy in Patients with Systemic Juvenile Idiopathic Arthritis in Clinical Remission: Results from a Phase IIIb/IV Open-Label, Randomized Study. *Arthritis Rheumatol.* **2021**, *73*, 336–346. [CrossRef]
30. Schieker, M.; Conaghan, P.G.; Mindeholm, L.; Praestgaard, J.; Solomon, D.H.; Scotti, C.; Gram, H.; Thuren, T.; Roubenoff, R.; Ridker, P.M. Effects of Interleukin-1β Inhibition on Incident Hip and Knee Replacement: Exploratory Analyses From a Randomized, Double-Blind, Placebo-Controlled Trial. *Ann. Intern. Med.* **2020**, *173*, 509–515. [CrossRef]
31. Caron, M.M.J.; Emans, P.J.; Cremers, A.; Surtel, D.A.M.; Coolsen, M.M.E.; van Rhijn, L.W.; Welting, T.J.M. Hypertrophic Differentiation during Chondrogenic Differentiation of Progenitor Cells Is Stimulated by BMP-2 but Suppressed by BMP-7. *Osteoarthr. Cartil.* **2013**, *21*, 604–613. [CrossRef]
32. Hunter, D.J.; Pike, M.C.; Jonas, B.L.; Kissin, E.; Krop, J.; McAlindon, T. Phase 1 Safety and Tolerability Study of BMP-7 in Symptomatic Knee Osteoarthritis. *BMC Musculoskelet. Disord.* **2010**, *11*, 232. [CrossRef] [PubMed]
33. Sreekumar, V.; Aspera-Werz, R.H.; Tendulkar, G.; Reumann, M.K.; Freude, T.; Breitkopf-Heinlein, K.; Dooley, S.; Pscherer, S.; Ochs, B.G.; Flesch, I.; et al. BMP9 a Possible Alternative Drug for the Recently Withdrawn BMP7? New Perspectives for (Re-)Implementation by Personalized Medicine. *Arch. Toxicol.* **2017**, *91*, 1353–1366. [CrossRef]

34. Ellsworth, J.L.; Berry, J.; Bukowski, T.; Claus, J.; Feldhaus, A.; Holderman, S.; Holdren, M.S.; Lum, K.D.; Moore, E.E.; Raymond, F.; et al. Fibroblast Growth Factor-18 Is a Trophic Factor for Mature Chondrocytes and Their Progenitors. *Osteoarthr. Cartil.* **2002**, *10*, 308–320. [CrossRef] [PubMed]

35. Mori, Y.; Saito, T.; Chang, S.H.; Kobayashi, H.; Ladel, C.H.; Guehring, H.; Chung, U.; Kawaguchi, H. Identification of Fibroblast Growth Factor-18 as a Molecule to Protect Adult Articular Cartilage by Gene Expression Profiling. *J. Biol. Chem.* **2014**, *289*, 10192–10200. [CrossRef]

36. Dahlberg, L.E.; Aydemir, A.; Muurahainen, N.; Gühring, H.; Fredberg Edebo, H.; Krarup-Jensen, N.; Ladel, C.H.; Jurvelin, J.S. A First-in-Human, Double-Blind, Randomised, Placebo-Controlled, Dose Ascending Study of Intra-Articular RhFGF18 (Sprifermin) in Patients with Advanced Knee Osteoarthritis. *Clin. Exp. Rheumatol.* **2016**, *34*, 445–450.

37. Hochberg, M.C.; Guermazi, A.; Guehring, H.; Aydemir, A.; Wax, S.; Fleuranceau-Morel, P.; Reinstrup Bihlet, A.; Byrjalsen, I.; Ragnar Andersen, J.; Eckstein, F. Effect of Intra-Articular Sprifermin vs Placebo on Femorotibial Joint Cartilage Thickness in Patients with Osteoarthritis: The FORWARD Randomized Clinical Trial. *JAMA* **2019**, *322*, 1360–1370. [CrossRef]

38. Brett, A.; Bowes, M.A.; Conaghan, P.G.; Ladel, C.; Guehring, H.; Moreau, F.; Eckstein, F. Automated MRI Assessment Confirms Cartilage Thickness Modification in Patients with Knee Osteoarthritis: Post-Hoc Analysis from a Phase II Sprifermin Study. *Osteoarthr. Cartil.* **2020**, *28*, 1432–1436. [CrossRef]

39. Eckstein, F.; Kraines, J.L.; Aydemir, A.; Wirth, W.; Maschek, S.; Hochberg, M.C. Intra-Articular Sprifermin Reduces Cartilage Loss in Addition to Increasing Cartilage Gain Independent of Location in the Femorotibial Joint: Post-Hoc Analysis of a Randomised, Placebo-Controlled Phase II Clinical Trial. *Ann. Rheum. Dis.* **2020**, *79*, 525–528. [CrossRef]

40. Roemer, F.W.; Kraines, J.; Aydemir, A.; Wax, S.; Hochberg, M.C.; Crema, M.D.; Guermazi, A. Evaluating the Structural Effects of Intra-Articular Sprifermin on Cartilage and Non-Cartilaginous Tissue Alterations, Based on SqMRI Assessment over 2 Years. *Osteoarthr. Cartil.* **2020**, *28*, 1229–1234. [CrossRef]

41. Scotti, C.; Gimbel, J.; Laurent, D.; Madar, A.; Peters, T.; Zhang, Y.; Polus, F.; Beste, M.; Vostiar, I.; Choudhury, S.; et al. LNA043, A Novel Cartilage Regenerative Treatment for Osteoarthritis. *Arthritis Rheumatol.* **2020**, *72* (Suppl. 19), 1485.

42. Candela, M.E.; Yasuhara, R.; Iwamoto, M.; Enomoto-Iwamoto, M. Resident Mesenchymal Progenitors of Articular Cartilage. *Matrix Biol.* **2014**, *39*, 44–49. [CrossRef]

43. De Bari, C.; Dell'Accio, F.; Tylzanowski, P.; Luyten, F.P. Multipotent Mesenchymal Stem Cells from Adult Human Synovial Membrane. *Arthritis Rheumatol.* **2001**, *44*, 1928–1942. [CrossRef]

44. Ok, S.M.; Kim, J.H.; Kim, J.S.; Jeong, E.G.; Park, Y.M.; Jeon, H.M.; Heo, J.Y.; Ahn, Y.W.; Yu, S.N.; Park, H.R.; et al. Local Injection of Growth Hormone for Temporomandibular Joint Osteoarthritis. *Yonsei Med. J.* **2020**, *61*, 331–340. [CrossRef]

45. Lubis, A.M.T.; Wonggokusuma, E.; Marsetio, A.F. Intra-Articular Recombinant Human Growth Hormone In-jection Compared with Hyaluronic Acid and Placebo for an Osteoarthritis Model of New Zealand Rabbits. *Knee Surg. Relat. Res.* **2019**, *31*, 44–53. [CrossRef]

46. Loffredo, F.S.; Pancoast, J.R.; Cai, L.; Vannelli, T.; Dong, J.Z.; Lee, R.T.; Patwari, P. Targeted Delivery to Carti-lage Is Critical for in Vivo Efficacy of Insulin-like Growth Factor 1 in a Rat Model of Osteoarthritis. *Arthritis Rheumatol.* **2014**, *66*, 1247–1255. [CrossRef]

47. Ding, Z.; Lu, W.; Dai, C.; Huang, W.; Liu, F.; Shan, W.; Cheng, C.; Xu, J.; Yin, Z.; He, W. The CRD of Frizzled 7 Exhibits Chondropro-tective Effects in Osteoarthritis via Inhibition of the Canonical Wnt3a/β-Catenin Signal-ing Pathway. *Int. Immunopharmacol.* **2020**, *82*, 106367. [CrossRef]

48. Xu, C.; Zhu, S.; Wu, M.; Zhao, Y.; Han, W.; Yu, Y. The Therapeutic Effect of RhMK on Osteoarthritis in Mice, Induced by Destabilization of the Medial Meniscus. *Biol. Pharm. Bull.* **2014**, *37*, 1803–1810. [CrossRef]

49. van Helvoort, E.M.; Popov-Celeketic, J.; Eijkelkamp, N.; Coeleveld, K.; Tryfonidou, M.A.; Wijne, C.D.; Hack, C.E.; Lafeber, F.P.J.G.; Mastbergen, S.C. Canine IL4-10 Fusion Protein Provides Disease Modifying Activity in a Canine Model of OA; an Exploratory Study. *PLoS ONE* **2019**, *14*, e0219587. [CrossRef]

50. Steen-Louws, C.; Popov-Celeketic, J.; Mastbergen, S.C.; Coeleveld, K.; Hack, C.E.; Eijkelkamp, N.; Tryfonidou, M.; Spruijt, S.; van Roon, J.A.G.; Lafeber, F.P.J.G. IL4-10 Fusion Protein Has Chondroprotective, An-ti-Inflammatory and Potentially Analgesic Effects in the Treatment of Osteoarthritis. *Osteoarthr. Cartil.* **2018**, *26*, 1127–1135. [CrossRef]

51. Chang, J.C.; Christiansen, B.A.; Murugesh, D.K.; Sebastian, A.; Hum, N.R.; Collette, N.M.; Hatsell, S.; Econo-mides, A.N.; Blanchette, C.D.; Loots, G.G. SOST/Sclerostin Improves Posttraumatic Osteoarthritis and Inhibits MMP2/3 Expression After Injury. *J. Bone Miner. Res.* **2018**, *33*, 1105–1113. [CrossRef]

52. Wei, J.-L.; Fu, W.; Ding, Y.-J.; Hettinghouse, A.; Lendhey, M.; Schwarzkopf, R.; Kennedy, O.D.; Liu, C.-J. Progranulin Derivative Atsttrin Protects against Early Osteoarthritis in Mouse and Rat Models. *Arthritis Res. Ther.* **2017**, *19*, 280. [CrossRef]

53. Parrish, W.R.; Byers, B.A.; Su, D.; Geesin, J.; Herzberg, U.; Wadsworth, S.; Bendele, A.; Story, B. Intra-Articular Therapy with Recombinant Human GDF5 Arrests Disease Progression and Stimulates Cartilage Repair in the Rat Medial Meniscus Transection (MMT) Model of Osteoarthritis. *Osteoarthr. Cartil.* **2017**, *25*, 554–560. [CrossRef]

54. Waller, K.A.; Chin, K.E.; Jay, G.D.; Zhang, L.X.; Teeple, E.; McAllister, S.; Badger, G.J.; Schmidt, T.A.; Fleming, B.C. Intra-Articular Recombinant Human Proteoglycan 4 Mitigates Cartilage Damage After Destabilization of the Medial Meniscus in the Yucatan Minipig. *Am. J. Sports Med.* **2017**, *45*, 1512–1521. [CrossRef]

55. Jay, G.D.; Fleming, B.C.; Watkins, B.A.; McHugh, K.A.; Anderson, S.C.; Zhang, L.X.; Teeple, E.; Waller, K.A.; Elsaid, K.A. Prevention of Cartilage Degeneration and Restoration of Chondroprotection by Lubricin Tribo-supplementation in the Rat Following Anterior Cruciate Ligament Transection. *Arthritis Rheumatol.* **2010**, *62*, 2382–2391. [CrossRef] [PubMed]

56. Chiusaroli, R.; Visentini, M.; Galimberti, C.; Casseler, C.; Mennuni, L.; Covaceuszach, S.; Lanza, M.; Ugolini, G.; Caselli, G.; Rovati, L.C.; et al. Targeting of ADAMTS5's Ancillary Domain with the Recombinant MAb CRB0017 Ameliorates Disease Progression in a Spontaneous Murine Model of Osteoarthritis. *Osteoarthr. Cartil.* **2013**, *21*, 1807–1810. [CrossRef] [PubMed]

57. Lee, H.-H.; Chang, C.-C.; Shieh, M.-J.; Wang, J.-P.; Chen, Y.-T.; Young, T.-H.; Hung, S.-C. Hypoxia Enhances Chondrogenesis and Prevents Terminal Differentiation through PI3K/Akt/FoxO Dependent Anti-Apoptotic Effect. *Sci. Rep.* **2013**, *3*, 2683. [CrossRef]

58. Ulrich-Vinther, M.; Duch, M.R.; Søballe, K.; O'Keefe, R.J.; Schwarz, E.M.; Pedersen, F.S. In Vivo Gene Delivery to Articular Chondrocytes Mediated by an Adeno-Associated Virus Vector. *J. Orthop. Res.* **2004**, *22*, 726–734. [CrossRef]

59. Watson, R.S.; Broome, T.A.; Levings, P.P.; Rice, B.L.; Kay, J.D.; Smith, A.D.; Gouze, E.; Gouze, J.-N.; Dacanay, E.A.; Hauswirth, W.W.; et al. ScAAV-Mediated Gene Transfer of Interleukin-1-Receptor Antagonist to Synovium and Articular Cartilage in Large Mammalian Joints. *Gene Ther.* **2013**, *20*, 670–677. [CrossRef]

60. Evans, C.H.; Ghivizzani, S.C.; Robbins, P.D. Gene Delivery to Joints by Intra-Articular Injection. *Hum. Gene Ther.* **2018**, *29*, 2–14. [CrossRef]

61. Clinicaltraials.gov. Available online: https://clinicaltrials.gov/ct2/show/NCT02790723 (accessed on 31 January 2021).

62. Clinicaltraials.gov. Available online: https://www.clinicaltrials.gov/ct2/show/NCT04119687 (accessed on 21 February 2021).

63. Clinicaltraials.gov. Available online: https://clinicaltrials.gov/ct2/show/NCT04124042 (accessed on 31 January 2021).

64. Clinicaltraials.gov. Available online: https://clinicaltrials.gov/ct2/show/NCT02727764 (accessed on 31 January 2021).

65. Nixon, A.J.; Grol, M.W.; Lang, H.M.; Ruan, M.Z.C.; Stone, A.; Begum, L.; Chen, Y.; Dawson, B.; Gannon, F.; Plutizki, S.; et al. Disease-Modifying Osteoarthritis Treatment with Interleukin-1 Receptor Antagonist Gene Therapy in Small and Large Animal Models. *Arthritis Rheumatol.* **2018**, *70*, 1757–1768. [CrossRef]

66. Stone, A.; Grol, M.W.; Ruan, M.Z.C.; Dawson, B.; Chen, Y.; Jiang, M.-M.; Song, I.-W.; Jayaram, P.; Cela, R.; Gannon, F.; et al. Combinatorial Prg4 and Il-1ra Gene Therapy Protects Against Hyperalgesia and Cartilage Degeneration in Post-Traumatic Osteoarthritis. *Hum. Gene Ther.* **2019**, *30*, 225–235. [CrossRef]

67. Nathwani, A.C.; Gray, J.T.; Ng, C.Y.C.; Zhou, J.; Spence, Y.; Waddington, S.N.; Tuddenham, E.G.D.; Kemball-Cook, G.; McIntosh, J.; Boon-Spijker, M.; et al. Self-Complementary Adeno-Associated Virus Vectors Containing a Novel Liver-Specific Human Factor IX Expression Cassette Enable Highly Efficient Transduction of Murine and Nonhuman Primate Liver. *Blood* **2006**, *107*, 2653–2661. [CrossRef]

68. Wang, G.; Evans, C.H.; Benson, J.M.; Hutt, J.A.; Seagrave, J.; Wilder, J.A.; Grieger, J.C.; Samulski, R.J.; Terse, P.S. Safety and Biodistribution Assessment of Sc-RAAV2.5IL-1Ra Administered via Intra-Articular Injection in a Mono-Iodoacetate-Induced Osteoarthritis Rat Model. *Mol. Ther. Methods Clin. Dev.* **2016**, *3*, 15052. [CrossRef]

69. Watson Levings, R.S.; Broome, T.A.; Smith, A.D.; Rice, B.L.; Gibbs, E.P.; Myara, D.A.; Hyddmark, E.V.; Nasri, E.; Zarezadeh, A.; Levings, P.P.; et al. Gene Therapy for Osteoarthritis: Pharmacokinetics of Intra-Articular Self-Complementary Adeno-Associated Virus Interleukin-1 Receptor Antagonist Delivery in an Equine Model. *Hum. Gene Ther. Clin. Dev.* **2018**, *29*, 90–100. [CrossRef]

70. Katsikis, P.D.; Chu, C.Q.; Brennan, F.M.; Maini, R.N.; Feldmann, M. Immunoregulatory Role of Interleukin 10 in Rheumatoid Arthritis. *J. Exp. Med.* **1994**, *179*, 1517–1527. [CrossRef]

71. Moroguchi, A.; Ishimura, K.; Okano, K.; Wakabayashi, H.; Maeba, T.; Maeta, H. Interleukin-10 Suppresses Proliferation and Remodeling of Extracellular Matrix of Cultured Human Skin Fibroblasts. *Eur. Surg. Res.* **2004**, *36*, 39–44. [CrossRef]

72. Broeren, M.G.A.; de Vries, M.; Bennink, M.B.; Arntz, O.J.; van Lent, P.L.E.M.; van der Kraan, P.M.; van den Berg, W.B.; van den Hoogen, F.H.J.; Koenders, M.I.; van de Loo, F.A.J. Suppression of the Inflammatory Response by Disease-Inducible Interleukin-10 Gene Therapy in a Three-Dimensional Micromass Model of the Human Synovial Membrane. *Arthritis Res. Ther.* **2016**, *18*. [CrossRef]

73. Watkins, L.R.; Chavez, R.A.; Landry, R.; Fry, M.; Green-Fulgham, S.M.; Coulson, J.D.; Collins, S.D.; Glover, D.K.; Rieger, J.; Forsayeth, J.R. Targeted Interleukin-10 Plasmid DNA Therapy in the Treatment of Osteoarthritis: Toxicology and Pain Efficacy Assessments. *Brain Behav. Immun.* **2020**, *90*, 155–166. [CrossRef]

74. Aalbers, C.J.; Bevaart, L.; Loiler, S.; de Cortie, K.; Wright, J.F.; Mingozzi, F.; Tak, P.P.; Vervoordeldonk, M.J. Preclinical Potency and Biodistribution Studies of an AAV 5 Vector Expressing Human Interferon-β (ART-I02) for Local Treatment of Patients with Rheumatoid Arthritis. *PLoS ONE* **2015**, *10*, e0130612. [CrossRef]

75. Bevaart, L.; Aalbers, C.J.; Vierboom, M.P.M.; Broekstra, N.; Kondova, I.; Breedveld, E.; Hauck, B.; Wright, J.F.; Tak, P.P.; Vervoordeldonk, M.J. Safety, Biodistribution, and Efficacy of an AAV-5 Vector Encoding Human Interferon-Beta (ART-I02) Delivered via Intra-Articular Injection in Rhesus Monkeys with Collagen-Induced Arthritis. *Hum. Gene Ther. Clin. Dev.* **2015**, *26*, 103–112. [CrossRef]

76. Ruan, M.Z.; Cerullo, V.; Cela, R.; Clarke, C.; Lundgren-Akerlund, E.; Barry, M.A.; Lee, B.H. Treatment of Osteo-arthritis Using a Helper-Dependent Adenoviral Vector Retargeted to Chondrocytes. *Mol. Ther. Methods Clin. Dev.* **2016**, *3*, 16008. [CrossRef]

77. Ortved, K.; Wagner, B.; Calcedo, R.; Wilson, J.; Schaefer, D.; Nixon, A. Humoral and Cell-Mediated Immune Response, and Growth Factor Synthesis after Direct Intraarticular Injection of RAAV2-IGF-I and RAAV5-IGF-I in the Equine Middle Carpal Joint. *Hum. Gene Ther.* **2015**, *26*, 161–171. [CrossRef]

78. Setten, R.L.; Rossi, J.J.; Han, S.-P. The Current State and Future Directions of RNAi-Based Therapeutics. *Nat. Rev. Drug Discov.* **2019**, *18*, 421–446. [CrossRef]

79. Sahin, U.; Karikó, K.; Türeci, Ö. MRNA-Based Therapeutics–Developing a New Class of Drugs. *Nat. Rev. Drug Discov.* **2014**, *13*, 759–780. [CrossRef]

80. Bennett, C.F.; Krainer, A.R.; Cleveland, D.W. Antisense Oligonucleotide Therapies for Neurodegenerative Diseases. *Annu. Rev. Neurosci.* **2019**, *42*, 385–406. [CrossRef]

81. Heo, Y.-A. Golodirsen: First Approval. *Drugs* **2020**, *80*, 329–333. [CrossRef]

82. Ashizawa, A.T.; Cortes, J. Liposomal Delivery of Nucleic Acid-Based Anticancer Therapeutics: BP-100-1.01. *Expert Opin. Drug Deliv.* **2015**, *12*, 1107–1120. [CrossRef]

83. Falzarano, M.S.; Passarelli, C.; Ferlini, A. Nanoparticle Delivery of Antisense Oligonucleotides and Their Application in the Exon Skipping Strategy for Duchenne Muscular Dystrophy. *Nucleic Acid. Ther.* **2014**, *24*, 87–100. [CrossRef]

84. Mäe, M.; Andaloussi, S.E.; Lehto, T.; Langel, U. Chemically Modified Cell-Penetrating Peptides for the Delivery of Nucleic Acids. *Expert Opin. Drug Deliv.* **2009**, *6*, 1195–1205. [CrossRef]

85. Vester, B.; Wengel, J. LNA (Locked Nucleic Acid): High-Affinity Targeting of Complementary RNA and DNA. *Biochemistry* **2004**, *43*, 13233–13241. [CrossRef]

86. Dhillon, S. Viltolarsen: First Approval. *Drugs* **2020**, *80*, 1027–1031. [CrossRef]

87. Hair, P.; Cameron, F.; McKeage, K. Mipomersen Sodium: First Global Approval. *Drugs* **2013**, *73*, 487–493. [CrossRef] [PubMed]

88. Hoy, S.M. Nusinersen: First Global Approval. *Drugs* **2017**, *77*, 473–479. [CrossRef]

89. Hoy, S.M. Patisiran: First Global Approval. *Drugs* **2018**, *78*, 1625–1631. [CrossRef]

90. Keam, S.J. Inotersen: First Global Approval. *Drugs* **2018**, *78*, 1371–1376. [CrossRef]

91. Paik, J.; Duggan, S. Volanesorsen: First Global Approval. *Drugs* **2019**, *79*, 1349–1354. [CrossRef] [PubMed]

92. Perry, C.M.; Balfour, J.A. Fomivirsen. *Drugs* **1999**, *57*, 375–380. [CrossRef]

93. Scott, L.J. Givosiran: First Approval. *Drugs* **2020**, *80*, 335–339. [CrossRef]

94. Scott, L.J.; Keam, S.J. Lumasiran: First Approval. *Drugs* **2021**. [CrossRef]

95. Syed, Y.Y. Eteplirsen: First Global Approval. *Drugs* **2016**, *76*, 1699–1704. [CrossRef]

96. Steinberg, J.; Ritchie, G.R.S.; Roumeliotis, T.I.; Jayasuriya, R.L.; Clark, M.J.; Brooks, R.A.; Binch, A.L.A.; Shah, K.M.; Coyle, R.; Pardo, M.; et al. Integrative Epigenomics, Transcriptomics and Proteomics of Patient Chon-drocytes Reveal Genes and Pathways Involved in Osteoarthritis. *Sci. Rep.* **2017**, *7*, 8935. [CrossRef] [PubMed]

97. Wang, X.; Ning, Y.; Zhou, B.; Yang, L.; Wang, Y.; Guo, X. Integrated Bioinformatics Analysis of the Osteoarthri-tis-associated MicroRNA Expression Signature. *Mol. Med. Rep.* **2018**, *17*, 1833–1838. [CrossRef]

98. Peffers, M.; Liu, X.; Clegg, P. Transcriptomic Signatures in Cartilage Ageing. *Arthritis Res. Ther.* **2013**, *15*, R98. [CrossRef]

99. Malemud, C.J. MicroRNAs and Osteoarthritis. *Cells* **2018**, *7*, 72. [CrossRef]

100. Sondag, G.R.; Haqqi, T.M. The Role of MicroRNAs and Their Targets in Osteoarthritis. *Curr. Rheumatol. Rep.* **2016**, *18*, 56. [CrossRef]

101. Wu, C.; Tian, B.; Qu, X.; Liu, F.; Tang, T.; Qin, A.; Zhu, Z.; Dai, K. MicroRNAs Play a Role in Chondrogenesis and Osteoarthritis (Review). *Int. J. Mol. Med.* **2014**, *34*, 13–23. [CrossRef]

102. Ji, M.-L.; Jiang, H.; Wu, F.; Geng, R.; Ya, L.K.; Lin, Y.C.; Xu, J.H.; Wu, X.T.; Lu, J. Precise Targeting of MiR-141/200c Cluster in Chondrocytes Attenuates Osteoarthritis Development. *Ann. Rheum. Dis.* **2020**. [CrossRef]

103. Tian, L.; Su, Z.; Ma, X.; Wang, F.; Guo, Y. Inhibition of MiR-203 Ameliorates Osteoarthritis Cartilage Degrada-tion in the Postmenopausal Rat Model: Involvement of Estrogen Receptor α. *Hum. Gene Ther. Clin. Dev.* **2019**, *30*, 160–168. [CrossRef]

104. Wang, X.-B.; Zhao, F.-C.; Yi, L.-H.; Tang, J.-L.; Zhu, Z.-Y.; Pang, Y.; Chen, Y.-S.; Li, D.-Y.; Guo, K.-J.; Zheng, X. MicroRNA-21-5p as a Novel Therapeutic Target for Osteoarthritis. *Rheumatology* **2019**. [CrossRef]

105. Endisha, H.; Datta, P.; Sharma, A.; Nakamura, S.; Rossomacha, E.; Younan, C.; Ali, S.A.; Tavallaee, G.; Lively, S.; Potla, P.; et al. MicroRNA-34a-5p Promotes Joint Destruction During Osteoarthritis. *Arthritis Rheumatol.* **2020**. [CrossRef]

106. Liu, X.; Liu, L.; Zhang, H.; Shao, Y.; Chen, Z.; Feng, X.; Fang, H.; Zhao, C.; Pan, J.; Zhang, H.; et al. MiR-146b Accelerates Osteoarthritis Progression by Targeting Alpha-2-Macroglobulin. *Aging* **2019**, *11*, 6014–6028. [CrossRef] [PubMed]

107. Nakamura, A.; Rampersaud, Y.R.; Nakamura, S.; Sharma, A.; Zeng, F.; Rossomacha, E.; Ali, S.A.; Krawetz, R.; Haroon, N.; Perruccio, A.V.; et al. MicroRNA-181a-5p Antisense Oligonucleotides Attenuate Osteoarthritis in Facet and Knee Joints. *Ann. Rheum. Dis.* **2019**, *78*, 111–121. [CrossRef]

108. Lian, W.-S.; Ko, J.-Y.; Wu, R.-W.; Sun, Y.-C.; Chen, Y.-S.; Wu, S.-L.; Weng, L.-H.; Jahr, H.; Wang, F.-S. Mi-croRNA-128a Represses Chondrocyte Autophagy and Exacerbates Knee Osteoarthritis by Disrupting Atg12. *Cell Death Dis.* **2018**, *9*, 919. [CrossRef]

109. Baek, D.; Lee, K.-M.; Park, K.W.; Suh, J.W.; Choi, S.M.; Park, K.H.; Lee, J.W.; Kim, S.-H. Inhibition of MiR-449a Promotes Cartilage Regeneration and Prevents Progression of Osteoarthritis in In Vivo Rat Models. *Mol. Ther. Nucleic Acids* **2018**, *13*, 322–333. [CrossRef] [PubMed]

110. Pi, Y.; Zhang, X.; Shao, Z.; Zhao, F.; Hu, X.; Ao, Y. Intra-Articular Delivery of Anti-Hif-2α SiRNA by Chondro-cyte-Homing Nanoparticles to Prevent Cartilage Degeneration in Arthritic Mice. *Gene Ther.* **2015**, *22*, 439–448. [CrossRef]

111. Hoshi, H.; Akagi, R.; Yamaguchi, S.; Muramatsu, Y.; Akatsu, Y.; Yamamoto, Y.; Sasaki, T.; Takahashi, K.; Sasho, T. Effect of Inhibiting MMP13 and ADAMTS5 by Intra-Articular Injection of Small Interfering RNA in a Sur-gically Induced Osteoarthritis Model of Mice. *Cell Tissue Res.* **2017**, *368*, 379–387. [CrossRef]

112. Nakagawa, R.; Akagi, R.; Yamaguchi, S.; Enomoto, T.; Sato, Y.; Kimura, S.; Ogawa, Y.; Sadamasu, A.; Ohtori, S.; Sasho, T. Single vs. Repeated Matrix Metalloproteinase-13 Knockdown with Intra-Articular Short Interfering RNA Administration in a Murine Osteoarthritis Model. *Connect. Tissue Res.* **2019**, *60*, 335–343. [CrossRef] [PubMed]

113. Gong, Y.; Li, S.-J.; Liu, R.; Zhan, J.-F.; Tan, C.; Fang, Y.-F.; Chen, Y.; Yu, B. Inhibition of YAP with SiRNA Pre-vents Cartilage Degradation and Ameliorates Osteoarthritis Development. *J. Mol. Med.* **2019**, *97*, 103–114. [CrossRef]

114. Li, L.; Li, M.; Pang, Y.; Wang, J.; Wan, Y.; Zhu, C.; Yin, Z. Abnormal Thyroid Hormone Receptor Signaling in Osteoarthritic Osteoblasts Regulates Microangiogenesis in Subchondral Bone. *Life Sci.* **2019**, *239*, 116975. [CrossRef] [PubMed]

115. Wang, J.; Wang, Y.; Zhang, H.; Gao, W.; Lu, M.; Liu, W.; Li, Y.; Yin, Z. Forkhead Box C1 Promotes the Pathology of Osteoarthritis by Upregulating β-Catenin in Synovial Fibroblasts. *FEBS J.* **2020**, *287*, 3065–3087. [CrossRef] [PubMed]

116. Kawanishi, Y.; Nakasa, T.; Shoji, T.; Hamanishi, M.; Shimizu, R.; Kamei, N.; Usman, M.A.; Ochi, M. In-tra-Articular Injection of Synthetic MicroRNA-210 Accelerates Avascular Meniscal Healing in Rat Medial Meniscal Injured Model. *Arthritis Res. Ther.* **2014**, *16*. [CrossRef]

117. Hu, J.; Wang, Z.; Pan, Y.; Ma, J.; Miao, X.; Qi, X.; Zhou, H.; Jia, L. MiR-26a and MiR-26b Mediate Osteoarthritis Progression by Targeting FUT4 via NF-KB Signaling Pathway. *Int. J. Biochem. Cell Biol.* **2018**, *94*, 79–88. [CrossRef]

118. Hu, G.; Zhao, X.; Wang, C.; Geng, Y.; Zhao, J.; Xu, J.; Zuo, B.; Zhao, C.; Wang, C.; Zhang, X. MicroRNA-145 At-tenuates TNF-α-Driven Cartilage Matrix Degradation in Osteoarthritis via Direct Suppression of MKK4. *Cell Death Dis.* **2017**, *8*, e3140. [CrossRef]

119. Si, H.-B.; Zeng, Y.; Liu, S.-Y.; Zhou, Z.-K.; Chen, Y.-N.; Cheng, J.-Q.; Lu, Y.-R.; Shen, B. Intra-Articular Injection of MicroRNA-140 (MiRNA-140) Alleviates Osteoarthritis (OA) Progression by Modulating Extracellular Ma-trix (ECM) Homeostasis in Rats. *Osteoarthr. Cartil.* **2017**, *25*, 1698–1707. [CrossRef]

120. Brown, S.; Kumar, S.; Sharma, B. Intra-Articular Targeting of Nanomaterials for the Treatment of Osteoarthri-tis. *Acta Biomater.* **2019**, *93*, 239–257. [CrossRef] [PubMed]

121. Almagro, J.C.; Daniels-Wells, T.R.; Perez-Tapia, S.M.; Penichet, M.L. Progress and Challenges in the Design and Clinical Development of Antibodies for Cancer Therapy. *Front. Immunol.* **2017**, *8*, 1751. [CrossRef]

122. Veronese, F.M.; Mero, A. The Impact of PEGylation on Biological Therapies. *BioDrugs* **2008**, *22*, 315–329. [CrossRef]

123. Liang, X.; Chen, Y.; Wu, L.; Maharjan, A.; Regmi, B.; Zhang, J.; Gui, S. In Situ Hexagonal Liquid Crystal for In-tra-Articular Delivery of Sinomenine Hydrochloride. *Biomed. Pharmacother.* **2019**, *117*, 108993. [CrossRef]

124. Roberts, T.C.; Langer, R.; Wood, M.J.A. Advances in Oligonucleotide Drug Delivery. *Nat. Rev. Drug Discov.* **2020**, *19*, 673–694. [CrossRef]

125. Geary, R.S.; Norris, D.; Yu, R.; Bennett, C.F. Pharmacokinetics, Biodistribution and Cell Uptake of Antisense Oligonucleotides. *Adv. Drug Deliv. Rev.* **2015**, *87*, 46–51. [CrossRef]

126. Zhang, X.; Prasadam, I.; Fang, W.; Crawford, R.; Xiao, Y. Chondromodulin-1 Ameliorates Osteoarthritis Pro-gression by Inhibiting HIF-2α Activity. *Osteoarthr. Cartil.* **2016**, *24*, 1970–1980. [CrossRef] [PubMed]

127. Chu, X.; You, H.; Yuan, X.; Zhao, W.; Li, W.; Guo, X. Protective Effect of Lentivirus-Mediated SiRNA Targeting ADAMTS-5 on Cartilage Degradation in a Rat Model of Osteoarthritis. *Int. J. Mol. Med.* **2013**, *31*, 1222–1228. [CrossRef]

128. Ji, Q.; Xu, X.; Kang, L.; Xu, Y.; Xiao, J.; Goodman, S.B.; Zhu, X.; Li, W.; Liu, J.; Gao, X.; et al. Hematopoietic PBX-Interacting Protein Mediates Cartilage Degeneration during the Pathogenesis of Osteoarthritis. *Nat. Commun.* **2019**, *10*, 313. [CrossRef] [PubMed]

129. Zhang, X.; Crawford, R.; Xiao, Y. Inhibition of Vascular Endothelial Growth Factor with ShRNA in Chondro-cytes Ameliorates Osteoarthritis. *J. Mol. Med. (Berl.)* **2016**, *94*, 787–798. [CrossRef]

130. Peng, J.-S.; Chen, S.-Y.; Wu, C.-L.; Chong, H.-E.; Ding, Y.-C.; Shiau, A.-L.; Wang, C.-R. Amelioration of Experi-mental Autoimmune Arthritis Through Targeting of Synovial Fibroblasts by Intraarticular Delivery of Mi-croRNAs 140-3p and 140-5p. *Arthritis Rheumatol.* **2016**, *68*, 370–381. [CrossRef]

131. Song, J.; Kim, D.; Chun, C.-H.; Jin, E.-J. MiR-370 and MiR-373 Regulate the Pathogenesis of Osteoarthritis by Modulating One-Carbon Metabolism via SHMT-2 and MECP-2, Respectively. *Aging Cell* **2015**, *14*, 826–837. [CrossRef]

132. Yan, S.; Wang, M.; Zhao, J.; Zhang, H.; Zhou, C.; Jin, L.; Zhang, Y.; Qiu, X.; Ma, B.; Fan, Q. MicroRNA-34a Af-fects Chondrocyte Apoptosis and Proliferation by Targeting the SIRT1/P53 Signaling Pathway during the Pathogenesis of Osteoarthritis. *Int. J. Mol. Med.* **2016**, *38*, 201–209. [CrossRef]

133. Anguela, X.M.; High, K.A. Entering the Modern Era of Gene Therapy. *Annu. Rev. Med.* **2019**, *70*, 273–288. [CrossRef]

134. Brown, T.D.; Johnston, R.C.; Saltzman, C.L.; Marsh, J.L.; Buckwalter, J.A. Posttraumatic Osteoarthritis: A First Estimate of Incidence, Prevalence, and Burden of Disease. *J. Orthop. Trauma* **2006**, *20*, 739–744. [CrossRef] [PubMed]

135. Kuyinu, E.L.; Narayanan, G.; Nair, L.S.; Laurencin, C.T. Animal Models of Osteoarthritis: Classification, Up-date, and Measure-ment of Outcomes. *J. Orthop. Surg. Res.* **2016**, *11*. [CrossRef]

136. Bijlsma, J.W.J.; Berenbaum, F.; Lafeber, F.P.J.G. Osteoarthritis: An Update with Relevance for Clinical Practice. *Lancet* **2011**, *377*, 2115–2126. [CrossRef]

137. Glyn-Jones, S.; Palmer, A.J.R.; Agricola, R.; Price, A.J.; Vincent, T.L.; Weinans, H.; Carr, A.J. Osteoarthritis. *Lancet* **2015**, *386*, 376–387. [CrossRef]

138. Juras, V.; Chang, G.; Regatte, R.R. Current Status of Functional MRI of Osteoarthritis for Diagnosis and Prog-nosis. *Curr. Opin. Rheumatol.* **2020**, *32*, 102–109. [CrossRef]

International Journal of
Molecular Sciences

Review

Potential of Exosomes for Diagnosis and Treatment of Joint Disease: Towards a Point-of-Care Therapy for Osteoarthritis of the Knee

Miki Maehara [1,2], Eriko Toyoda [1,2], Takumi Takahashi [1,2], Masahiko Watanabe [1,2] and Masato Sato [1,2,*]

1   Department of Orthopaedic Surgery, Surgical Science, Tokai University School of Medicine, 143 Shimokasuya, Isehara 259-1193, Kanagawa, Japan; m-maehara@tsc.u-tokai.ac.jp (M.M.); etoyoda@tokai-u.jp (E.T.); ttakahashi@tokai-u.jp (T.T.); masahiko@is.icc.u-tokai.ac.jp (M.W.)
2   Center for Musculoskeletal innovative Research and Advancement (C-MiRA), Tokai University Graduate School of Medicine, 143 Shimokasuya, Isehara 259-1193, Kanagawa, Japan
*   Correspondence: sato-m@is.icc.u-tokai.ac.jp; Tel.: +81-463-93-1121 or +81-463-96-4404

Abstract: In the knee joint, articular cartilage injury can often lead to osteoarthritis of the knee (OAK). Currently, no point-of-care treatment can completely address OAK symptoms and regenerate articular cartilage to restore original functions. While various cell-based therapies are being developed to address OAK, exosomes containing various components derived from their cells of origin have attracted attention as a cell-free alternative. The potential for exosomes as a novel point-of-care treatment for OAK has been studied extensively, especially in the context of intra-articular treatments. Specific exosomal microRNAs have been identified as possibly effective in treating cartilage defects. Additionally, exosomes have been studied as biomarkers through their differences in body fluid composition between joint disease patients and healthy subjects. Exosomes themselves can be utilized as a drug delivery system through their manipulation and encapsulation of specific contents to be delivered to specific cells. Through the combination of exosomes with tissue engineering, novel sustained release drug delivery systems are being developed. On the other hand, many of the functions and activities of exosomes are unknown and challenges remain for clinical applications. In this review, the possibilities of intra-articular treatments utilizing exosomes and the challenges in using exosomes in therapy are discussed.

Keywords: exosome; point-of-care therapy; osteoarthritis of the knee; cartilage regeneration

Citation: Maehara, M.; Toyoda, E.; Takahashi, T.; Watanabe, M.; Sato, M. Potential of Exosomes for Diagnosis and Treatment of Joint Disease: Towards a Point-of-Care Therapy for Osteoarthritis of the Knee. *Int. J. Mol. Sci.* **2021**, *22*, 2666. https://doi.org/10.3390/ijms22052666

Academic Editor: Kurt A. Jellinger

Received: 2 February 2021
Accepted: 3 March 2021
Published: 6 March 2021

Publisher's Note: MDPI stays neutral with regard to jurisdictional claims in published maps and institutional affiliations.

1. Introduction

Cartilage has poor self-repair and regeneration ability once injured, as first reported in 1743 by William Hunter [1]. Thus, the regeneration of articular cartilage has been a long-standing challenge for physicians and researchers in the field of orthopedics. Articular cartilage consists of highly elastic hyaline cartilage, which is mainly composed of water and extracellular matrix (ECM), such as proteoglycans and collagen fibers. The cellular component of articular cartilage makes up only about 5% of its wet weight and <10% of its tissue volume. Articular cartilage has poor self-renewal ability due to the extremely low infiltration of oxygen, nutrients, and progenitor cells, as it is avascular [2,3].

Osteoarthritis of the knee (OAK) is one of the most common joint diseases. It is a chronic, progressive disease in which the articular cartilage gradually degenerates due to complex interactions between various factors including trauma due to accidents and sports, obesity, skeletal structure, heredity, and age-related overload on the joints. The global prevalence of OAK is estimated to be approximately 3.8%, and it is observed more in women (mean, 4.8%) than in men (mean, 2.8%) [4]. Initially, partial-thickness articular cartilage defects with wear and tear on the cartilage surface are observed. As the pathological condition progresses, the subchondral bone becomes exposed, leading to full-thickness articular cartilage defects. One of the most serious symptoms is pain during daily activities such as walking and climbing the stairs, leading to decreased physical activity and quality of life.

Non-surgical treatments for articular cartilage injury include non-pharmacological treatments such as exercise therapy, weight loss therapy, and orthotic therapy. Pharmacological treatments include utilization of analgesics or non-steroidal anti-inflammatory drugs (NSAID) and intra-articular injection of corticosteroids or hyaluronic acid. Currently, these are the limited options that are widely available for point-of-care treatment for OAK, and they fail to reverse the progress of cartilage degeneration. For patients who show no improvements with non-pharmacological and pharmacological treatments, surgical treatment options are considered.

Surgical options for injuries with small defects include subchondral drilling [5], microfracture [6,7], and mosaicplasty [8,9]. However, the cartilage regenerated through these methods is often fibrocartilage, which is mechanically inferior to the hyaline cartilage that constitutes the original articular cartilage. For larger cartilage defects, autologous chondrocyte implantation (ACI) [10] has been utilized. Currently, products such as JACC® [11,12], CaRes® [13,14], Hyalograft®C [15], and BioSeed®C [16] are on the market and used worldwide. However, the regenerated cartilage is reported to be either fibrocartilage or a composite of fibrocartilage and hyaline cartilage, and these methods are not yet applicable to OAK [17,18]. In Japan, cell sheet transplantation with high tibial osteotomy [19–21] is a cell therapy that has been approved for the treatment of OAK as an advanced medical care. However, the treatment is costly and time consuming. After other options are exhausted, total knee arthroplasty (TKA) is applied. Although the treatment outcomes of TKA are remarkable [22], artificial joints may need revision after 15 to 20 years. Thus TKA is often applied to people over the age of 65.

To bridge the gap between traditional pharmacological and surgical treatments, cell therapies such as mesenchymal stem cells (MSCs) [23,24] and platelet rich plasmas (PRPs) [25–27] have been in use as a point-of-care treatment for OAK. Patients can receive these treatments through intra-articular injections at relatively low costs in select places. However, therapeutic effects attributed to various modes of action including cytokines and growth factors can be variable [28–31]. In addition, risks of inappropriate transformation of the transplanted cells and inflammatory reactions exist [32–36]. Thus, a cell-free therapy that maximizes the therapeutic effect through the selected administration of identified factors is ideal.

To address such issues, extracellular vesicles (EVs), especially exosomes, are attracting attention as a novel tool that can be applied to treat cartilage and bone disease [37,38]. The therapeutic effects of MSCs have also been partially attributed to exosomes carrying specific cargos [39–43]. With recent advancements, exosomes are used as efficient drug delivery systems that can be manipulated to carry specific cargo such as micro ribonucleic acids (miRNAs). As such, exosomes may be used as a point-of-care treatment for joint disease, especially through intra-articular injection. In addition, exosomes may be useful as biomarkers to detect joint disease such as OAK at early onset. More recently, a combination of exosomes with tissue engineering methods has been investigated to prolong and improve the efficacy of exosomes. In this review, we present the recent concepts of exosomes that can be applied to the development of point-of-care treatment for joint disease such as OAK.

2. Extracellular Vesicles

Historically, membrane vesicles released by cells have been thought of as non-functional, inactive "debris" resulting from cell damage or turnover of the cellular membrane. However, in 1983, the existence of a 100-nm functional endoplasmic reticulum comprised of lipid bilayer membranes secreted from cells was confirmed [44]. Subsequent research identified membrane vesicles secreted by various cells, which were named according to their features, such as ectosomes, microvesicles, oncosomes, exosomes, and apoptotic bodies [45–52]. To resolve confusion, the International Society for Extracellular Vesicles (ISEV) suggested using the term "extracellular vesicles" (EVs) as "the generic term for particles naturally released from the cell that are delimited by a lipid bilayer membrane and cannot replicate, i.e., do not contain a functional nucleus" [53]. Various subtypes of EVs share similar characteristics, and a method to accurately classify EVs and their fundamental biological roles is unclear [54]. Therefore, the ISEV has suggested a classification of the subtypes of EVs according to the following characteristics: (a) physical characteristics, such as size ("small EVs" (sEVs, <100–200 nm) and "medium/large EVs" (m/lEVs, >200 nm)) or density (low, middle, high); (b) biochemical composition (cluster of differentiation (CD)63+/CD81+ EVs, Annexin A5-stained EVs, etc.); or (c) descriptions of conditions or cell of origin (podocyte EVs, hypoxic EVs, large oncosomes, apoptotic bodies) [53].

3. Biogenesis, Composition, and Isolation of Exosomes

The nomenclature of exosomes was suggested in 1987 [46]. Exosomes are a subtype of EVs that have a small diameter of 50–150 nm. They are secreted from most of the cells in the body and are observed in body fluids and cell culture media [47,52].

Exosomes are known to be formed by endocytosis (Figure 1). Early endosomes transition to late endosomes. Then a large number of intraluminal membrane vesicles (ILVs) are formed by internal budding of the endosome membrane. Late endosomes are called multivesicular bodies (MVBs). When the MVBs fuse with the cell membrane, the internal ILVs are released into the extracellular space as exosomes [55] (Figure 1). Endosomal sorting complexes required for transport (ESCRT) and tetraspanins are thought to be involved in the formation of endosomes [47,56,57]. The mechanisms for uptake of the exosomes into target cells and the delivery of the exosome contents are unclear. Many studies have suggested that endosomes are the putative location of EV-content delivery, and contents are released into the cytoplasm of the target cells by fusing with endosome membranes [47,55]. Other mechanisms of content delivery to target cells, such as interaction with the receptors present on the cell membrane and direct fusion with the cell membrane, have also been proposed [47,55].

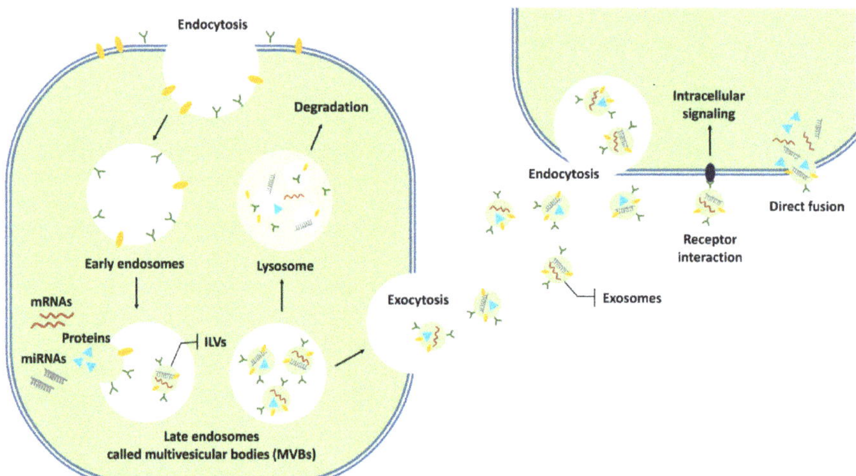

Figure 1. Formation of exosomes and uptake into target cells. Early endosomes are formed by endocytosis and they transition to late endosomes. These are called multivesicular bodies (MVBs), with intraluminal membrane vesicles (ILVs). ILVs that are released from MVBs into the extracellular space are called exosomes. Released exosomes are taken up by the target cells through endocytosis, interactions with receptors on the cell membrane, or direct fusion with the cell membrane.

Exosomes are comprised of a lipid bilayer membrane containing lipid raft constituents, such as cholesterol, sphingomyelin, and ceramide, and contain messenger RNAs (mRNAs), miRNAs, deoxyribonucleic acid (DNA), and various proteins (Figure 2) [58–61]. As of

January 2021, 9769 proteins, 1116 lipids, 3408 mRNAs, and 2838 miRNAs have been registered in the ExoCarta database (http://www.exocarta.org/, accessed on 2 January 2021), which catalogues exosome contents. Although various proteins are associated with the lipid membrane of exosomes, no clear biomarker has been identified. Proteins, such as major histocompatibility complex (MHC) class I and class II, heat shock proteins (e.g., heat shock protein (Hsp) 70, Hsp90), flottilin-1, and actin, have been used as "exosome markers" in the past and have been shown to be present in all types of EVs [62]. In addition, the co-expression of tetraspanins (e.g., CD9, CD63, CD81) and proteins associated with MVB formation (e.g., ALG-2-interacting protein X (ALIX), tumor susceptibility gene 101 (TSG101)) have been suggested (Figure 2) [62]. It is not yet clear whether exosomes have specific functions not found in other EVs or whether they can be reliably distinguished and separated from other EVs.

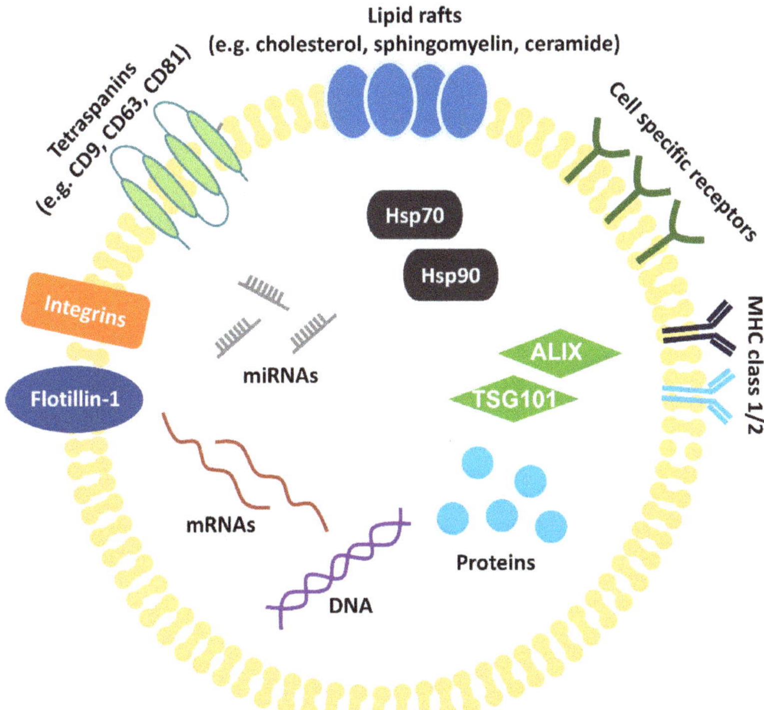

Figure 2. Structure of exosomes. Exosomes have a lipid bilayer membrane composed of lipid raft constituents, tetraspanins (such as CD9, 63 and 81), and proteins (such as integrin, cell specific receptors, MHC class I and class II, and flottilin-1). Exosomes contain cellular components such as proteins, DNA, mRNAs, miRNAs, proteins associated with MVB formation (ALIX, TSG101), and heat shock proteins (Hsp70, Hsp90).

EVs are a heterogeneous population with a mixture of cell vesicles of various fractions. In order to use exosomes for research and treatment, a technique for isolating only "exosome fractions" in large quantities and with high purity is required. Various exosome isolation methods have been developed to this day [63–65]. The most classic method is ultracentrifugation. This method is still the most widely used [66] and has become the gold standard for exosome isolation. In addition, there are various methods such as ultrafiltration, polymer precipitation, size exclusion chromatography, immunoaffinity purification, and microfluidics-based techniques. Furthermore, various exosome isolation kits such as exoEasy Maxi kit (QIAGEN, Venlo, Nederland), ExoQuick (System Biosciences, Palo

Alto, CA, USA), qEV size exclusion columns (Izon Science, Oxford, United Kingdom), and MagCapture™ Exosome Isolation Kit PS (Wako Pure Chemical corporation, Osaka, Japan) are on the market. Each isolation method has advantages and disadvantages, and the quality and amount of exosomes obtained vary greatly depending on the isolation method selected [63–65]. In order to isolate the exosome fraction with the desired activity, it is important to select the optimal exosome isolation method for each objective.

4. Potential of Exosomes for Diagnosis and Treatment of Joint Diseases

The potential for exosomes and their derivatives to be delivered through intra-articular injection opens up new possibilities for the treatment of joint diseases such as OAK. Exosomes contain specific information about the cells from which they are released, and may have the ability to deliver molecules to specific organs or tissues at a distance [61,67,68]. Studies have shown that exosomal miRNAs play an important role in joint homeostasis. Furthermore, the imbalances in exosomes created during joint disease can be utilized as diagnostic biomarkers. Studies have also shown the potential for exosomes to be utilized as a drug delivery system (DDS) to deliver specific cargo to localized areas. In combination with tissue engineering, studies have utilized exosomes in a sustained release drug delivery system (SRDDS) to further improve the efficacy. Here, we discuss such topics in regards to their applications to the diagnosis and treatment of joint diseases.

4.1. Exosomal miRNAs

MiRNAs are small non-coding RNAs that have approximately 22 nucleotides. These induce gene silencing by sequence-complementary binding to target sites located within the three prime untranslated regions (3′UTR) of the target mRNA and regulate multiple biochemical pathways [69]. Many miRNAs associated with chondrocyte function and homeostasis have been reported to contribute to cartilage repair [70,71]. For example, miR-140, a promising miRNA for OAK treatment, is expressed specifically in articular cartilage and plays an important role in cartilage development and metabolic balance of the cartilage matrix by inhibiting a disintegrin and metalloproteinase with thrombospondin motifs 5 (ADAMTS5) and matrix metalloproteinase 13 (MMP13) [72,73]. Ukai et al. examined cartilage tissues from pediatric patients with polydactylism, young patients with anterior cruciate ligament injury, and elderly patients with OAK and identified miRNA-199a-3p, 193b, and 320c as markers that suppress age-related cartilage metabolism [74]. The discovery that mRNAs and miRNAs are packaged in exosomes that are transported between cells and function in target cells was a breakthrough [59,75–77]. Exosomal miRNAs are protected from endogenous RNase activity by the exosome membrane and can exist stably in body fluids and cell culture media. Therefore, exosomal miRNAs can be collected and administered through intra-articular injections. This section introduces the current research on intra-articular treatments with miRNAs (i.e., exosomal miRNAs) secreted extracellularly via exosomes (Table 1).

Studies have shown that exosomal miRNAs may influence cartilage differentiation and homeostasis. Sun et al. performed a microarray analysis of exosomes secreted from both human bone marrow-derived MSCs (hBMSCs) undifferentiated and hBMSCs differentiated into cartilage [78]. Through the comparative analysis, they identified 35 miRNAs, including miR-1246, miR-1290, miR-193a-5p, miR-320c, and miR-92a, that were upregulated in exosomes derived from hBMSCs differentiated into cartilage. In addition, they demonstrated that exosomes derived from miR-320c-overexpressing hBMSCs upregulate SRY-box transcription factor 9 (SOX9) and downregulate MMP13 in OA chondrocytes, and enhance chondrogenesis of hBMSCs [78]. Mao et al. demonstrated that exosomes secreted by OA cartilage express significantly less miR-95-5p compared to normal cartilage [79]. In addition, they showed that exosomes derived from miR-95-5p-overexpressing primary chondrocytes promote cartilage development and cartilage matrix expression by hBMSCs and chondrocytes, and that miR-95-5p may be effective in the treatment of OA by directly targeting histone deacetylase (HDAC) 2/8. Li et al. showed that chondrocyte-derived exosomal miR-8485 promotes the chondrocyte differentiation of BMSCs [80]. The study suggested that miR-8485 may suppress the production of glycogen synthase kinase 3 beta (GSK-3β) by regulating the expression and activity of GSK-3β, activate the Wnt/β-catenin pathway through phosphorylation of GSK-3β by targeting disheveled binding antagonist of beta catenin 1 (DACT1), and promote chondrocyte differentiation.

Table 1. Studies on exosomal miRNAs regulating chondrocyte function and homeostasis for the purpose of intra-articular treatments.

Exosome Origin Cell Source	Study Design	Animal Models	miRNA	Result(s)	Target(s)	Reference
Human BMSCs	in vitro	-	miR-320c	Upregulates SOX9 and downregulates MMP13 expression in OA chondrocyte.	Not mentioned	[78]
Human BMSCs	in vitro	-	miR-95-5p	Enhances histone H3 acetylation and maintains the function of articular chondrocytes. Promotes SOX9, COL2A1 and Aggrecan expression and enhances cartilage development.	HDAC2/8	[79]
Human chondrocytes	in vitro	-	miR-8485	Activates Wnt/β-catenin pathways. Promotes chondrogenic differentiation of hBMSCs.	GSK3B, DACT1	[80]
Human BMSCs	in vitro and in vivo	Mice	miR-92a-3p	Promotes cartilage proliferation. In both MSCs and PHCs, promotes matrix genes expression and inhibits WNT5A expression.	WNT5A	[81]
Human BMSCs	in vitro and in vivo	Rats	miR-26a-5p	It was shown that the damage of synovial fibroblasts is suppressed in vitro, and the OA damage is alleviated in vivo.	PTGS2	[82]
Human IPFP MSCs	in vitro and in vivo	Mice	miR-100-5p	Protects cartilage from damage and ameliorate gait patterns of DMM-induced OA mice.	mTOR	[83]
Human SMSCs	in vitro and in vivo	Rats	miR-140-5p	Enhances the proliferation and migration of ACs, and the progression of early OA and prevents severe damage to knee articular cartilage in the OA rats.	RalA	[84]

hBMSC: human bone marrow-derived MSC, IPFP: infrapatellar fat pad, SMSC: synovium-derived MSC, MSC: mesenchymal stem cell, PHC: primary human chondrocytes, WNT5A: Wnt family member 5A, OA: osteoarthritis, SOX9: SRY-box transcription factor 9, MMP13: matrix metalloproteinase 13, COL2A1: Collagen type II alpha 1, DMM: destabilization of the medial meniscus, AC: articular cartilage, PTGS2: Prostaglandin-endoperoxide synthase 2, HDAC: histone deacetylase, mTOR: mammalian target of rapamycin, RalA: Ras-related protein, GSK-3β: glycogen synthase kinase 3 beta, DACT1: dishevelled binding antagonist of beta catenin 1.

Other studies have shown the potential of exosomal miRNAs in various animal models. Mao et al. showed that miR-92a-3p directly targets Wnt family member 5A (WNT5A) and promotes proliferation and migration of chondrocytes and the differentiation of MSCs into chondrocytes [81]. In addition, they showed that exosomes derived from miR-92a-3p-overexpressing hBMSCs inhibit the progression of early osteoarthritis (OA) and prevent the severe damage of cartilage in an OA model mouse induced with collagenase VII. Jin et al. found that the expression of miR-26a-5p decreases and the expression of prostaglandin-endoperoxide synthase 2 (PTGS2) increases in OA patients and in synovial fibroblasts treated with interleukin-1 beta (IL-1β) [82]. In addition, they showed that hBMSC-derived exosomes provide miR-26a-5p to synovial fibroblasts and alleviate the damage of OA synovial fibroblasts by controlling PTGS2 expression. Moreover, articular injection of exosomes derived from miR-26a-5p-overexpressing hBMSC into an OA rat model prevents OA damage by alleviating the inflammation of synovial tissues and decreasing apoptosis. Wu et al. demonstrated that infrapatellar fat pad (IPFP) MSC-derived exosomes

(MSC^{IPFP}-Exos) can deliver miR-100-5p to chondrocytes [83]. The miR-100-5p specifically targets the 3'UTR region of mammalian target of rapamycin (mTOR) and significantly enhances autophagy levels in chondrocytes, inhibits cell apoptosis, enhances anabolism, and represses catabolism in IL-1β-treated OA chondrocytes. In addition, MSC^{IPFP}-Exos improved the pathological degree of severity and foot gait patterns in a destabilization of the medial meniscus (DMM)-induced OA mouse model. Tao et al. showed that exosomes derived from synovium-derived MSCs (SMSCs) activate Yes-associated protein (YAP) and promote the proliferation and migration of chondrocytes, but decrease ECM secretion [84]. Therefore, exosomes derived from miR-140-5p-overexpressing SMSCs (SMSC-140-Exos) affected proliferation and migration of chondrocytes without ECM secretion being influenced. Moreover, the administration of SMSC-140-Exos to an anterior cruciate ligament transection (ACLT) OA rat model demonstrated the inhibition of severe damage of the articular cartilage and the progression of OA.

In addition, exosomes are thought to be deeply involved in the development of various physiological phenomena and pathological conditions, and further research on exosomal miRNAs may lead to the elucidation of the mechanism of onset and progression of joint diseases such as OAK. Ni et al. showed that exosome-like vesicles from OA chondrocytes inhibit the expression of autophagy related 4B cysteine peptidase (ATG4B) via miR-449a-5p [85]. This result suggests that inflammation of the synovial membrane may be promoted by the inhibition of autophagy of macrophages and the increase in production of mature IL-1β, leading to progression of OA diseases.

4.2. Exosomes as Biomarkers

As OAK patients experience few symptoms during early onset and because of the difficulty in providing treatment as the disease progresses, early diagnostic methods using tools such as biomarkers are desired. Exosomes are anticipated to be useful as novel biomarkers as they contain specific information from the released cells and can stably carry encapsulated substances. In the diagnosis of lung cancer, Exo Dx™ Lung (ALK) (Exosome Diagnostics, Inc., Waltham, MA, USA) is the first Clinical Laboratory Improvement Amendments-validated exosome-based clinical liquid biopsy test that isolates exosomal mRNAs in blood to diagnose cancer [86,87]. For the diagnosis of OAK, exosomes isolated from blood or joint fluid have been identified as useful to date.

For instance, several studies have shown that the expression of specific exosomal miRNAs is altered in patients with joint disease. Kolhe et al. showed that synovial fluid exosomal miRNAs may be altered in OAK patients [88]. In particular, exosomal miRNAs expressed specifically in female OAK patients are responsive to estrogen and target the Toll-like receptor (TLR) signal pathway. Meng et al. confirmed that the expression of exosomal miRNA-193b-3p was decreased in the plasma of OAK patients compared to that of healthy subjects [89].

In addition, Zhao et al. mentioned the possibility that long noncoding RNAs (lncRNAs) can be used as biomarkers for OAK. They conducted an analysis of plasma-derived and synovial fluid-derived exosomal lncRNAs to assess the progression of OAK [90]. As a result, the expression of plasma-derived exosomal lncRNAs showed no significant difference; however, the expression of synovial fluid-derived exosomal lncRNAs was higher in early OA and late-stage OA than in healthy subjects. Specifically, exosomal lncRNA prostate-specific transcript 1 (PCGEM1) was much higher in late OA than in early OA and much higher in early OA than in healthy subjects.

Other studies have examined the use of exosomes as biomarkers to differentiate joint diseases. Chen et al. reported that specific plasma-derived exosomal miRNAs were connected to the common pathogenesis of psoriatic arthritis, psoriasis vulgaris, rheumatoid arthritis (RA), and gouty arthritis. An analysis of the potential target genes revealed that these miRNAs are related to immune disorders and bone injury [91]. In addition, Tsuno et al. demonstrated that serum exosomes of patients with active RA possess different protein profiles compared to patients with inactive RA, patients with OA, and healthy donors [92].

Together, these results indicate that differences in exosome profiles in patients with joint diseases and healthy subjects can be investigated using body fluids such as synovial fluid and plasma, which may lead to the development of various tests and diagnostic methods for OAK and other joint diseases in the future.

4.3. Exosomes as a Drug Delivery System (DDS)

If specific miRNAs and other molecules are identified as having a therapeutic effect for the treatment of OAK, they can be packaged in exosomes that can act as a DDS to transport specific contents to target tissues or organs [68,93]. There are two types of methods to enclose the target substances in exosomes: pre-loading and post-loading methods [94] (Figure 3).

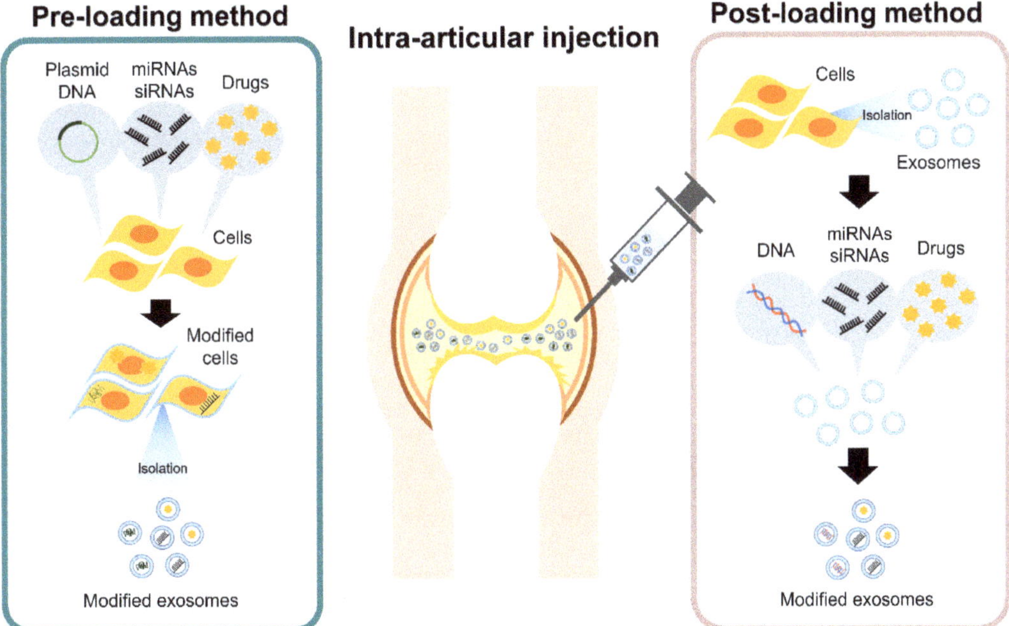

Figure 3. Intra-articular treatments utilizing modified exosomes as a drug delivery system. The loading of specific components into exosomes can be conducted either before exosome isolation (pre-loading method) or after exosome isolation (post-loading method).

Pre-loading methods collect exosomes released by the modified cells after cell modification is performed by miRNA or small interfering RNA (siRNA) transfection or by the uptake of target substances from the culture media during cell culture (Figure 3). Many studies have confirmed the ability of cartilage repair by exosomes collected from cells overexpressing specific miRNAs [71,74,76]. In addition, Wang et al. showed that the expression of miR-135b within the released exosomes can be upregulated by stimulating MSCs using transforming growth factor beta 1 (TGF-β1) [95]. The pre-loading method is a relatively simple procedure that relies on cells to package exosomes. However, it can be difficult to control the loading of specific target molecules in exosomes. A further understanding of the mechanisms by which intracellular miRNAs, mRNAs, and proteins load exosomes efficiently with specific target molecules is necessary.

In contrast, post-loading methods extract exosomes from cells and body fluids, and the target molecules are encapsulated in the exosomes through disruption of the exosome membrane (Figure 3). Post-loading methods include physical induction by electroporation, freeze/thaw method, sonication, etc. and chemical induction by saponin treatment, transfection reagents, etc. In the post-loading method, the target molecules can be directly loaded into the exosomes, so the loading efficiency is expected to be higher than that of pre-loading methods. However, physical disruption may affect the size and shape of exosomes. Chemical induction has been shown to be more efficient in loading target molecules compared to physical induction methods [96]. However, chemicals used for the loading of molecules may be toxic, so removing them in the process is essential.

In terms of applications to joint therapy, post-loading methods may allow the targeting of specific tissues, such as cartilage, subchondral bone, or synovium, to treat the various symptoms. Liang et al. reported on the loading of specific miRNAs through electroporation into engineered exosomes [97]. They engineered chondrocyte-affinity peptide (CAP) exosomes by fusing CAP with lysosome-associated membrane glycoprotein 2b (Lamp2b) on the surface of exosomes. Furthermore, they loaded miR-140 into the CAP exosomes using electroporation (CAP exosome/miR-140). The CAP exosomes/miR-140 allowed for the specific delivery of chondrocytes. Moreover, in a DMM-induced OA rat model, they demonstrated that the progression of OA can be reduced by CAP-exosome/miR-140. Through modification of exosomes and post-loading methods, they were able to simultaneously achieve tissue-specific delivery and inclusion of target molecules into exosomes. These improvements in exosome modification technology could enhance the effectiveness of exosome-based intra-articular treatments.

4.4. Potential of Exosomes in a Sustained Release Drug Delivery System (SRDDS)

Exosomes are relatively stable in vivo, but multiple administrations may be required to overcome the natural elimination of exosomes in the joint. As such, to prolong and enhance the effect of joint treatment methods, studies have combined exosomes with various biomaterials to be used in a SRDDS [98–101]. Chen et al. produced a 3D-printed cartilage ECM/gelatin methacrylate (GelMA)/exosome scaffold with radially oriented channels through a desktop-stereolithography technology [102]. This 3D printed scaffold was shown to continuously release exosomes for 14 days in vitro. When the 3D printed scaffold was transplanted through an invasive procedure in a rabbit osteochondral model, the scaffold retained exosomes for at least 7 days in vivo and significantly promoted cartilage repair.

Other studies have combined exosomes with injectable materials that can be administered through intra-articular injection. Liu et al. utilized stem cell-derived exosomes in combination with photoinduced imine crosslinking hydrogel glue to produce a cell-free exosome tissue patch [103]. The tissue patch continuously released exosomes for 14 days. In vivo studies showed that the new cartilage tissue regenerated by the tissue patch was hyaline cartilage-like in a rabbit articular cartilage defect model. Alternatively, Hu et al. produced a Gelma/nanoclay hydrogel (Gel-nano-sEVs) with embedded human umbilical cord MSC-derived small extracellular vesicles (hUC-MSCs-sEVs) [104]. In this study, they showed that miR-23a-3p, which is highly expressed in hUC-MSCs-sEVs, activates the phosphatase and tensin homolog deleted from chromosome 10 (PTEN)/AKT serine/threonine kinase (AKT) signaling pathway and promotes cartilage regeneration. The Gel-nano-sEVs were demonstrated to release exosomes continuously for approximately 30 days and contribute to cartilage regeneration. Further studies are expected in this field.

5. Advantages and Challenges of Developing Treatments Utilizing Exosomes

Treatments utilizing exosomes have remarkable potential to target and affect specific tissues. With proper control, specific cargos such as miRNAs can be packaged to be delivered to various tissues affected by OAK, such as cartilage, synovium, and subchondral bone. Unlike cell-based therapies, exosomes have a lower immunogenicity and an inability to directly form tumors [105,106]. Cell-based therapies rely on a more complicated mode of action such as the production of humoral factors that may be influenced by local conditions [34,107]. While exosomes may also be influenced by local conditions as well, with proper manufacturing they can carry and deliver a consistent set of cargos at desired concentrations.

On the other hand, many challenges remain, such as the large number of cells required for the manufacturing of exosomes, the multiple steps and time required for collection, and the unclear risks in ensuring safety and efficacy [39,41,43,55]. In addition, when exosomes are used clinically, the possible transmission of infectious diseases, due to contamination with viruses, bacteria, and fungi, harmful effects of components other than exosomes that are simultaneously administered, and potential risks of inconsistent efficacy and quality must be considered.

Exosomes possess complicated characteristics such as their size, the cargos they hold, and their destinations, which must be properly characterized in addition to the origins of the exosomes. As such, setting quality standards in the manufacturing process will be a challenge. When manufacturing exosomes for treatments, the quality of cells used as raw materials and their constituents and the management of the manufacturing process will be important additions to the quality control of the exosomes themselves. The ISEV proposed Minimal Information for Studies of Extracellular Vesicles (MISEV) guidelines in 2014, which was updated in 2018, suggesting a comprehensive yet evolving guideline for studies using extracellular vehicles, including exosomes [53,108]. To secure the quality, efficacy, and safety of exosomes, strict quality control and manufacturing control must be implemented.

6. Conclusions

EVs, including exosomes, present new possibilities for treatment, diagnosis, and drug development for patients with joint diseases including OAK. Exosomes from body fluids such as joint fluid and plasma can provide important prognoses of disease to administer appropriate treatments and other preventive measures. The use of exosomes for treating OAK through intra-articular injection provides a novel treatment method to bridge the gap between pharmacological and surgical procedures. However, many unclear aspects of the functions and mechanisms of EVs, including exosomes, and their roles in treatments of joint diseases such as OAK, remain to be elucidated. For the realization of exosome therapy, it is essential to clarify the risks of the treatment methods, implement strict manufacturing controls, and prepare appropriate guidelines and laws.

Further research in this field and scientific evidence of safety and efficacy may greatly contribute to the point-of-care treatment and cell-free therapy of joint disease.

Author Contributions: Conceptualization, M.M. and M.S.; writing—original draft preparation, M.M., E.T., T.T. and M.S.; writing—review and editing, M.M. and T.T.; supervision, M.W. and M.S.; funding acquisition, M.M. and M.S. All authors have read and agreed to the published version of the manuscript.

Funding: Authors received financial support for the authorship and publication of this article by JSPS KAKENHI Grant Number JP19K18509 and TERUMO LIFE SCIENCE FOUNDATION.

Conflicts of Interest: The authors declare no conflict of interest.

Abbreviations

3′UTR	three prime untranslated region
ACI	autologous chondrocyte implantation
ACLT	anterior cruciate ligament transection
ADAMTS5	a disintegrin and metalloproteinase with thrombospondin motifs 5
AKT	AKT serine/threonine kinase 1
ALIX	ALG-2-interacting protein X
ATG4B	autophagy related 4B cysteine peptidase
CAP	chondrocyte-affinity peptide
CD	cluster of differentiation
DACT1	dishevelled binding antagonist of beta catenin 1
DDS	drug delivery system
DMM	destabilization of the medial meniscus
DMOAD	disease-modifying OA drugs
DNA	deoxyribonucleic acid
ECM	extracellular matrix
ESCRT	endosomal sorting complexes required for transport
EV	extracellular vesicle
GelMA	gelatin methacrylate
GSK-3β	glycogen synthase kinase 3 beta
hBMSC	human bone marrow-derived MSC
HDAC	histone deacetylase
Hsp	heat shock protein
hUC	human umbilical cord
IL-1β	interleukin-1 beta
ILV	intraluminal membrane vesicle
IPFP	infrapatellar fat pad
ISEV	International Society for Extracellular Vesicles
Lamp2b	lysosome-associated membrane glycoprotein 2b
lncRNA	long noncoding RNA
MCH	major histocompatibility complex
MCS	mesenchymal stem cell
miRNA	microRNA
MISEV	Minimal Information for Studies of Extracellular Vesicles
MMP13	matrix metalloproteinase 13
mTOR	mammalian target of rapamycin
MVB	multivesicular body
NSAID	non-steroidal anti-inflammatory drugs
OAK	osteoarthritis of knee
PCGEM1	prostate-specific transcript 1
PHC	primary human chondrocytes
PTEN	phosphatase and tensin homolog deleted from chromosome 10
PTGS2	prostaglandin-endoperoxide synthase 2
RA	rheumatoid arthritis
RNA	ribonucleic acid
siRNA	small interfering RNA
SMSC	synovium-derived MSC
SOX9	SRY-box transcription factor 9
SRDDS	sustained release drug delivery system
TGF-β1	transforming growth factor beta 1
TKA	total knee arthroplasty
TLR	Toll-like receptor
TSG101	tumor susceptibility gene 101
WNT5A	Wnt family member 5A
YAP	Yes-associated protein

References

1. Hunter, W. Of the Structure and Diseases of Articu-lating Cartilages. *Philos. Trans. R. Soc. Lond.* **1743**, *42*, 514–521. [CrossRef]
2. Ulrich-Vinther, M.; Maloney, M.D.; Schwarz, E.M.; Rosier, R.; O'Keefe, R.J. Articular Cartilage Biology. *JAAOS J. Am. Acad. Orthop. Surg.* **2003**, *11*, 421–430. [CrossRef]
3. Aspden, R.; Saunders, F. Osteoarthritis as an Organ Disease: From the Cradle to the Grave. *Eur. Cells Mater.* **2019**, *37*, 74–87. [CrossRef]
4. Cross, M.; Smith, E.; Hoy, D.; Nolte, S.; Ackerman, I.; Fransen, M.; Bridgett, L.; Williams, S.; Guillemin, F.; Hill, C.L.; et al. The Global Burden of Hip and Knee Osteoarthritis: Estimates from the Global Burden of Disease 2010 Study. *Ann. Rheum. Dis.* **2014**, *73*, 1323–1330. [CrossRef] [PubMed]
5. Pridie, K.; Gordon-Strachan, G. A Method of Resurfacing Osteoarthritis Knee Joints. *J. Bone Jt. Surg. Br.* **1959**, *41*, 618–619.
6. Steadman, J.R.; Rodkey, W.G.; Rodrigo, J.J. Microfracture: Surgical Technique and Rehabilitation to Treat Chondral Defects. *Clin. Orthop.* **2001**, *391*, S362–S369. [CrossRef]
7. Mithoefer, K.; Williams, R.J.; Warren, R.F.; Potter, H.G.; Spock, C.R.; Jones, E.C.; Wickiewicz, T.L.; Marx, R.G. Chondral Resurfacing of Articular Cartilage Defects in the Knee with the Microfracture Technique: Surgical Technique. *JBJS Essent. Surg. Tech.* **2006**, *88*, 294–304. [CrossRef]
8. Hangody, L.; Vásárhelyi, G.; Hangody, L.R.; Sükösd, Z.; Tibay, G.; Bartha, L.; Bodó, G. Autologous Osteochondral Grafting—Technique and Long-Term Results. *Injury* **2008**, *39*, 32–39. [CrossRef] [PubMed]
9. Szerb, I.; Hangody, L.; Duska, Z.; Kaposi, N.P. Mosaicplasty: Long-Term Follow-Up. *Bull. Hosp. Jt. Dis.* **2005**, *63*, 54–62.
10. Brittberg, M.; Lindahl, A.; Nilsson, A.; Ohlsson, C.; Isaksson, O.; Peterson, L. Treatment of Deep Cartilage Defects in the Knee with Autologous Chondrocyte Transplantation. *N. Engl. J. Med.* **1994**, *331*, 889–895. [CrossRef] [PubMed]
11. Uchio, Y.; Ochi, M.; Matsusaki, M.; Kurioka, H.; Katsube, K. Human Chondrocyte Proliferation and Matrix Synthesis Cultured in Atelocollagen Gel. *J. Biomed. Mater. Res.* **2000**, *50*, 138–143. [CrossRef]
12. Ochi, M.; Uchio, Y.; Kawasaki, K.; Wakitani, S.; Iwasa, J. Transplantation of Cartilage-like Tissue Made by Tissue Engineering in the Treatment of Cartilage Defects of the Knee. *J. Bone Jt. Surg. Br.* **2002**, *84*, 571–578. [CrossRef]
13. Nehrer, S.; Halbwirth, F.; Luksch, T. CaReS®, Cartilage Regeneration System: Autologous Chondrocyte Transplantation in a Collagen Gel. In *Techniques in Cartilage Repair Surgery*; Shetty, A.A., Kim, S.-J., Nakamura, N., Brittberg, M., Eds.; Springer: Berlin/Heidelberg, Germany, 2014; pp. 245–250. ISBN 978-3-642-41921-8.
14. Schneider, U.; Rackwitz, L.; Andereya, S.; Siebenlist, S.; Fensky, F.; Reichert, J.; Löer, I.; Barthel, T.; Rudert, M.; Nöth, U. A Prospective Multicenter Study on the Outcome of Type I Collagen Hydrogel-Based Autologous Chondrocyte Implantation (CaReS) for the Repair of Articular Cartilage Defects in the Knee. *Am. J. Sports Med.* **2011**, *39*, 2558–2565. [CrossRef]
15. Tognana, E.; Borrione, A.; De Luca, C.; Pavesio, A. Hyalograft C: Hyaluronan-Based Scaffolds in Tissue-Engineered Cartilage. *Cells Tissues Organs* **2007**, *186*, 97–103. [CrossRef] [PubMed]
16. Kreuz, P.C.; Müller, S.; Ossendorf, C.; Kaps, C.; Erggelet, C. Treatment of Focal Degenerative Cartilage Defects with Polymer-Based Autologous Chondrocyte Grafts: Four-Year Clinical Results. *Arthritis Res. Ther.* **2009**, *11*, R33. [CrossRef]
17. Roberts, S.; McCall, I.W.; Darby, A.J.; Menage, J.; Evans, H.; Harrison, P.E.; Richardson, J.B. Autologous Chondrocyte Implantation for Cartilage Repair: Monitoring Its Success by Magnetic Resonance Imaging and Histology. *Arthritis Res. Ther.* **2003**, *5*, R60–R73. [CrossRef]
18. Bartlett, W.; Skinner, J.A.; Gooding, C.R.; Carrington, R.W.J.; Flanagan, A.M.; Briggs, T.W.R.; Bentley, G. Autologous Chondrocyte Implantation versus Matrix-Induced Autologous Chondrocyte Implantation for Osteochondral Defects of the Knee: A Prospective, Randomised Study. *J. Bone Jt. Surg. Br.* **2005**, *87*, 640–645. [CrossRef] [PubMed]
19. Maehara, M.; Sato, M.; Toyoda, E.; Takahashi, T.; Okada, E.; Kotoku, T.; Watanabe, M. Characterization of Polydactyly-Derived Chondrocyte Sheets versus Adult Chondrocyte Sheets for Articular Cartilage Repair. *Inflamm. Regen.* **2017**, *37*, 22. [CrossRef] [PubMed]
20. Sato, M.; Yamato, M.; Mitani, G.; Takagaki, T.; Hamahashi, K.; Nakamura, Y.; Ishihara, M.; Matoba, R.; Kobayashi, H.; Okano, T.; et al. Combined Surgery and Chondrocyte Cell-Sheet Transplantation Improves Clinical and Structural Outcomes in Knee Osteoarthritis. *NPJ Regen. Med.* **2019**, *4*, 4. [CrossRef]
21. Takahashi, T.; Sato, M.; Toyoda, E.; Maehara, M.; Takizawa, D.; Maruki, H.; Tominaga, A.; Okada, E.; Okazaki, K.; Watanabe, M. Rabbit Xenogeneic Transplantation Model for Evaluating Human Chondrocyte Sheets Used in Articular Cartilage Repair. *J. Tissue Eng. Regen. Med.* **2018**, *12*, 2067–2076. [CrossRef] [PubMed]
22. Varacallo, M.; Luo, T.D.; Johanson, N.A. Total Knee Arthroplasty (TKA) Techniques. In *StatPearls*; StatPearls Publishing: Treasure Island, FL, USA, 2020.
23. Colombini, A.; Perucca Orfei, C.; Kouroupis, D.; Ragni, E.; De Luca, P.; ViganÒ, M.; Correa, D.; de Girolamo, L. Mesenchymal Stem Cells in the Treatment of Articular Cartilage Degeneration: New Biological Insights for an Old-Timer Cell. *Cytotherapy* **2019**, *21*, 1179–1197. [CrossRef]
24. Mancuso, P.; Raman, S.; Glynn, A.; Barry, F.; Murphy, J.M. Mesenchymal Stem Cell Therapy for Osteoarthritis: The Critical Role of the Cell Secretome. *Front. Bioeng. Biotechnol.* **2019**, *7*, 9. [CrossRef] [PubMed]
25. Mascarenhas, R.; Saltzman, B.M.; Fortier, L.A.; Cole, B.J. Role of Platelet-Rich Plasma in Articular Cartilage Injury and Disease. *J. Knee Surg.* **2015**, *28*, 3–10. [CrossRef] [PubMed]

26. Dhillon, M.S.; Patel, S.; John, R. PRP in OA Knee—Update, Current Confusions and Future Options. *SICOT J.* **2017**, *3*, 27. [CrossRef]
27. Wasai, S.; Sato, M.; Maehara, M.; Toyoda, E.; Uchiyama, R.; Takahashi, T.; Okada, E.; Iwasaki, Y.; Suzuki, S.; Watanabe, M. Characteristics of Autologous Protein Solution and Leucocyte-Poor Platelet-Rich Plasma for the Treatment of Osteoarthritis of the Knee. *Sci. Rep.* **2020**, *10*, 10572. [CrossRef] [PubMed]
28. Kuffler, D.P. Variables Affecting the Potential Efficacy of PRP in Providing Chronic Pain Relief. *J. Pain Res.* **2018**, *12*, 109–116. [CrossRef] [PubMed]
29. Kretlow, J.D.; Jin, Y.-Q.; Liu, W.; Zhang, W.J.; Hong, T.-H.; Zhou, G.; Baggett, L.S.; Mikos, A.G.; Cao, Y. Donor Age and Cell Passage Affects Differentiation Potential of Murine Bone Marrow-Derived Stem Cells. *BMC Cell Biol.* **2008**, *9*, 60. [CrossRef]
30. Zaim, M.; Karaman, S.; Cetin, G.; Isik, S. Donor Age and Long-Term Culture Affect Differentiation and Proliferation of Human Bone Marrow Mesenchymal Stem Cells. *Ann. Hematol.* **2012**, *91*, 1175–1186. [CrossRef]
31. Mazzocca, A.D.; McCarthy, M.B.R.; Chowaniec, D.M.; Dugdale, E.M.; Hansen, D.; Cote, M.P.; Bradley, J.P.; Romeo, A.A.; Arciero, R.A.; Beitzel, K. The Positive Effects of Different Platelet-Rich Plasma Methods on Human Muscle, Bone, and Tendon Cells. *Am. J. Sports Med.* **2012**, *40*, 1742–1749. [CrossRef]
32. Breitbach, M.; Bostani, T.; Roell, W.; Xia, Y.; Dewald, O.; Nygren, J.M.; Fries, J.W.U.; Tiemann, K.; Bohlen, H.; Hescheler, J.; et al. Potential Risks of Bone Marrow Cell Transplantation into Infarcted Hearts. *Blood* **2007**, *110*, 1362–1369. [CrossRef]
33. Lalu, M.M.; McIntyre, L.; Pugliese, C.; Fergusson, D.; Winston, B.W.; Marshall, J.C.; Granton, J.; Stewart, D.J. Safety of Cell Therapy with Mesenchymal Stromal Cells (SafeCell): A Systematic Review and Meta-Analysis of Clinical Trials. *PLoS ONE* **2012**, *7*, e47559. [CrossRef]
34. Kan, C.; Chen, L.; Hu, Y.; Lu, H.; Li, Y.; Kessler, J.A.; Kan, L. Microenvironmental Factors That Regulate Mesenchymal Stem Cells: Lessons Learned from the Study of Heterotopic Ossification. *Histol. Histopathol.* **2017**, *32*, 977–985. [CrossRef] [PubMed]
35. Caplan, H.; Olson, S.D.; Kumar, A.; George, M.; Prabhakara, K.S.; Wenzel, P.; Bedi, S.; Toledano-Furman, N.E.; Triolo, F.; Kamhieh-Milz, J.; et al. Mesenchymal Stromal Cell Therapeutic Delivery: Translational Challenges to Clinical Application. *Front. Immunol.* **2019**, *10*, 1645. [CrossRef]
36. Drela, K.; Stanaszek, L.; Nowakowski, A.; Kuczynska, Z.; Lukomska, B. Experimental Strategies of Mesenchymal Stem Cell Propagation: Adverse Events and Potential Risk of Functional Changes. Available online: https://www.hindawi.com/journals/sci/2019/7012692/ (accessed on 1 March 2021).
37. Liu, Y.; Wang, Y.; Lv, Q.; Li, X. Exosomes: From Garbage Bins to Translational Medicine. *Int. J. Pharm.* **2020**, *583*, 119333. [CrossRef]
38. Pourakbari, R.; Khodadadi, M.; Aghebati-Maleki, A.; Aghebati-Maleki, L.; Yousefi, M. The Potential of Exosomes in the Therapy of the Cartilage and Bone Complications; Emphasis on Osteoarthritis. *Life Sci.* **2019**, *236*, 116861. [CrossRef] [PubMed]
39. Allan, D.; Tieu, A.; Lalu, M.; Burger, D. Mesenchymal Stromal Cell-derived Extracellular Vesicles for Regenerative Therapy and Immune Modulation: Progress and Challenges toward Clinical Application. *Stem Cells Transl. Med.* **2020**, *9*, 39–46. [CrossRef]
40. Witwer, K.W.; Balkom, B.W.M.V.; Bruno, S.; Choo, A.; Dominici, M.; Gimona, M.; Hill, A.F.; Kleijn, D.D.; Koh, M.; Lai, R.C.; et al. Defining Mesenchymal Stromal Cell (MSC)-Derived Small Extracellular Vesicles for Therapeutic Applications. *J. Extracell. Vesicles* **2019**, *8*, 1609206. [CrossRef]
41. Elahi, F.M.; Farawell, D.G.; Nolta, J.A.; Anderson, J.D. Preclinical Translation of Exosomes Derived from Mesenchymal Stem/Stromal Cells. *Stem Cells* **2020**, *38*, 15–21. [CrossRef]
42. Phinney, D.G.; Pittenger, M.F. Concise Review: MSC-Derived Exosomes for Cell-Free Therapy. *Stem Cells* **2017**, *35*, 851–858. [CrossRef]
43. Rohde, E.; Pachler, K.; Gimona, M. Manufacturing and Characterization of Extracellular Vesicles from Umbilical Cord–Derived Mesenchymal Stromal Cells for Clinical Testing. *Cytotherapy* **2019**, *21*, 581–592. [CrossRef]
44. Pan, B.T.; Johnstone, R.M. Fate of the Transferrin Receptor during Maturation of Sheep Reticulocytes in Vitro: Selective Externalization of the Receptor. *Cell* **1983**, *33*, 967–978. [CrossRef]
45. Stahl, P.D.; Raposo, G. Extracellular Vesicles: Exosomes and Microvesicles, Integrators of Homeostasis. *Physiology* **2019**, *34*, 169–177. [CrossRef]
46. Johnstone, R.M.; Adam, M.; Hammond, J.R.; Orr, L.; Turbide, C. Vesicle Formation during Reticulocyte Maturation. Association of Plasma Membrane Activities with Released Vesicles (Exosomes). *J. Biol. Chem.* **1987**, *262*, 9412–9420. [CrossRef]
47. Mathieu, M.; Martin-Jaular, L.; Lavieu, G.; Théry, C. Specificities of Secretion and Uptake of Exosomes and Other Extracellular Vesicles for Cell-to-Cell Communication. *Nat. Cell Biol.* **2019**, *21*, 9–17. [CrossRef]
48. Stein, J.M.; Luzio, J.P. Ectocytosis Caused by Sublytic Autologous Complement Attack on Human Neutrophils. The Sorting of Endogenous Plasma-Membrane Proteins and Lipids into Shed Vesicles. *Biochem. J.* **1991**, *274*, 381–386. [CrossRef]
49. Cocucci, E.; Racchetti, G.; Meldolesi, J. Shedding Microvesicles: Artefacts No More. *Trends Cell Biol.* **2009**, *19*, 43–51. [CrossRef] [PubMed]
50. Di Vizio, D.; Kim, J.; Hager, M.H.; Morello, M.; Yang, W.; Lafargue, C.J.; True, L.; Rubin, M.A.; Adam, R.M.; Beroukhim, R.; et al. Oncosome Formation in Prostate Cancer: Association with a Region of Frequent Chromosomal Deletion in Metastatic Disease. *Cancer Res.* **2009**, *69*, 5601–5609. [CrossRef]
51. Hristov, M.; Erl, W.; Linder, S.; Weber, P.C. Apoptotic Bodies from Endothelial Cells Enhance the Number and Initiate the Differentiation of Human Endothelial Progenitor Cells in Vitro. *Blood* **2004**, *104*, 2761–2766. [CrossRef] [PubMed]

52. Van Niel, G.; D'Angelo, G.; Raposo, G. Shedding Light on the Cell Biology of Extracellular Vesicles. *Nat. Rev. Mol. Cell Biol.* **2018**, *19*, 213–228. [CrossRef] [PubMed]
53. Théry, C.; Witwer, K.W.; Aikawa, E.; Alcaraz, M.J.; Anderson, J.D.; Andriantsitohaina, R.; Antoniou, A.; Arab, T.; Archer, F.; Atkin-Smith, G.K.; et al. Minimal Information for Studies of Extracellular Vesicles 2018 (MISEV2018): A Position Statement of the International Society for Extracellular Vesicles and Update of the MISEV2014 Guidelines. *J. Extracell. Vesicles* **2018**, *7*, 1535750. [CrossRef] [PubMed]
54. Tkach, M.; Kowal, J.; Théry, C. Why the Need and How to Approach the Functional Diversity of Extracellular Vesicles. *Philos. Trans. R. Soc. B Biol. Sci.* **2018**, *373*, 20160479. [CrossRef]
55. Kalluri, R.; LeBleu, V.S. The Biology, Function, and Biomedical Applications of Exosomes. *Science* **2020**, *367*, eaau6977. [CrossRef]
56. Colombo, M.; Raposo, G.; Théry, C. Biogenesis, Secretion, and Intercellular Interactions of Exosomes and Other Extracellular Vesicles. *Annu. Rev. Cell Dev. Biol.* **2014**, *30*, 255–289. [CrossRef]
57. Villarroya-Beltri, C.; Baixauli, F.; Gutiérrez-Vázquez, C.; Sánchez-Madrid, F.; Mittelbrunn, M. SORTING IT OUT: REGULATION OF EXOSOME LOADING. *Semin. Cancer Biol.* **2014**, *28*, 3–13. [CrossRef]
58. Guescini, M.; Genedani, S.; Stocchi, V.; Agnati, L.F. Astrocytes and Glioblastoma Cells Release Exosomes Carrying MtDNA. *J. Neural Transm.* **2010**, *117*, 1–4. [CrossRef]
59. Valadi, H.; Ekström, K.; Bossios, A.; Sjöstrand, M.; Lee, J.J.; Lötvall, J.O. Exosome-Mediated Transfer of MRNAs and MicroRNAs Is a Novel Mechanism of Genetic Exchange between Cells. *Nat. Cell Biol.* **2007**, *9*, 654–659. [CrossRef]
60. Turturici, G.; Tinnirello, R.; Sconzo, G.; Geraci, F. Extracellular Membrane Vesicles as a Mechanism of Cell-to-Cell Communication: Advantages and Disadvantages. *Am. J. Physiol. Cell Physiol.* **2014**, *306*, C621–C633. [CrossRef] [PubMed]
61. Raposo, G.; Stoorvogel, W. Extracellular Vesicles: Exosomes, Microvesicles, and Friends. *J. Cell Biol.* **2013**, *200*, 373–383. [CrossRef] [PubMed]
62. Kowal, J.; Arras, G.; Colombo, M.; Jouve, M.; Morath, J.P.; Primdal-Bengtson, B.; Dingli, F.; Loew, D.; Tkach, M.; Théry, C. Proteomic Comparison Defines Novel Markers to Characterize Heterogeneous Populations of Extracellular Vesicle Subtypes. *Proc. Natl. Acad. Sci. USA* **2016**, *113*, E968–E977. [CrossRef] [PubMed]
63. Gurunathan, S.; Kang, M.-H.; Jeyaraj, M.; Qasim, M.; Kim, J.-H. Review of the Isolation, Characterization, Biological Function, and Multifarious Therapeutic Approaches of Exosomes. *Cells* **2019**, *8*, 307. [CrossRef]
64. Konoshenko, M.Y.; Lekchnov, E.A.; Vlassov, A.V.; Laktionov, P.P. Isolation of Extracellular Vesicles: General Methodologies and Latest Trends. *BioMed. Res. Int.* **2018**, *2018*, 1–27. [CrossRef]
65. Li, P.; Kaslan, M.; Lee, S.H.; Yao, J.; Gao, Z. Progress in Exosome Isolation Techniques. *Theranostics* **2017**, *7*, 789–804. [CrossRef]
66. Gardiner, C.; Vizio, D.D.; Sahoo, S.; Théry, C.; Witwer, K.W.; Wauben, M.; Hill, A.F. Techniques Used for the Isolation and Characterization of Extracellular Vesicles: Results of a Worldwide Survey. *J. Extracell. Vesicles* **2016**, *5*, 32945. [CrossRef] [PubMed]
67. Kosaka, N.; Yoshioka, Y.; Fujita, Y.; Ochiya, T. Versatile Roles of Extracellular Vesicles in Cancer. *J. Clin. Investig.* **2016**, *126*, 1163–1172. [CrossRef] [PubMed]
68. Tominaga, N.; Yoshioka, Y.; Ochiya, T. A Novel Platform for Cancer Therapy Using Extracellular Vesicles. *Adv. Drug Deliv. Rev.* **2015**, *1*, 50–55. [CrossRef]
69. He, L.; Hannon, G.J. MicroRNAs: Small RNAs with a Big Role in Gene Regulation. *Nat. Rev. Genet.* **2004**, *5*, 522–531. [CrossRef]
70. Malemud, C.J. MicroRNAs and Osteoarthritis. *Cells* **2018**, *7*, 92. [CrossRef]
71. Hong, E.; Reddi, A.H. MicroRNAs in Chondrogenesis, Articular Cartilage, and Osteoarthritis: Implications for Tissue Engineering. *Tissue Eng. Part B Rev.* **2012**, *18*, 445–453. [CrossRef] [PubMed]
72. Grigelioniene, G.; Suzuki, H.I.; Taylan, F.; Mirzamohammadi, F.; Borochowitz, Z.U.; Ayturk, U.M.; Tzur, S.; Horemuzova, E.; Lindstrand, A.; Weis, M.A.; et al. Gain-of-Function Mutation of MicroRNA-140 in Human Skeletal Dysplasia. *Nat. Med.* **2019**, *25*, 583–590. [CrossRef]
73. Miyaki, S.; Nakasa, T.; Otsuki, S.; Grogan, S.P.; Higashiyama, R.; Inoue, A.; Kato, Y.; Sato, T.; Lotz, M.K.; Asahara, H. MicroRNA-140 Is Expressed in Differentiated Human Articular Chondrocytes and Modulates Interleukin-1 Responses. *Arthritis Rheum.* **2009**, *60*, 2723–2730. [CrossRef]
74. Ukai, T.; Sato, M.; Akutsu, H.; Umezawa, A.; Mochida, J. MicroRNA-199a-3p, MicroRNA-193b, and MicroRNA-320c Are Correlated to Aging and Regulate Human Cartilage Metabolism. *J. Orthop. Res.* **2012**, *30*, 1915–1922. [CrossRef] [PubMed]
75. Kosaka, N.; Iguchi, H.; Yoshioka, Y.; Takeshita, F.; Matsuki, Y.; Ochiya, T. Secretory Mechanisms and Intercellular Transfer of MicroRNAs in Living Cells. *J. Biol. Chem.* **2010**, *285*, 17442–17452. [CrossRef] [PubMed]
76. Pegtel, D.M.; Cosmopoulos, K.; Thorley-Lawson, D.A.; Van Eijndhoven, M.A.J.; Hopmans, E.S.; Lindenberg, J.L.; De Gruijl, T.D.; Würdinger, T.; Middeldorp, J.M. Functional Delivery of Viral MiRNAs via Exosomes. *Proc. Natl. Acad. Sci. USA* **2010**, *107*, 6328–6333. [CrossRef]
77. Zhang, J.; Li, S.; Li, L.; Li, M.; Guo, C.; Yao, J.; Mi, S. Exosome and Exosomal MicroRNA: Trafficking, Sorting, and Function. *Genom. Proteom. Bioinform.* **2015**, *13*, 17–24. [CrossRef]
78. Sun, H.; Hu, S.; Zhang, Z.; Lun, J.; Liao, W.; Zhang, Z. Expression of Exosomal MicroRNAs during Chondrogenic Differentiation of Human Bone Mesenchymal Stem Cells. *J. Cell. Biochem.* **2019**, *120*, 171–181. [CrossRef]
79. Mao, G.; Hu, S.; Zhang, Z.; Wu, P.; Zhao, X.; Lin, R.; Liao, W.; Kang, Y. Exosomal MiR-95-5p Regulates Chondrogenesis and Cartilage Degradation via Histone Deacetylase 2/8. *J. Cell. Mol. Med.* **2018**, *22*, 5354–5366. [CrossRef] [PubMed]

80. Li, Z.; Wang, Y.; Xiang, S.; Zheng, Z.; Bian, Y.; Feng, B.; Weng, X. Chondrocytes-Derived Exosomal MiR-8485 Regulated the Wnt/β-Catenin Pathways to Promote Chondrogenic Differentiation of BMSCs. *Biochem. Biophys. Res. Commun.* **2020**, *523*, 506–513. [CrossRef] [PubMed]

81. Mao, G.; Zhang, Z.; Hu, S.; Zhang, Z.; Chang, Z.; Huang, Z.; Liao, W.; Kang, Y. Exosomes Derived from MiR-92a-3p-Overexpressing Human Mesenchymal Stem Cells Enhance Chondrogenesis and Suppress Cartilage Degradation via Targeting WNT5A. *Stem Cell Res. Ther.* **2018**, *9*, 247. [CrossRef] [PubMed]

82. Jin, Z.; Ren, J.; Qi, S. Human Bone Mesenchymal Stem Cells-Derived Exosomes Overexpressing MicroRNA-26a-5p Alleviate Osteoarthritis via down-Regulation of PTGS2. *Int. Immunopharmacol.* **2020**, *78*, 105946. [CrossRef] [PubMed]

83. Wu, J.; Kuang, L.; Chen, C.; Yang, J.; Zeng, W.-N.; Li, T.; Chen, H.; Huang, S.; Fu, Z.; Li, J.; et al. MiR-100-5p-Abundant Exosomes Derived from Infrapatellar Fat Pad MSCs Protect Articular Cartilage and Ameliorate Gait Abnormalities via Inhibition of MTOR in Osteoarthritis. *Biomaterials* **2019**, *206*, 87–100. [CrossRef]

84. Tao, S.C.; Yuan, T.; Zhang, Y.-L.; Yin, W.-J.; Guo, S.-C.; Zhang, C.-Q. Exosomes Derived from MiR-140-5p-Overexpressing Human Synovial Mesenchymal Stem Cells Enhance Cartilage Tissue Regeneration and Prevent Osteoarthritis of the Knee in a Rat Model. *Theranostics* **2017**, *7*, 180–195. [CrossRef]

85. Ni, Z.; Kuang, L.; Chen, H.; Xie, Y.; Zhang, B.; Ouyang, J.; Wu, J.; Zhou, S.; Chen, L.; Nan, S.; et al. The Exosome-like Vesicles from Osteoarthritic Chondrocyte Enhanced Mature IL-1β Production of Macrophages and Aggravated Synovitis in Osteoarthritis. *Cell Death Dis.* **2019**, *10*, 522. [CrossRef]

86. Sheridan, C. Exosome Cancer Diagnostic Reaches Market. *Nat. Biotechnol.* **2016**, *34*, 359–360. [CrossRef] [PubMed]

87. Brinkmann, K.; Enderle, D.; Flinspach, C.; Meyer, L.; Skog, J.; Noerholm, M. Exosome Liquid Biopsies of NSCLC Patients for Longitudinal Monitoring of ALK Fusions and Resistance Mutations. *J. Clin. Oncol.* **2018**, *36*, e24090. [CrossRef]

88. Kolhe, R.; Hunter, M.; Liu, S.; Jadeja, R.N.; Pundkar, C.; Mondal, A.K.; Mendhe, B.; Drewry, M.; Rojiani, M.V.; Liu, Y.; et al. Gender-Specific Differential Expression of Exosomal MiRNA in Synovial Fluid of Patients with Osteoarthritis. *Sci. Rep.* **2017**, *7*, 2029. [CrossRef]

89. Meng, F.; Li, Z.; Zhang, Z.; Yang, Z.; Kang, Y.; Zhao, X.; Long, D.; Hu, S.; Gu, M.; He, S.; et al. MicroRNA-193b-3p Regulates Chondrogenesis and Chondrocyte Metabolism by Targeting HDAC3. *Theranostics* **2018**, *8*, 2862–2883. [CrossRef]

90. Zhao, Y.; Xu, J. Synovial Fluid-Derived Exosomal LncRNA PCGEM1 as Biomarker for the Different Stages of Osteoarthritis. *Int. Orthop.* **2018**, *42*, 2865–2872. [CrossRef] [PubMed]

91. Chen, X.M.; Zhao, Y.; Wu, X.D.; Wang, M.J.; Yu, H.; Lu, J.J.; Hu, Y.J.; Huang, Q.C.; Huang, R.Y.; Lu, C.J. Novel Findings from Determination of Common Expressed Plasma Exosomal MicroRNAs in Patients with Psoriatic Arthritis, Psoriasis Vulgaris, Rheumatoid Arthritis, and Gouty Arthritis. *Discov. Med.* **2019**, *28*, 47–68. [PubMed]

92. Tsuno, H.; Arito, M.; Suematsu, N.; Sato, T.; Hashimoto, A.; Matsui, T.; Omoteyama, K.; Sato, M.; Okamoto, K.; Tohma, S.; et al. A Proteomic Analysis of Serum-Derived Exosomes in Rheumatoid Arthritis. *BMC Rheumatol.* **2018**, *2*, 35. [CrossRef]

93. Somiya, M.; Yoshioka, Y.; Ochiya, T. Drug Delivery Application of Extracellular Vesicles; Insight into Production, Drug Loading, Targeting, and Pharmacokinetics. *AIMS Bioeng.* **2017**, *4*, 73. [CrossRef]

94. Antimisiaris, S.G.; Mourtas, S.; Marazioti, A. Exosomes and Exosome-Inspired Vesicles for Targeted Drug Delivery. *Pharmaceutics* **2018**, *10*, 218. [CrossRef]

95. Wang, R.; Xu, B.; Xu, H. TGF-B1 Promoted Chondrocyte Proliferation by Regulating Sp1 through MSC-Exosomes Derived MiR-135b. *Cell Cycle* **2018**, *17*, 2756–2765. [CrossRef]

96. Haney, M.J.; Klyachko, N.L.; Zhao, Y.; Gupta, R.; Plotnikova, E.G.; He, Z.; Patel, T.; Piroyan, A.; Sokolsky, M.; Kabanov, A.V.; et al. Exosomes as Drug Delivery Vehicles for Parkinson's Disease Therapy. *J. Control. Release* **2015**, *207*, 18–30. [CrossRef] [PubMed]

97. Liang, Y.; Xu, X.; Li, X.; Xiong, J.; Li, B.; Duan, L.; Wang, D.; Xia, J. Chondrocyte-Targeted MicroRNA Delivery by Engineered Exosomes toward a Cell-Free Osteoarthritis Therapy. *ACS Appl. Mater. Interfaces* **2020**, *12*, 36938–36947. [CrossRef] [PubMed]

98. Xie, H.; Wang, Z.; Zhang, L.; Lei, Q.; Zhao, A.; Wang, H.; Li, Q.; Cao, Y.; Zhang, W.J.; Chen, Z. Extracellular Vesicle-Functionalized Decalcified Bone Matrix Scaffolds with Enhanced Pro-Angiogenic and Pro-Bone Regeneration Activities. *Sci. Rep.* **2017**, *7*, 45622. [CrossRef] [PubMed]

99. Shi, Q.; Qian, Z.; Liu, D.; Sun, J.; Wang, X.; Liu, H.; Xu, J.; Guo, X. GMSC-Derived Exosomes Combined with a Chitosan/Silk Hydrogel Sponge Accelerates Wound Healing in a Diabetic Rat Skin Defect Model. *Front. Physiol.* **2017**, *8*, 904. [CrossRef]

100. Yang, J.; Chen, Z.; Pan, D.; Li, H.; Shen, J. Umbilical Cord-Derived Mesenchymal Stem Cell-Derived Exosomes Combined Pluronic F127 Hydrogel Promote Chronic Diabetic Wound Healing and Complete Skin Regeneration. *Int. J. Nanomed.* **2020**, *11*, 5911–5926. [CrossRef]

101. Li, L.; Zhang, Y.; Mu, J.; Chen, J.; Zhang, C.; Cao, H.; Gao, J. Transplantation of Human Mesenchymal Stem-Cell-Derived Exosomes Immobilized in an Adhesive Hydrogel for Effective Treatment of Spinal Cord Injury. *Nano Lett.* **2020**, *20*, 4298–4305. [CrossRef] [PubMed]

102. Chen, P.; Zheng, L.; Wang, Y.; Tao, M.; Xie, Z.; Xia, C.; Gu, C.; Chen, J.; Qiu, P.; Mei, S.; et al. Desktop-Stereolithography 3D Printing of a Radially Oriented Extracellular Matrix/Mesenchymal Stem Cell Exosome Bioink for Osteochondral Defect Regeneration. *Theranostics* **2019**, *9*, 2439–2459. [CrossRef] [PubMed]

103. Liu, X.; Yang, Y.; Li, Y.; Niu, X.; Zhao, B.; Wang, Y.; Bao, C.; Xie, Z.; Lin, Q.; Zhu, L. Integration of Stem Cell-Derived Exosomes with in Situ Hydrogel Glue as a Promising Tissue Patch for Articular Cartilage Regeneration. *Nanoscale* **2017**, *9*, 4430–4438. [CrossRef] [PubMed]

104. Hu, H.; Dong, L.; Bu, Z.; Shen, Y.; Luo, J.; Zhang, H.; Zhao, S.; Lv, F.; Liu, Z. MiR-23a-3p-Abundant Small Extracellular Vesicles Released from Gelma/Nanoclay Hydrogel for Cartilage Regeneration. *J. Extracell. Vesicles* **2020**, *9*, 1778883. [CrossRef] [PubMed]
105. Liew, L.C.; Katsuda, T.; Gailhouste, L.; Nakagama, H.; Ochiya, T. Mesenchymal Stem Cell-Derived Extracellular Vesicles: A Glimmer of Hope in Treating Alzheimer's Disease. *Int. Immunol.* **2017**, *29*, 11–19. [CrossRef] [PubMed]
106. Mendt, M.; Rezvani, K.; Shpall, E. Mesenchymal Stem Cell-Derived Exosomes for Clinical Use. *Bone Marrow Transpl.* **2019**, *54*, 789–792. [CrossRef] [PubMed]
107. Murphy, M.B.; Moncivais, K.; Caplan, A.I. Mesenchymal Stem Cells: Environmentally Responsive Therapeutics for Regenerative Medicine. *Exp. Mol. Med.* **2013**, *45*, e54. [CrossRef]
108. Lener, T.; Gimona, M.; Aigner, L.; Börger, V.; Buzas, E.; Camussi, G.; Chaput, N.; Chatterjee, D.; Court, F.A.; del Portillo, H.A.; et al. Applying Extracellular Vesicles Based Therapeutics in Clinical Trials—An ISEV Position Paper. *J. Extracell. Vesicles* **2015**, *4*, 30087. [CrossRef] [PubMed]

International Journal of
Molecular Sciences

Review

Potential Mechanism of Action of Current Point-of-Care Autologous Therapy Treatments for Osteoarthritis of the Knee—A Narrative Review

Jennifer Woodell-May [1,*], Kathleen Steckbeck [1] and William King [2]

1   Zimmer Biomet, 56 East Bell Drive, Warsaw, IN 46580, USA; kathleen.steckbeck@zimmerbiomet.com
2   Owl Manor, 720 East Winona Avenue, Warsaw, IN 46580, USA; william.king@owlmanormedical.com
*   Correspondence: jennifer.woodell-may@zimmerbiomet.com

Abstract: Osteoarthritis (OA) is a progressive degenerative disease that manifests as pain and inflammation and often results in total joint replacement. There is significant interest in understanding how intra-articular injections made from autologous blood or bone marrow could alleviate symptoms and potentially intervene in the progression of the disease. There is in vitro an in vivo evidence that suggests that these therapies, including platelet-rich plasma (PRP), autologous anti-inflammatories (AAIs), and concentrated bone marrow aspirate (cBMA), can interrupt cartilage matrix degradation driven by pro-inflammatory cytokines. This review analyzes the evidence for and against inclusion of white blood cells, the potential role of platelets, and the less studied potential role of blood plasma when combining these components to create an autologous point-of-care therapy to treat OA. There has been significant focus on the differences between the various autologous therapies. However, evidence suggests that there may be more in common between groups and perhaps we should be thinking of these therapies on a spectrum of the same technology, each providing significant levels of anti-inflammatory cytokines that can be antagonists against the inflammatory cytokines driving OA symptoms and progression. While clinical data have demonstrated symptom alleviation, more studies will need to be conducted to determine whether these preclinical disease-modifying findings translate into clinical practice.

Keywords: platelet-rich plasma; autologous anti-inflammatory; concentrated bone marrow aspirate; osteoarthritis; intra-articular injection; mechanism of action

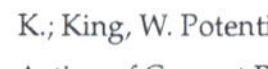
check for
updates

Citation: Woodell-May, J.; Steckbeck, K.; King, W. Potential Mechanism of Action of Current Point-of-Care Autologous Therapy Treatments for Osteoarthritis of the Knee—A Narrative Review. *Int. J. Mol. Sci.* **2021**, *22*, 2726. https://doi.org/10.3390/ijms22052726

Academic Editor: Masato Sato

Received: 1 February 2021
Accepted: 23 February 2021
Published: 8 March 2021

Publisher's Note: MDPI stays neutral with regard to jurisdictional claims in published maps and institutional affiliations.

1. Introduction

Osteoarthritis (OA) is a degenerative and disabling articulating joint disease that affects both younger, more active patients (e.g., patients with trauma or who have prolonged participation in highly demanding sports) and the elderly [1,2]. The disease is progressive and debilitating, eventually resulting in pain that may be so severe that restive sleep is impossible, along with life-altering loss of function.

Surgical intervention is clinically successful, and widely used, in treating severe degenerative OA; however, treatment modalities for less advanced OA are associated with varying rates of success. Current treatment options include non-steroidal anti-inflammatory drugs (NSAIDs), corticosteroid injections, and hyaluronic acid (HA) injections. Although these treatments can relieve pain temporarily for some OA patients, they may not address the biological mechanisms causing the disease [3].

Although OA is classified as a non-inflammatory disease, inflammation is implicated in many symptoms and in OA progression. Pro-inflammatory cytokines involved in OA development include interleukin-1 (IL-1), tumor necrosis factor alpha (TNFα), interleukin-6 (IL-6), and interleukin-8 (IL-8) [4]. The cytokines associated with inflammation in OA, primarily interleukin-1 beta (IL-1β) and TNFα, are also implicated in cartilage matrix breakdown [5]. These cytokines induce chondrocytes to produce matrix metalloproteinases

(MMPs) that in turn are responsible for cartilage matrix degradation [6]. These cartilage matrix breakdown products in synovial fluid are thought to increase synovial inflammation [7], creating a positive feed-forward loop that increases inflammation and cartilage breakdown (Figure 1).

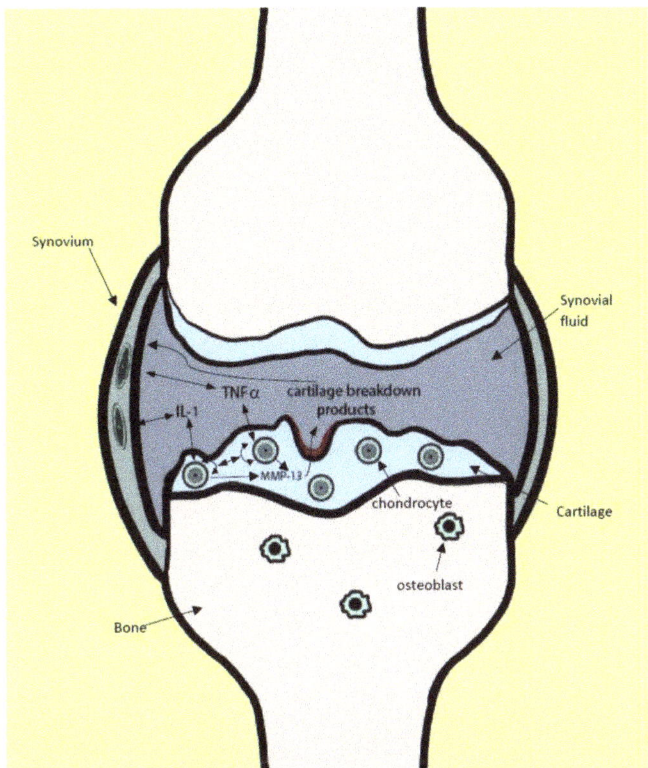

Figure 1. Model of feed-forward loop of pro-inflammatory cytokine-driven OA progression. IL-1β and TNFα bind to chondrocytes and induce expression of MMP-13. MMP-13 causes cartilage matrix breakdown. Cartilage breakdown products inflame cells in the synovium, inducing more production of IL-1β and TNFα.

As IL-1β and TNFα play important roles in inflammation and cartilage breakdown, inhibition of these cytokines may limit inflammation and matrix degradation. Consequently, inhibition of these proteins may constitute an effective OA therapy. The anti-inflammatory cytokine interleukin-1 receptor antagonist (IL-1ra), a competitive IL-1 receptor antagonist, blocks the signaling activity of IL-1 while exhibiting no signal-inducing activity itself [8]. Soluble forms of the IL-1 receptor (sIL-1R) can also bind with IL-1, reducing IL-1 biologic activity by preventing it from binding to surface receptors on the cells [9]. Moreover, soluble forms of the cell receptors for TNFα, known as sTNF-RI and sTNF-RII, can bind to TNFα, preventing TNFα surface receptor binding and thus inhibiting cell signaling [10].

Over the past two decades, preparations from autologous blood or bone marrow to create platelet-rich plasma (PRP), autologous anti-inflammatories (AAIs), and concentrated bone marrow (cBMA or BMAC) have been studied in order to treat osteoarthritis. Each of these formulations are created from similar starting materials by combining plasma, platelets, white blood cells (including early lineages) in various combinations. Each formulation is in on a spectrum of similar mechanistic approaches to treating OA. Regulatory approval to treat OA vary for these formulations depending on the region in the world.

The purpose of this review is to summarize the data that exist for each of these classes of autologous therapies and how each may intervene in the OA disease process.

2. Methods

Search terms in both PubMed database and Google Scholar included "platelet-rich plasma", PRP, osteoarthritis, "cell culture", preclinical, and "mechanism of action" in various combinations. The same combinations were repeated for "concentrated bone marrow aspirate", "cBMA", "BMAC", "autologous anti-inflammatory", and "AAI". Additionally, the authors' own knowledge and work in the field was included when relevant. A focus was made on the specific role that these therapies play in the treatment of OA, and therefore cartilage-specific studies were only included to expand on the potential role that cartilage repair has during treatment of OA.

3. Results

3.1. Platelet-Rich Plasma

Platelet-rich plasma (PRP), or platelet concentrate, is defined as an autologous therapy prepared point of care that contains a concentration of platelets higher than whole blood. This review will focus on PRP prepared by centrifugation that takes advantage of the density differences between the elements in peripheral blood allowing for separation of components. In general, PRP can be divided broadly into two classifications. The first preparation class is created with "slow" centrifugation speeds (e.g., 1500 rpm for 5 min) and have platelet concentrations 2–3 times above baseline collected in blood plasma (platelet-poor plasma; PPP), and with very few red blood cells or white blood cells. These PRPs are yellow in color. The second classification is prepared from "hard" spin cycles (e.g., 3200 rpm for 15 min) resulting in PRPs with platelet concentrations 5–9 above whole blood and typically with white blood cells also above baseline levels. This method typically contains some RBC and are red in color. One popular classification method to discern these two PRPs has focused on the differences in white blood cell concentrations, calling the products with low white blood cell concentrations leukocyte-poor platelet-rich plasma (LP-PRP) and those with higher WBC concentrations as leucocyte-rich platelet-rich plasma (LR-PRP). An evaluation of blood components can be compared by complete blood counts (CBC) (Table 1). Platelet concentrates can also be activated to form a fibrin clot, often referred to as platelet-rich fibrin (PRF). These gel products are often utilized in direct application to a surgical or wound site [11], but are not included in this OA review.

Table 1. Complete blood counts (CBC) of common autologous therapies. (PPP: platelet-poor plasma; LP-PRP:leucocyte-poor platelet-rich plasma; LR-PRP: leucocyte-rich platelet-rich plasma; APS: autologous protein solution; ACS: autologous conditioned serum; BMA: bone marrow aspirate; cBMA: concentrated bone marrow aspirate; BL: below lower limit; NC: not calculated, NM: not measured).

	WBC (k/ul)	PLT (k/ul)	RBC (M/ul)	IL-1ra (pg/mL)	sIL-1RII (pg/mL)	sTNF-RII (pg/mL)	IL-1β (pg/mL)	IL-1ra:IL-1 ratio
Whole blood [12]	5.4 ± 1.8	175 ± 70	5.5 ± 1.1	5665 ± 2318	7135 ± 1766	1125 ± 253	3.4 ± 2.0	4842 ± 2756
PPP [13]	0.1 ±0.0	28 ± 9.3	0.00 ± 0.00	296 ± 141	19,922 ± 2938	3080 ± 635	BL	NC
LP-PRP [13]	1.5 ± 2.0	399 ± 108	0.04 ± 0.06	673 ± 741	15,596 ± 2159	2894 ± 689	BL	NC
LR-PRP [12]	28.1 ± 6.9	1745 ± 439	0.9 ± 0.3	22,395 ± 12,900	NM	NM	3.5 ± 1.0	6369 ± 2321
APS [12]	46.5 ± 14.0	707 ± 444	1.5 ± 1.1	30,853 ± 16,734	20,483 ± 5819	9492 ± 1387	3.8 ± 0.8	8535 ± 3999
ACS [12]	0.0 ± 0.0	14 ± 6	0.0 ± 0.0	1618 ± 675	15,678 ± 2356	2696 ± 679	14.7 ± 14.8	291 ± 256
BMA [14]	22 ± 10	116 ± 30	4.1 ± 0.3	18,110 ± 6681	6768 ± 1995	1292 ± 153	3.0 ± 1.1	6154 ± 1357
cBMA [14]	133 ± 91	885 ± 201	1.3 ± 0.2	73,978 ± 39,464	9814 ± 3199	3932 ± 1301	14.5 ± 11.4	5856 ± 2745

As PRP first became popular in orthopedics, clinical use was focused on both hard (bone) and soft tissues (tendon, muscle, etc.) because the platelets concentrated contain growth factors (PDGF, VEGF, TGF-β1, EGF) that are chemoattractive, proliferative, and angiogenic, therefore, playing a key role in signaling the wound healing cascade. When these mechanisms were translated into treating OA, the initial belief was that the growth factors in platelets would stimulate chondrocyte cell growth in the cartilage, thereby healing cartilage defects. There is ample evidence that PRPs can induce chondrocyte proliferation [15,16] in cell culture. However, chondrocyte proliferation effects have not yet translated into a clinical benefit when treating OA. PRP OA mechanism has evolved to include intervening in the inflammatory process. Interestingly, the factors that most likely contribute anti-inflammatory cytokines in PRP are found in the blood plasma and white blood cells and not in the platelets (Figure 2).

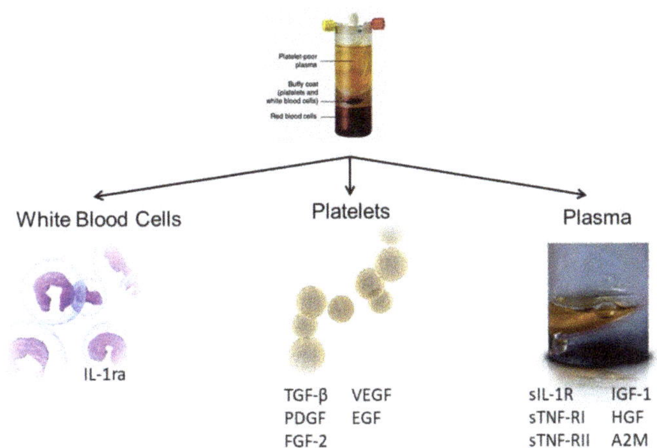

Figure 2. Cytokine contribution from components of blood used to make autologous therapies.

PRP in vitro mechanistic data in OA have focused on the differences between LP-PRP and LR-PRP. The cytokines found in a LP-PRP and LR-PRP reflect their platelet and white blood cell concentration (Figure 1). LP-PRP has lower concentrations of platelet-derived growth factors (PDGF, TGF- β1) than LR-PRP due to lower platelet concentrations (Table 1). LR-PRP has more white blood cell-derived cytokines over LP-PRP that include both pro-inflammatory cytokines (IL-1β, MMP-9) [17] and anti-inflammatory cytokines (IL-1ra) [12]. IL-1ra is produced from monocytes [18] and neutrophils [19]. Therefore, in order for a PRP to have high concentrations of IL-1ra, WBCs need to be included. In order to ensure that the PRP is not pro-inflammatory, the ratio of IL-1β and IL-1ra is important. In a common LR-PRP, the amount of IL-1β is 3.5 ± 1.0 pg/mL while the IL-1ra is 22,395 ± 12,900 pg/mL, giving a ratio between the two of 6369 ± 2321 [12]. In order to evaluate how much IL-1ra is required to block the activity of IL-1β, investigators intravenously injected IL-1β to induce fever in rabbits and then increasing concentrations of the antagonist IL-1ra to determine how much in excess was required. Symptoms were blocked 50% when IL-1ra was injected at 100-fold increase over IL-1β (100 IL-1ra:IL-1β) and were completely blocked by 1000-fold excess of IL-1ra (1000 IL-1ra:IL-1β) [20].

In addition to production of pro-inflammatory cytokines, in cell culture LR-PRP has demonstrated higher levels of synoviocyte cell death and pro-inflammatory cytokine production compared to LP-PRP or PPP [21]. As LR-PRP has higher concentrations of RBC, it has been observed that layers of RBC deposit on the top of a monolayer cell culture, inhibiting nutrient transfer, and could lead to increased cell death and stimulation of pro-inflammatory cytokines. Cell culture techniques with autologous therapies that contain higher red blood cell content can be improved by either suspending the treatment above the monolayer culture with a cell culture insert (Nunc™ Polycarbonate Membrane Inserts, Thermo Scientific™) or by creating a thrombin-activated releasate. Using these techniques, both LP-PRP and LR-PRP can stimulate chondrocyte cell proliferation [15,16].

Additionally, human intra-articular injections of LR-PRP to treat OA has not resulted in increased pro-inflammatory cytokines in synovial fluid or blood plasma [22].

A less lauded component of PRP that contributes anti-inflammatory cytokines is the platelet-poor plasma. All autologous therapies are collected in some volume of blood plasma (Figure 2). Blood plasma contains sIL-1R1, sTNF-RI, sTNF-RII, and A2M [12,23,24]. Interestingly, no matter what autologous therapy, these factors contributed from blood plasma will be present at least in concentrations similar to whole blood, and in some formulations increased levels over whole blood. When examining the anti-inflammatory mechanism of action in OA, the PPP contributes more anti-inflammatory cytokines than platelets. This is a novel approach to understanding the activity of a LP-PRP product in the treatment of OA.

The in vitro positive effects on chondrocytes has translated into demonstrated benefit in animal models of chondral defects [25,26]. PRP has also demonstrated improved histology scores compared to saline in collagenase-induced OA rabbit model [27] and ACL transection rabbit model [28]. As treatment of OA is often focused on the alleviation of pain, in a randomized control trial in dogs with naturally occurring OA, PRP treated dogs had significant decrease in lameness grades compared to saline control subjects [29].

While there has been evidence of intra-articular injections of both LR-PRP and LP-PRP [30,31] providing clinical benefits for patients with knee OA pain, to date, there have been no correlation between any component of the PRP to clinical outcomes. The question still remains whether platelets, white blood cells, or, in fact, the blood plasma, contains the necessary ingredients to treat knee OA pain or to have any disease-modifying effect.

3.2. Autologous Anti-Inflammatory

Autologous anti-inflammatories (AAIs) are autologous blood-derived technologies that focus on concentrating high levels of anti-inflammatory cytokines [32,33]. As the focus on platelet-rich plasma is to concentrate platelets, the AAI designation was created to differentiate from products focused on platelets to those focused on the concentration of anti-inflammatory cytokines. As mentioned earlier, IL-1ra is produced from monocytes and [18] and neutrophils [19]. IL-1ra correlates to the number of WBCs contained in the output [34]. Therefore, AAIs typically utilize strategies to maximize WBC or to collect cytokines produced from WBCs.

Two examples of AAIs are autologous protein solution (APS) and autologous conditioned serum (ACS) [12]. APS is processed through a two-step centrifugation method where anticoagulated peripheral blood is centrifuged to isolate a leukocyte-rich buffy coat suspended in plasma. The buffy coat is mixed with polyacrylamide beads to absorb water; the output is then centrifuged a second time creating a WBC concentrate output suspended in a concentrated PPP [33,35]. APS is designed as a point-of-care single injection OA therapy [36,37]. ACS is produced by incubating venous blood at 37 °C in borosilicate glass-bead containing tubes for 6–24 h. The blood clots and the WBCs expel IL-1ra and other cytokines. Following incubation, tubes are centrifuged and serum is removed, aliquoted into syringes, and stored frozen −20 °C for future multiple injections [32]. The major difference between APS and ACS is that APS output contains WBCs and ACS is cell-free serum collected from incubated WBCs.

The composition of AAIs is designed to inhibit the synergistic feed-forward progression of IL-1β and TNF-α by concentrating their respective antagonists (IL-1ra, sIL-1RII, sTNF-RI, and sTNF-RII) as well as multiple other cytokines and growth factors [32,38] with the goal of intervening in the pro-inflammatory cytokines (IL-1β and TNFα) that cause the degradation of OA [39] (Table 1).

Since AAIs have been specifically created with OA treatment in mind, there is an extensive amount of preclinical data supporting their mechanism of action of intervening in the IL-β and TNFα catabolic pathways described earlier. For example, APS has been shown to reduce MMP-13 production from IL-1β and TNFα-stimulated chondrocytes [35] and decreased IL-8 production from IL-1β-stimulated activated macrophages [40]. AAIs have also demonstrated ability to reduce matrix degradation. In a stimulated equine cartilage and synovium explant co-culture, AAIs significantly downregulated IL-1β expression in cartilage, reduced catabolic cartilage production of PGE2, and downregulated MMP-1 in the synovium [41]. AAIs also upregulated type II collagen and aggrecan expression [41]. AAIs provided chondroprotective effects and decreased matrix degradation in cartilage explants treated with IL-1 and TNFα [38,41].

When translating cell culture experiments into animal models, an intra-articular injection of APS demonstrated cartilage matrix protection by improved cartilage histology scoring compared to animals treated with saline in both a meniscal-tear OA model [42] and IL-1β-induced OA model [43] in athymic rats. Single injections of APS have also demonstrated significant improvement in lameness compared to a single injection of saline in both equine [44] and canine [45] studies.

AAIs have demonstrated early clinical evidence linking cytokine and cellular content and clinical response in patients with OA. Wasai et al. found that the IL-1ra concentration in APS positively correlated with changes in the clinical outcome scores in subjects with knee OA [46]. In a clinical study of OA subjects treated with a single injection of APS, characterization analysis showed 85.7 % of subject's APS had an IL-1ra:IL-1 ratio greater than 1000 or a WBC count greater than 30 k/μL. These subjects were also high OMERACT-OARSI clinical responders six months post-injection (Figure 3) [34,47]. Interestingly, this is the same ratio (1000 IL-1ra:IL-1β) that demonstrated complete blockage of an IL-1β fever-induction model in rabbits [20]. These early results suggest that WBC, and consequently IL-1ra concentrations, can improve outcomes in OA subjects.

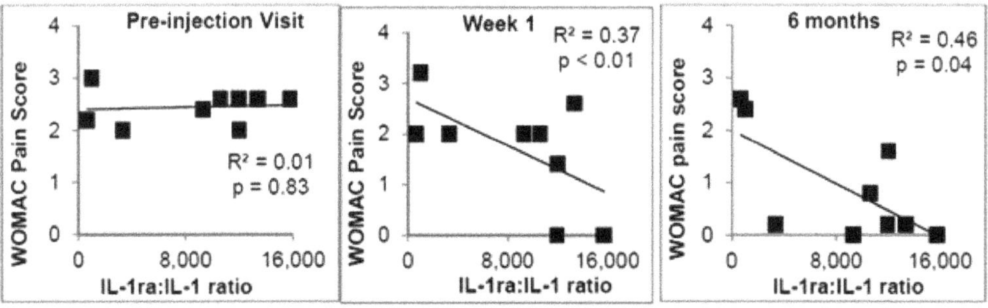

Figure 3. Correlation of WOMAC pain score and ratio of IL-1ra:IL-1β injection in OA subjects [34,47].

Multiple clinical trials in the management of knee OA have been completed using AAIs as they have proven capable of producing high concentrations of anti-inflammatory cytokines from healthy subjects as well as subjects diagnosed with OA [32,33,46] Angadi et al. has recently reviewed and compared the current landscape of AAI clinical evidence and concluded their safety profiles satisfactory for clinical use and presented similar risk profiles of general intra-articular injections [48]. Two knee OA randomized controlled trials of ACS intra-articular injections, Yang et al. and Baltzer et al., demonstrated OA symptom improvements of ACS patients over placebo injections [49,50]. Several studies of a single injection of APS in knee OA have also demonstrated clinical benefits [36,37,51–53].

3.3. Concentrated Bone Marrow Aspirate

Devices that concentrate bone marrow aspirate have been explored to address osteoarthritis in preclinical models [54–58] and clinical trials [59–62]. Understanding what is in cBMA could inform how it might address osteoarthritis. The first proposed MOA for cBMA in treating OA was attributed to "stem cells [63]." However, cBMA contains a low number of stem cells in comparison to culture-expanded stem cell approaches [64]. An emerging understanding of cBMA is that it is a WBC, progenitor cell, and platelet-rich product which enables it to contain a high concentration of anti-inflammatory cytokines and anabolic growth factors [65]. Understanding that cBMA is in the same spectrum of options with PRPs and AAIs will help scientists and clinicians better understand the potential of this option.

The output of cBMA devices is determined by the components of bone marrow, the surgeon's aspiration technique, and the devices used to concentrate the cells and platelets. Bone marrow is a complex tissue made of hematopoietic stem cells (HSCs) which form blood [66], mesenchymal stem cells (MSCs) which form connective tissues [67], endothelial progenitor cells (EPCs) which form blood vessels [68], WBCs of mixed phenotypes, and platelets. Ultimately, the function of bone marrow is to create new blood cells. Stem cells are rare in bone marrow: Caplan has estimated that 1:10,000 cells in bone marrow is an MSC in newborns and that percentage declines with age [69]. It is important to note that the same white blood cells and platelets that are in bone marrow migrate to the blood stream. They are connected and part of the same system [70]. A surgeon's bone marrow aspiration technique impacts what cells are obtained for processing. It is not possible to obtain pure bone marrow without some peripheral blood dilution. Hernigou and colleagues have shown that smaller volume bone marrow draws, smaller volume syringes, and changing aspirate locations can minimize the dilution of bone marrow cells with peripheral blood [71]. Furthermore, different locations from aspiration could vary the cellular content of bone marrow aspirate [72]. While these techniques have been shown to be "best practices", these approaches are considered by some surgeons to be not pragmatic. Therefore, it is likely that many surgeons may be delivering a product that is closer to a blood-derived PRP than cBMA.

Providing bone marrow stem cells (BMSCs) have been one of the rationales proposed as for the intra-articular injection of cBMA for OA. Cultured BMSCs have had beneficial effects in cell-culture models of osteoarthritis including their differentiation into chondrocytes [73] and inhibiting inflammation [74]. BMSCs have had positive effects in a large animal model of OA that was used in support of a larger human program. Specifically, culture-expanded BMSCs stimulated the regeneration of meniscal tissue and slowed the progression of OA in a medial meniscus excision/anterior cruciate resection goat model [75]. However, these large animal positive results did not translate into regulatory approvals in humans. In a Phase I/II clinical study, 55 patients who underwent a partial medial meniscectomy received two doses of allogeneic mesenchymal stem cells and a vehicle control. While the cells showed safety and there was some new meniscal tissue evident in a portion of subjects twelve months post-meniscectomy [76], the study sponsor did not proceed to a confirmatory trial. Currently, there are no approved cultured stem cell therapies for treatment of osteoarthritis in the United States. Potential gaps in translation of these stem cells include a lack of clear phenotypic parameters for the cells, no set mechanism of action to optimize, and a complex manufacturing process [77]. Together, these data and history suggest that if cBMA has a role in addressing OA, then other non-stem cell components must play a significant role.

There has been significant characterization work performed on cBMA. cBMA contains a high concentration of WBC and therefore has a high concentration of IL-1ra [78]. Indeed, there is a strong and significant correlation between the WBC concentration in cBMA with IL-1ra (R^2 = 0.92) [14]. In addition to IL-1ra, cBMA contains high concentrations of other anti-inflammatory cytokines like sIL-1RII, sTNF-RI, and sTNF-RII and low concentrations of inflammatory cytokines like IL-1β and TNFα (Table 1). Interestingly, cBMA is the only autologous therapy where the ratio of IL-1ra:IL-1β decreases in the concentrated version over the starting material (Table 1). While still very low, this is due to an increase in IL-1β by almost 4-fold (3.0 ± 1.1 pg/mL in BMA and 14.5 ± 11.4 pg/mL in cBMA). However, cBMA has by far the most IL-1ra concentration of any autologous blood product with an average of 73,978± 39,464 pg/mL.

Just as with PRP, cBMA has significant concentrations of anabolic and angiogenic growth factors such as TGF-β1, PDGF-AB, PDGF-BB, and EGF [14]. These trophic factors are also secreted from cultured mesenchymal stem cells in culture through extracellular vesicles (EVs). EVs could be a method to utilize the signal factors from allogenic cell sources such as umbilical cord or Wharton's jelly. EVs from MSCs have been shown to be chondroprotective and anti-inflammatory in cell culture, improve histology scores in both RA and OA animal models, and demonstrates early promise in Phase I/II clinical studies [79].

In addition to IL-1ra, the WBC concentration in cBMA is also significantly correlated with colony forming units–fibroblasts (CFU-F) [80], which is a surrogate marker for BM-SCs [81]. cBMA has other progenitor cells including concentrated HSCs and EPCs [82], whose intra-articular role in osteoarthritis has not been extensively characterized. These separate anti-inflammatory and pro-angiogenic properties of cBMA could have differential effects when its injected in different locations.

cBMA is delivered to an OA joint via intra-articular injection. However, cBMA has been used in cartilage repair techniques in combination with varying biomaterials and drug delivery systems including hydrogels and microspheres, and in combination with microfracture surgery. These delivery techniques are intended to increase their residence time and bioavailability of the bioactive factors released from the cells [83]. These approaches have produced durable pain relief in several single-arm studies [84–86]. The intra-articular injection of cBMA for treatment of OA has produced mixed results in randomized and controlled trials. For example, in one study 25 patients with bilateral knee pain received injections of saline in one knee and cBMA in the contralateral knee. Clinical improvements were seen in both knees, but there wasn't a statistical difference between saline and cBMA [59]. In another study, 90 patients with symptomatic knee OA received intra-articular injection of either PRP or cBMA. Both groups had clinically improved symptoms 12 months post-injection, but there were not significant differences between groups. To date, clinical evidence has demonstrated relief of OA symptoms but has not demonstrated reliably that they regrow lost cartilage tissue in OA.

A recent model of OA has been put forward which conceptualizes the "joint as an organ." Subchondral bone plays an important role in OA. Indeed, changes in subchondral bone can alter the function and pain of the whole joint [87–89]. Bone marrow lesions are hypothesized to originate from a traumatic event and have been shown to correlate with pain [90–92]. The proposed mechanism of action for cBMA in these lesions could be its angiogenic molecules and cells recruiting new blood vessels and bone turnover in the lesion, restoring bone health [93]. In a canine subchondral bone lesion model, cBMA and calcium phosphate-injected lesions enhanced knee range of motion and limb loading through improved trabecular bone remodeling [94]. Direct injection of cBMA into an avascular necrosis defect along with decompression has also demonstrated some promise [95]. This general approach of delivering cBMA with or without a carrier has been explored in clinical case series [96–98]. However, further clinical evidence would likely be required.

4. Conclusions

There is significant interest in autologous therapies to treat OA patients, both to alleviate symptoms and to potentially delay progression of the disease. While there has been a focus on the differences between these therapies, particularly the presence or absences of white blood cells, the cell culture and animal studies suggest more in common than not. As we examine the mechanistic evidence of each of the therapies, they all have in common the production of anti-inflammatory cytokines at varying concentrations that can be shown both in cell culture and in animal models to delay disease progression. If autologous therapies' mechanism of action is via modifying the local inflammatory environment in the joint, and PRP, AAIs, and cBMA have significant concentrations of anti-inflammatory cytokines, this could explain why all three therapies have demonstrated some level of clinical efficacy. It may be true that the proliferative cytokines derived from platelets, and found in all three therapies, may also play a role in OA, though beneficial effects have been demonstrated in preclinical models but not yet in a clinical setting.

To date, clinical evidence suggests that they can all alleviate symptoms from patients with OA, but have not yet definitively demonstrated disease modification. Correlations between autologous therapy content and clinical outcomes have been the holy grail in autologous therapies. There is early evidence that potentially the IL-1ra:IL-1b ratio in the output of the therapy may play a role. This fact challenges the dogma that white blood cells, whether sourced from peripheral blood or from bone marrow, should be eliminated from autologous therapies altogether in the treatment of OA. Further studies will be needed to confirm these hypotheses explored in this review. Ultimately, convincing clinical evidence of disease modification will require long-term studies that utilize imaging or other surrogate markers including biomarkers or delay of total joint replacement.

Author Contributions: J.W.-M., K.S., and W.K. drafted, reviewed, and edited this manuscript. All authors have read and agreed to the published version of the manuscript.

Funding: This review received no external funding.

Institutional Review Board Statement: Not applicable.

Informed Consent Statement: Not applicable.

Data Availability Statement: The data presented in this study are available on request from the corresponding author.

Conflicts of Interest: J.W.-M. and K.S. are paid employees of Zimmer Biomet and W.K. is a paid employee of Owl Manor.

References

1. Aigner, T.; Schmitz, N. Pathogenesis and pathology of osteoarthritis. *Rheumatology* **2011**, *7*, 1741–1759.
2. Lequesne, M.G.; Dang, N.; Lane, N.E. Sport practice and osteoarthritis of the limbs. *Osteoarthr. Cartil.* **1997**, *5*, 75–86. [CrossRef]
3. Brzusek, D.; Petron, D. Treating knee osteoarthritis with intra-articular hyaluronans. *Curr. Med. Res. Opin.* **2008**, *24*, 3307–3322. [CrossRef]
4. Fernandes, J.C.; Martel-Pelletier, J.; Pelletier, J.P. The role of cytokines in osteoarthritis pathophysiology. *Biorheology* **2002**, *39*, 237–246. [PubMed]
5. Goldring, M.B.; Otero, M.; Tsuchimochi, K.; Ijiri, K.; Li, Y. Defining the roles of inflammatory and anabolic cytokines in cartilage metabolism. *Ann. Rheum. Dis.* **2008**, *67* (Suppl. 3), iii75–iii82. [CrossRef] [PubMed]
6. Shlopov, B.V.; Gumanovskaya, M.L.; Hasty, K.A. Autocrine regulation of collagenase 3 (matrix metalloproteinase 13) during osteoarthritis. *Arthritis Rheumatol.* **2000**, *43*, 195–205. [CrossRef]
7. Goldring, S.R.; Goldring, M.B. The role of cytokines in cartilage matrix degeneration in osteoarthritis. *Clin. Orthop. Relat. Res.* **2004**, *427*, S27–S36. [CrossRef] [PubMed]
8. Dinarello, C.A. The role of the interleukin-1-receptor antagonist in blocking inflammation mediated by interleukin-1. *N. Eng. J. Med.* **2000**, *343*, 732–734. [CrossRef]
9. Symons, J.A.; Young, P.R.; Duff, G.W. Soluble type II interleukin 1 (IL-1) receptor binds and blocks processing of IL-1 beta precursor and loses affinity for IL-1 receptor antagonist. *Proc. Natl. Acad. Sci. USA* **1995**, *92*, 1714–1718. [CrossRef]
10. Van Zee, K.J.; Kohno, T.; Fischer, E.; Rock, C.S.; Moldawer, L.L.; Lowry, S.F. Tumor necrosis factor soluble receptors circulate during experimental and clinical inflammation and can protect against excessive tumor necrosis factor alpha in vitro and in vivo. *Proc. Natl. Acad. Sci. USA* **1992**, *89*, 4845–4849. [CrossRef]
11. Giudice, A.; Esposito, M.; Bennardo, F.; Brancaccio, Y.; Buti, J.; Fortunato, L. Dental extractions for patients on oral antiplatelet: A within-person randomised controlled trial comparing haemostatic plugs, advanced-platelet-rich fibrin (A-PRF+) plugs, leukocyte- and platelet-rich fibrin (L-PRF) plugs and suturing alone. *Int. J. Oral Implant.* **2019**, *12*, 77–87.
12. King, W.J.; Steckbeck, K.; O'Shaughnessey, K.M.; Woodell-May, J. *Effect of Preparation Technique on Anti-Inflammatory Properties of Autologous Therapies*; Orthopaedic Research Society: Las Vegas, NV, USA, 2015.
13. Steckbeck, K.; Woodell-May, J. *Characterization Comparison of Patient-Matched Autologous Therapy Product Concentrations*; Orthopaedic Research Society: Phoenix, AZ, USA, 2020.
14. King, W.; Tan, M.; Ponticiello, M.; Woodell-May, J. *Anti-Inflammatory Properties of the Output of an Autologous Bone Marrow Concentrating Device*; Orthopaedic Research Society: Orlando, FL, USA, 2016.
15. Akeda, K.; An, H.S.; Okuma, M.; Attawia, M.; Miyamoto, K.; Thonar, E.J.; Lenz, M.E.; Sah, R.L.; Masuda, K. Platelet-Rich Plasma Stimulates Porcine Articular Chondrocyte Proliferation and Matrix Biosynthesis. *Osteoarthr. Cartil.* **2006**, *14*, 1272–1280. [CrossRef]
16. Cavallo, C.; Filardo, G.; Mariani, E.; Kon, E.; Marcacci, M.; Pereira Ruiz, M.T.; Facchini, A.; Grigolo, B. Comparison of platelet-rich plasma formulations for cartilage healing: An in vitro study. *J. Bone Jt. Surg. Am.* **2014**, *96*, 423–429. [CrossRef]

17. Sundman, E.A.; Cole, B.J.; Fortier, L.A. Growth factor and catabolic cytokine concentrations are influenced by the cellular composition of platelet-rich plasma. *Am. J. Sports Med.* **2011**, *39*, 2135–2140. [CrossRef] [PubMed]
18. Arend, W.P.; Malyak, M.; Guthridge, C.J.; Gabay, C. Interleukin-1 receptor antagonist: Role in biology. *Annu. Rev. Immunol.* **1998**, *16*, 27–55. [CrossRef]
19. McColl, S.R.; Paquin, R.; Menard, C.; Beaulieu, A.D. Human neutrophils produce high levels of the interleukin 1 receptor antagonist in response to granulocyte/macrophage colony-stimulating factor and tumor necrosis factor alpha. *J. Exp. Med.* **1992**, *176*, 593–598. [CrossRef]
20. Dinarello, C.A.; Thompson, R.C. Blocking IL-1: Interleukin 1 receptor antagonist in vivo and in vitro. *Immunol. Today* **1991**, *12*, 404–410. [CrossRef]
21. Braun, H.J.; Kim, H.J.; Chu, C.R.; Dragoo, J.L. The effect of platelet-rich plasma formulations and blood products on human synoviocytes implications for intra-articular injury and therapy. *Am. J. Sports Med.* **2014**, *42*, 1204–1210. [CrossRef]
22. Mariani, E.; Canella, V.; Cattini, L.; Kon, E.; Marcacci, M.; Di Matteo, B.; Pulsatelli, L.; Filardo, G. Leukocyte-Rich Platelet-Rich Plasma Injections Do Not. Up-Modulate Intra-Articular Pro-Inflammatory Cytokines in the Osteoarthritic Knee. *PLoS ONE* **2016**, *11*, e0156137. [CrossRef]
23. Arend, W.P. The mode of action of cytokine inhibitors. *J. Rheumatol. Suppl.* **2002**, *65*, 16–21. [PubMed]
24. Cuéllar, J.M.; Cuéllar, V.G.; Scuderi, G.J. α(2)-Macroglobulin: Autologous Protease Inhibition Technology. *Phys. Med. Rehabil. Clin. N. Am.* **2016**, *27*, 909–918. [CrossRef]
25. Milano, G.; Deriu, L.; Sanna, P.E.; Masala, G.; Manunta, A.; Postacchini, R.; Saccomanno, M.F.; Fabbriciani, C. Repeated platelet concentrate injections enhance reparative response of microfractures in the treatment of chondral defects of the knee: An experimental study in an animal model. *Arthroscopy* **2012**, *28*, 688–701. [CrossRef]
26. Sun, Y.; Feng, Y.; Zhang, C.Q.; Chen, S.B.; Cheng, X.G. The regenerative effect of platelet-rich plasma on healing in large osteochondral defects. *Int. Orthop.* **2010**, *34*, 589–597. [CrossRef]
27. Kwon, D.R.; Park, G.Y.; Lee, S.U. The effects of intra-articular platelet-rich plasma injection according to the severity of collagenase-induced knee osteoarthritis in a rabbit model. *Ann. Rehabil. Med.* **2012**, *36*, 458–465. [CrossRef]
28. Saito, M.; Takahashi, K.A.; Arai, Y.; Inoue, A.; Sakao, K.; Tonomura, H.; Honjo, K.; Nakagawa, S.; Inoue, H.; Tabata, Y.; et al. Intraarticular administration of platelet-rich plasma with biodegradable gelatin hydrogel microspheres prevents osteoarthritis progression in the rabbit knee. *Clin. Exp. Rheumatol.* **2009**, *27*, 201–207.
29. Fahie, M.A.; Ortolano, G.A.; Guercio, V.; Schaffer, J.A.; Johnston, G.; Au, J.; Hettlich, B.A.; Phillips, T.; Allen, M.J.; Bertone, A.L. A randomized controlled trial of the efficacy of autologous platelet therapy for the treatment of osteoarthritis in dogs. *J. Am. Vet. Med. Assoc.* **2013**, *243*, 1291–1297. [CrossRef] [PubMed]
30. Cole, B.J.; Karas, V.; Hussey, K.; Merkow, D.B.; Pilz, K.; Fortier, L.A. Hyaluronic Acid Versus Platelet-Rich Plasma: A Prospective, Double-Blind. Randomized Controlled Trial Comparing Clinical Outcomes and Effects on Intra-articular Biology for the Treatment of Knee Osteoarthritis. *Am. J. Sports Med.* **2017**, *45*, 339–346. [CrossRef] [PubMed]
31. Filardo, G.; Kon, E.; Pereira Ruiz, M.T.; Vaccaro, F.; Guitaldi, R.; Di, M.A.; Cenacchi, A.; Fornasari, P.M.; Marcacci, M. Platelet-rich plasma intra-articular injections for cartilage degeneration and osteoarthritis: Single- versus double-spinning approach. *Knee Surg. Sports Traumatol. Arthrosc.* **2011**, *20*, 2082–2091. [CrossRef]
32. Wehling, P.; Moser, C.; Frisbie, D.; McIlwraith, C.W.; Kawcak, C.E.; Krauspe, R.; Reinecke, J.A. Autologous conditioned serum in the treatment of orthopedic diseases: The orthokine therapy. *BioDrugs* **2007**, *21*, 323–332. [CrossRef]
33. O'Shaughnessey, K.; Matuska, A.; Hoeppner, J.; Farr, J.; Klaassen, M.; Kaeding, C.; Lattermann, C.; King, W.; Woodell-May, J. Autologous protein solution prepared from the blood of osteoarthritic patients contains an enhanced profile of anti-inflammatory cytokines and anabolic growth factors. *J. Orthop. Res.* **2014**, *32*, 1349–1355. [CrossRef]
34. King, W.J.; Toler, K.; Woodell-May, J. White Blood Cell Concentration Correlates with Increased Concentrations of IL-1ra and Changes in WOMAC Pain Scores in an Open-Label. Safety Study of Autologous Protein Solution. *Exp. Orthop.* **2016**, *3*, 9. [CrossRef] [PubMed]
35. Woodell-May, J.E.; Matuska, A.; Oyster, M.; Welch, Z.; O'Shaughnessey, K.M.; Hoeppner, J. Autologous protein solution inhibits MMP-13 production by IL-1beta and TNFalpha-stimulated human articular chondrocytes. *J. Orthop. Res.* **2011**, *29*, 1320–1326. [CrossRef] [PubMed]
36. Kon, E.; Engebretsen, L.; Verdonk, P.; Nehrer, S.; Filardo, G. Clinical outcomes of knee osteoarthritis treated with an autologous protein solution injection: A 1-year pilot double-blinded randomized controlled trial. *Am. J. Sports Med.* **2018**, *46*, 171–180. [CrossRef]
37. Hix, J.; Klaassen, M.; Foreman, R.; Cullen, E.; Toler, K.; King, W.; Woodell-May, J.A. Autologous Anti-Inflammatory Protein Solution Yielded a Favorable Safety Profile and Significant Pain Relief in an Open-Label. Pilot Study of Patients with Osteoarthritis. *Biores. Open Access.* **2017**, *6*, 151–158. [CrossRef] [PubMed]
38. Matuska, A.; O'Shaughnessey, K.; King, W.; Woodell-May, J. Autologous solution protects bovine cartilage explants from IL-1α-and TNFα-induced cartilage degradation. *J. Orthop. Res.* **2013**, *31*, 1929–1935. [CrossRef] [PubMed]
39. Wojdasiewicz, P.A.; Poniatowski, Ł.; Szukiewicz, D. The role of inflammatory and anti-inflammatory cytokines in the pathogenesis of osteoarthritis. *Mediat. Inflamm.* **2014**, *2014*, 561459. [CrossRef]
40. O'Shaughnessey, K.M.; Panitch, A.; Woodell-May, J.E. Blood-derived anti-inflammatory protein solution blocks the effect of IL-1b on human macrophages in vitro. *Inflamm. Res.* **2011**, *60*, 929–936. [CrossRef]

41. Velloso Alvarez, A.; Boone, L.H.; Pondugula, S.R.; Caldwell, F.; Wooldridge, A.A. Effects of Autologous Conditioned Serum, Autologous Protein Solution, and Triamcinolone on Inflammatory and Catabolic Gene Expression in Equine Cartilage and Synovial Explants Treated With IL-1β in Co-culture. *Front. Vet. Sci.* **2020**, *7*, 323. [CrossRef]

42. King, W.; Bendele, A.; Marohl, T.; Woodell-May, J. Human blood-based anti-inflammatory solution inhibits osteoarthritis progression in a meniscal-tear rat study. *J. Orthop. Res.* **2017**, *35*, 2260–2268. [CrossRef]

43. King, W.J.; Han, B.; Woodell-May, J. *Autologous Protein Solution Inhibits Osteoarthritis Disease Progression in a IL-1β–Induced Animal Model*; Orthopaedic Research Society Meeting: San Diego, CA, USA, 2017.

44. Bertone, A.L.; Ishihara, A.; Zekas, L.J.; Wellman, M.L.; Lewis, K.B.; R, S.; Barnaba, A.; Schmall, M.L.; Kanter, P.M.; Genovese, R.L. Evaluation of a single intra-articular injection of autologous protein solution for treatment of osteoarthritis in horses. *Am. J. Vet. Res.* **2014**, *75*, 141–151. [CrossRef]

45. Wanstrath, A.; Hettlich, B.A.; Su, L.; Smith, A.; Zekas, L.J.; Allen, A.J.; Bertone, A.L. Evaluation of a single intra-articular injection of autologous protein solution for treatment of osteoarthritis in a canine population. *Vet. Surg.* **2016**, *45*, 764–774. [CrossRef] [PubMed]

46. Wasai, S.; Sato, M.; Maehara, M.; Toyoda, E.; Uchiyama, R.; Takahashi, T.; Okada, E.; Iwasaki, Y.; Suzuki, S.; Watanabe, M. Characteristics of autologous protein solution and leucocyte-poor platelet-rich plasma for the treatment of osteoarthritis of the knee. *Sci. Rep.* **2020**, *10*, 10572. [CrossRef] [PubMed]

47. King, W.J.; VanDerWeegen, W.; Van Drumpt, R.; Soons, H.; Toler, K.; Woodell-May, J.E. Characterizing the Relationship between White Blood Cell and IL-1ra Concentration in Whole Blood and Decreased Osteoarthritis Pain in an Open-Label. Study of Autologous Protein Solution. In Proceedings of the European Federation of National Associations of Orthopaedics & Traumatology Congress, 16th Annual Meeting, Prague, Czech Republic, 27–29 May 2015.

48. Darshan, S.A.; Navraj Atwal, H.M. Autologous cell-free serum preparations in the management of knee osteoarthritis: What is the current clinical evidence? *Knee Surg. Relat. Res.* **2020**, *32*, 1–10.

49. Yang, K.G.; Raijmakers, N.J.; van Arkel, E.R.; Caron, J.J.; Rijk, P.C.; Willems, W.J.; Zijl, J.A.; Verbout, A.J.; Dhert, W.J.; Saris, D.B. Autologous interleukin-1 receptor antagonist improves function and symptoms in osteoarthritis when compared to placebo in a prospective randomized controlled trial. *Osteoarthr. Cartil.* **2008**, *16*, 498–505. [CrossRef] [PubMed]

50. Baltzer, A.W.; Moser, C.; Jansen, S.A.; Krauspe, R. Autologous conditioned serum (Orthokine) is an effective treatment for knee osteoarthritis. *Osteoarthr. Cartil.* **2008**, *17*, 152–160. [CrossRef] [PubMed]

51. Van Drumpt, R.A.M.; van Der Weegen, W.; King, W.J.; Toler, K.; Macenski, M.M. Safety and Treatment Effectiveness of a Single Autologous Protein Solution Injection in Patients with Knee Osteoarthritis. *Biores. Open Access* **2016**, *5*, 261–268. [CrossRef] [PubMed]

52. Van Genechten, W.; Vuylsteke, K.; Swinnen, L.; Martinez, P.R.; Verdonk, P. Autologous protein solution as selective treatment for advanced patellofemoral osteoarthritis in the middle-aged female patient: 54% response rate at 1 year follow-up. *Knee Surg. Sports Traumatol. Arthrosc.* **2021**, *29*, 988–997. [CrossRef]

53. Kon, E.; Engebretsen, L.; Verdonk, P.; Nehrer, S.; Filardo, G. Autologous Protein Solution Injections for the Treatment of Knee Osteoarthritis: 3-Year Results. *Am. J. Sports Med.* **2020**, *48*, 2703–2710. [CrossRef] [PubMed]

54. Jagodzinski, M.; Liu, C.; Guenther, D.; Burssens, A.; Petri, M.; Abedian, R.; Willbold, E.; Krettek, C.; Haasper, C.; Witte, F. Bone marrow-derived cell concentrates have limited effects on osteochondral reconstructions in the mini pig. *Tissue Eng. Part C Methods* **2014**, *20*, 215–226. [CrossRef]

55. Desando, G.; Bartolotti, I.; Cavallo, C.; Schiavinato, A.; Secchieri, C.; Kon, E.; Filardo, G.; Paro, M.; Grigolo, B. Short-term homing of hyaluronan-primed cells: Therapeutic implications for osteoarthritis treatment. *Tissue Eng. Part C Methods* **2018**, *24*, 121–133. [CrossRef]

56. Wang, Z.; Zhai, C.; Fei, H.; Hu, J.; Cui, W.; Wang, Z.; Li, Z.; Fan, W. Intraarticular injection autologous platelet-rich plasma and bone marrow concentrate in a goat osteoarthritis model. *J. Orthop. Res.* **2018**, *36*, 2140–2146. [CrossRef]

57. Song, F.; Tang, J.; Geng, R.; Hu, H.; Zhu, C.; Cui, W.; Fan, W. Comparison of the efficacy of bone marrow mononuclear cells and bone mesenchymal stem cells in the treatment of osteoarthritis in a sheep model. *Int. J. Clin. Exp. Pathol.* **2014**, *7*, 1415. [PubMed]

58. Singh, A.; Goel, S.; Gupta, K.; Kumar, M.; Arun, G.; Patil, H.; Kumaraswamy, V.; Jha, S. The role of stem cells in osteoarthritis: An experimental study in rabbits. *Bone Jt. Res.* **2014**, *3*, 32–37. [CrossRef] [PubMed]

59. Shapiro, S.A.; Kazmerchak, S.E.; Heckman, M.G.; Zubair, A.C.; O'Connor, M.I. A prospective, single-blind, placebo-controlled trial of bone marrow aspirate concentrate for knee osteoarthritis. *Am. J. Sports Med.* **2017**, *45*, 82–90. [CrossRef] [PubMed]

60. Chahla, J.; Dean, C.S.; Moatshe, G.; Pascual-Garrido, C.; Serra Cruz, R.; LaPrade, R.F. Concentrated bone marrow aspirate for the treatment of chondral injuries and osteoarthritis of the knee: A systematic review of outcomes. *Orthop. J. Sports Med.* **2016**, *4*, 2325967115625481. [CrossRef]

61. Mautner, K.; Bowers, R.; Easley, K.; Fausel, Z.; Robinson, R. Functional Outcomes Following Microfragmented Adipose Tissue Versus Bone Marrow Aspirate Concentrate Injections for Symptomatic Knee Osteoarthritis. *Stem Cells Transl. Med.* **2019**, *8*, 1149–1156. [CrossRef] [PubMed]

62. Themistocleous, G.S.; Chloros, G.D.; Kyrantzoulis, I.M.; Georgokostas, I.A.; Themistocleous, M.S.; Papagelopoulos, P.J.; Savvidou, O.D. Effectiveness of a single intra-articular bone marrow aspirate concentrate (BMAC) injection in patients with grade 3 and 4 knee osteoarthritis. *Heliyon* **2018**, *4*, e00871. [CrossRef]

63. Hernigou, P.; Bouthors, C.; Bastard, C.; Lachaniette, C.H.F.; Rouard, H.; Dubory, A. Subchondral bone or intra-articular injection of bone marrow concentrate mesenchymal stem cells in bilateral knee osteoarthritis: What better postpone knee arthroplasty at fifteen years? A randomized study. *Int. Orthop.* **2020**, *45*, 391–399. [CrossRef] [PubMed]

64. Ha, C.-W.; Park, Y.-B. Editorial Commentary: Considering Clinical Application of Bone Marrow Aspirate Concentrate for Restoration of Cartilage Defects in the Knee? Is It a Kind of Stem Cell Therapy? *Arthroscopy* **2019**, *35*, 1878–1879. [CrossRef]

65. King, W.; Toler, K.; Woodell-May, J. Role of White Blood Cells in Blood-and Bone Marrow-Based Autologous Therapies. *Biomed. Res. Int.* **2018**, *2018*. [CrossRef]

66. Cheshier, S.H.; Morrison, S.J.; Liao, X.; Weissman, I.L. In vivo proliferation and cell cycle kinetics of long-term self-renewing hematopoietic stem cells. *Proc. Natl. Acad. Sci. USA* **1999**, *96*, 3120–3125. [CrossRef]

67. Pittenger, M.F.; Mackay, A.M.; Beck, S.C.; Jaiswal, R.K.; Douglas, R.; Mosca, J.D.; Moorman, M.A.; Simonetti, D.W.; Craig, S.; Marshak, D.R. Multilineage potential of adult human mesenchymal stem cells. *Science* **1999**, *284*, 143–147. [CrossRef]

68. Asahara, T.; Masuda, H.; Takahashi, T.; Kalka, C.; Pastore, C.; Silver, M.; Kearne, M.; Magner, M.; Isner, J.M. Bone marrow origin of endothelial progenitor cells responsible for postnatal vasculogenesis in physiological and pathological neovascularization. *Circ. Res.* **1999**, *85*, 221–228. [CrossRef] [PubMed]

69. Caplan, A.I. Mesenchymal stem cells: Cell–based reconstructive therapy in orthopedics. *Tissue Eng.* **2005**, *11*, 1198–1211. [CrossRef] [PubMed]

70. Sacchetti, B.; Funari, A.; Michienzi, S.; Di Cesare, S.; Piersanti, S.; Saggio, I.; Tagliafico, E.; Ferrari, S.; Robey, P.G.; Riminucci, M. Self-renewing osteoprogenitors in bone marrow sinusoids can organize a hematopoietic microenvironment. *Cell* **2007**, *131*, 324–336. [CrossRef]

71. Hernigou, P.; Homma, Y.; Lachaniette, C.H.F.; Poignard, A.; Allain, J.; Chevallier, N.; Rouard, H. Benefits of small volume and small syringe for bone marrow aspirations of mesenchymal stem cells. *Int. Orthop.* **2013**, *37*, 2279–2287. [CrossRef]

72. Marx, R.E.; Tursun, R. A qualitative and quantitative analysis of autologous human multipotent adult stem cells derived from three anatomic areas by marrow aspiration: Tibia, anterior ilium, and posterior ilium. *Int. J. Oral Maxillofac. Implant.* **2013**, *28*, e290–e294. [CrossRef]

73. Richardson, S.M.; Curran, J.M.; Chen, R.; Vaughan-Thomas, A.; Hunt, J.A.; Freemont, A.J.; Hoyland, J.A. The differentiation of bone marrow mesenchymal stem cells into chondrocyte-like cells on poly-L-lactic acid (PLLA) scaffolds. *Biomaterials* **2006**, *27*, 4069–4078. [CrossRef] [PubMed]

74. Van Buul, G.; Villafuertes, E.; Bos, P.; Waarsing, J.; Kops, N.; Narcisi, R.; Weinans, H.; Verhaar, J.; Bernsen, M.; Van Osch, G. Mesenchymal stem cells secrete factors that inhibit inflammatory processes in short-term osteoarthritic synovium and cartilage explant culture. *Osteoarthr. Cartil.* **2012**, *20*, 1186–1196. [CrossRef] [PubMed]

75. Murphy, J.M.; Fink, D.J.; Hunziker, E.B.; Barry, F.P. Stem cell therapy in a caprine model of osteoarthritis. *Arthritis Rheum. Off. J. Am. Coll. Rheumatol.* **2003**, *48*, 3464–3474. [CrossRef]

76. Vangsness, C.T., Jr.; Farr, J.I.; Boyd, J.; Dellaero, D.T.; Mills, C.R.; LeRoux-Williams, M. Adult human mesenchymal stem cells delivered via intra-articular injection to the knee following partial medial meniscectomy: A randomized, double-blind, controlled study. *JBJS* **2014**, *96*, 90–98. [CrossRef] [PubMed]

77. Barry, F. MSC therapy for osteoarthritis: An unfinished story. *J. Orthop. Res.* **2019**, *37*, 1229–1235. [CrossRef]

78. Cassano, J.M.; Kennedy, J.G.; Ross, K.A.; Fraser, E.J.; Goodale, M.B.; Fortier, L.A. Bone marrow concentrate and platelet-rich plasma differ in cell distribution and interleukin 1 receptor antagonist protein concentration. *Knee Surg. Sports Traumatol. Arthrosc.* **2018**, *26*, 333–342. [CrossRef] [PubMed]

79. Arrigoni, C.; D'Arrigo, D.; Rossella, V.; Candrian, C.; Albertini, V.; Moretti, M. Umbilical Cord MSCs and Their Secretome in the Therapy of Arthritic Diseases: A Research and Industrial Perspective. *Cells* **2020**, *9*, 1343. [CrossRef]

80. Han, B.; Tan, M.; King, W.; Woodell-May, J. White blood cell concentration correlates with CFU-F concentration in the output of a point-of-care bone marrow concentrating device. In *TERMIS*; European Cells and Materials: Uppsala, Sweden, 2016.

81. Friedenstein, A.J.; Gorskaja, J.; Kulagina, N. Fibroblast precursors in normal and irradiated mouse hematopoietic organs. *Exp. Hematol.* **1976**, *4*, 267.

82. Woodell-May, J.E.; Tan, M.L.; King, W.J.; Swift, M.J.; Welch, Z.R.; Murphy, M.P.; McKale, J.M. Characterization of the cellular output of a point-of-care device and the implications for addressing critical limb ischemia. *Biores. Open Access* **2015**, *4*, 417–424. [CrossRef] [PubMed]

83. Patel, J.M.; Saleh, K.S.; Burdick, J.A.; Mauck, R.L. Bioactive factors for cartilage repair and regeneration: Improving delivery, retention, and activity. *Acta Biomater.* **2019**, *93*, 222–238. [CrossRef]

84. Skowroński, J.; Rutka, M. Osteochondral lesions of the knee reconstructed with mesenchymal stem cells-results. *Ortop. Traumatol. Rehabil.* **2013**, *15*, 195–204. [CrossRef]

85. De Girolamo, L.; Schönhuber, H.; Viganò, M.; Bait, C.; Quaglia, A.; Thiebat, G.; Volpi, P. Autologous matrix-induced chondrogenesis (amic) and amic enhanced by autologous concentrated bone marrow aspirate (bmac) allow for stable clinical and functional improvements at up to 9 years follow-up: Results from a randomized controlled study. *J. Clin. Med.* **2019**, *8*, 392. [CrossRef]

86. Enea, D.; Cecconi, S.; Calcagno, S.; Busilacchi, A.; Manzotti, S.; Gigante, A. One-step cartilage repair in the knee: Collagen-covered microfracture and autologous bone marrow concentrate. A pilot study. *Knee* **2015**, *22*, 30–35. [CrossRef]

87. Radin, E.L.; Rose, R.M. Role of subchondral bone in the initiation and progression of cartilage damage. *Clin. Orthop. Relat. Res.* **1986**, *213*, 34–40. [CrossRef]

88. Madry, H.; van Dijk, C.N.; Mueller-Gerbl, M. The basic science of the subchondral bone. *Knee Surg. Sports Traumatol. Arthrosc.* **2010**, *18*, 419–433. [CrossRef] [PubMed]
89. Buckwalter, J. Articular cartilage injuries. *Clin. Orthop. Relat. Res.* **2002**, *402*, 21–37. [CrossRef]
90. Koo, K.-H.; Ahn, I.-O.; Kim, R.; Song, H.-R.; Jeong, S.-T.; Na, J.-B.; Kim, Y.-S.; Cho, S.-H. Bone marrow edema and associated pain in early stage osteonecrosis of the femoral head: Prospective study with serial MR images. *Radiology* **1999**, *213*, 715–722. [CrossRef]
91. Felson, D.T.; Chaisson, C.E.; Hill, C.L.; Totterman, S.M.; Gale, M.E.; Skinner, K.M.; Kazis, L.; Gale, D.R. The association of bone marrow lesions with pain in knee osteoarthritis. *Ann. Intern. Med.* **2001**, *134*, 541–549. [CrossRef] [PubMed]
92. Felson, D.T.; Niu, J.; Guermazi, A.; Roemer, F.; Aliabadi, P.; Clancy, M.; Torner, J.; Lewis, C.E.; Nevitt, M.C. Correlation of the development of knee pain with enlarging bone marrow lesions on magnetic resonance imaging. *Arthritis Rheum. Off. J. Am. Coll. Rheumatol.* **2007**, *56*, 2986–2992. [CrossRef] [PubMed]
93. Stoker, A.M.; Baumann, C.A.; Stannard, J.P.; Cook, J.L. Bone marrow aspirate concentrate versus platelet rich plasma to enhance osseous integration potential for osteochondral allografts. *J. Knee Surg.* **2018**, *31*, 314–320.
94. Oliver, H.A.; Bozynski, C.C.; Cook, C.R.; Kuroki, K.; Sherman, S.L.; Stoker, A.M.; Cook, J.L. Enhanced Subchondroplasty Treatment for Post-Traumatic Cartilage and Subchondral Bone Marrow Lesions in a Canine Model. *J. Orthop. Res.* **2020**, *38*, 740–746. [CrossRef] [PubMed]
95. Martin, J.R.; Houdek, M.T.; Sierra, R.J. Use of concentrated bone marrow aspirate and platelet rich plasma during minimally invasive decompression of the femoral head in the treatment of osteonecrosis. *Croat. Med. J.* **2013**, *54*, 219–224. [CrossRef] [PubMed]
96. Kasik, C.S.; Martinkovich, S.; Mosier, B.; Akhavan, S. Short-Term Outcomes for the Biologic Treatment of Bone Marrow Edema of the Knee Using Bone Marrow Aspirate Concentrate and Injectable Demineralized Bone Matrix. *Arthrosc. Sports Med. Rehabil.* **2019**, *1*, e7–e14. [CrossRef] [PubMed]
97. Ankem, H.K.; Diulus, S.C.; Maldonado, D.R.; Ortiz-Declet, V.; Rosinsky, P.J.; Meghpara, M.B.; Shapira, J.; Lall, A.C.; Domb, B.G. Arthroscopic-Assisted Intraosseous Bioplasty of the Acetabulum. *Arthrosc. Tech.* **2020**, *9*, e1531–e1539. [CrossRef] [PubMed]
98. Hood, J.R.C.R.; Miller, J.R. The triad of osteobiology–Rehydrating calcium phosphate with bone marrow aspirate concentrate for the treatment of bone marrow lesions. *Bone* **2016**, *5*, 8.

International Journal of
Molecular Sciences

Article

Effect of Platelet-Rich Plasma on M1/M2 Macrophage Polarization

Ryoka Uchiyama [1,2], Eriko Toyoda [1,2], Miki Maehara [1,2], Shiho Wasai [1,2], Haruka Omura [1,2], Masahiko Watanabe [1,2] and Masato Sato [1,2,*]

[1] Department of Orthopaedic Surgery, Surgical Science, Tokai University School of Medicine, 143 Shimokasuya, Isehara, Kanagawa 259-1193, Japan; 9bmrm002@mail.u-tokai.ac.jp (R.U.); etoyoda@tokai-u.jp (E.T.); m-maehara@tsc.u-tokai.ac.jp (M.M.); s4.1x2.5s@gmail.com (S.W.); hrk.omr@gmail.com (H.O.); masahiko@is.icc.u-tokai.ac.jp (M.W.)

[2] Center for Musculoskeletal innovative Research and Advancement (C-MiRA), Tokai University Graduate School of Medicine, 143 Shimokasuya, Isehara, Kanagawa 259-1193, Japan

* Correspondence: sato-m@is.icc.u-tokai.ac.jp; Tel.: +81-463-93-1121; Fax: +81-463-96-4404

Abstract: Osteoarthritis of the knee (OAK) is a chronic degenerative disease and progresses with an imbalance of cytokines and macrophages in the joint. Studies regarding the use of platelet-rich plasma (PRP) as a point-of-care treatment for OAK have reported on its effect on tissue repair and suppression of inflammation but few have reported on its effect on macrophages and macrophage polarization. Based on our clinical experience with two types of PRP kits Cellaid Serum Collection Set P type kit (leukocyte-poor-PRP) and an Autologous Protein Solution kit (APS leukocyte-rich-PRP), we investigated the concentrations of humoral factors in PRPs prepared from the two kits and the effect of humoral factors on macrophage phenotypes. We found that the concentrations of cell components and humoral factors differed between PRPs purified using the two kits; APS had a higher concentration of M1 and M2 macrophage related factors. The addition of PRP supernatants to the culture media of monocyte-derived macrophages and M1 polarized macrophages revealed that PRPs suppressed M1 macrophage polarization and promoted M2 macrophage polarization. This research is the first to report the effect of PRPs purified using commercial kits on macrophage polarization.

Keywords: platelet rich plasma; autologous protein solution; macrophage

check for updates

Citation: Uchiyama, R.; Toyoda, E.; Maehara, M.; Wasai, S.; Omura, H.; Watanabe, M.; Sato, M. Effect of Platelet-Rich Plasma on M1/M2 Macrophage Polarization. *Int. J. Mol. Sci.* **2021**, *22*, 2336. https://doi.org/10.3390/ijms22052336

Academic Editor: Malgorzata Kloc

Received: 27 January 2021
Accepted: 23 February 2021
Published: 26 February 2021

Publisher's Note: MDPI stays neutral with regard to jurisdictional claims in published maps and institutional affiliations.

1. Introduction

Osteoarthritis of the knee (OAK) is a chronic degenerative disease that damages articular cartilage. OAK is characterized by aberrant cartilage metabolism, osteophyte formation, synovial tissue thickening, and macrophage infiltration. These changes contribute to the progress of OAK, resulting in joint pain and dysfunction [1,2]. There are about 25 million patients with OAK in Japan, and the number is expected to increase in tandem with Japan's hyper-aging population [3]. No treatment currently exists that lead to a complete recovery from OAK. Conservative treatments (i.e., exercise, oral treatment, intra-articular injection of hyaluronic acids) are performed for mild cases, and surgical treatments (i.e., arthroscopic debridement of the knee joint, high tibial osteotomy, and total knee arthroplasty) for severe cases [4,5]. Joint treatment with platelet-rich plasma (PRP) has recently attracted attention as a point-of-care treatment option to bridge the gap between conservative and surgical treatments.

PRP is an autologous blood product generated by centrifuging whole blood and concentrating platelets, containing diverse growth factors and cytokines [6]. The high concentration of anti-inflammatory factors in PRPs may suppress inflammation, growth factors may promote tissue repair, and in total these factors may improve symptoms [7,8]. Efficacy of platelet concentrates in promoting wound healing and tissue regeneration has been at the center of scientific debate over the past few decades [9]. Various purification kits have recently been marketed as this treatment has become more widespread. However, the blood cell components and humoral factors in PRPs depend on the purification method. Leukocyte-poor PRP (LP-PRP) lacks leukocytes whereas leukocyte-rich PRP (LR-PRP) do contain leukocytes. We have clinical experience with both types of kits: Cellaid Serum Collection Set P type kit (LP-PRP) and an Autologous Protein Solution kit (APS, LR-PRP). APS, with an added dehydration step using polyacrylamide beads during purification, has been reported to achieve a high concentration of not only platelets and leukocytes but also humoral factors such as IGF-1 [6]. However, there is no consensus on how differences in the components of PRP products affect their therapeutic efficacies.

One possibility is the effect of PRPs on macrophages. Macrophages are immune cells that play a crucial role in innate immunity and also participate in tissue repair and remodeling [10]. Macrophages can be polarized into two phenotypes in response to stimuli from their microenvironment. Classically activated macrophages (M1 macrophages) are induced by, for example, interferon-gamma (IFN-γ), tumor necrosis factor-alpha (TNF-α), and bacterial-derived lipopolysaccharides (LPS), and are known to release inflammatory cytokines that promote tissue damage and inflammation [11–13]. Alternatively activated macrophages (M2 macrophages) are induced by, for example, interleukin-4 (IL-4) and IL-13, and are known to release anti-inflammatory cytokines that promote tissue remodeling and suppress inflammation [14,15].

Recent studies have shown that M1 macrophages are present in the synovium and synovial fluid in OAK patients and involved in disease progression, suggesting they may be targeted for treatment [16–19]. Many studies have reported PRPs suppressing inflammation and inducing articular cartilage repair, but few studies have reported the effect of PRPs on macrophage phenotypes [20–22].

In this study, LP-PRP and APS were purified from peripheral blood of healthy subjects using two PRP preparation kits in which we have clinical experience. We then compared the concentrations of humoral factors related to M1/M2 macrophage polarization in the respective PRPs, and then added the respective PRP supernatants to the culture media of monocyte-derived macrophages (MDMs) to investigate their effect on macrophage phenotypes.

2. Results

2.1. Hematological Analysis of PRPs

From each healthy donor, 120 mL of peripheral blood was collected and 60 mL per kit was used to prepare both LP-PRP and APS. Immediately after preparation, hematological analysis was performed for whole blood, LP-PRP, and APS (Figure 1A). Erythrocyte concentration was negligible for LP-PRP while it was lower for APS compared to whole blood. Leukocyte concentration was negligible for LP-PRP while it was higher for APS compared to whole blood. Platelet concentrations were higher for both LP-PRP and APS compared to whole blood and did not significantly differ between LP-PRP and APS (Figure 1B). A comparison of the ratio of leukocyte types contained in whole blood and APS revealed that in APS, the ratios of neutrophils and eosinophils were higher while the ratios of monocytes and lymphocytes were lower (Figure 1C). According to the coding system presented by Kon et. al., the PRP codes for LP-PRP and APS are 210-00-00 and 214-15-10, respectively [23].

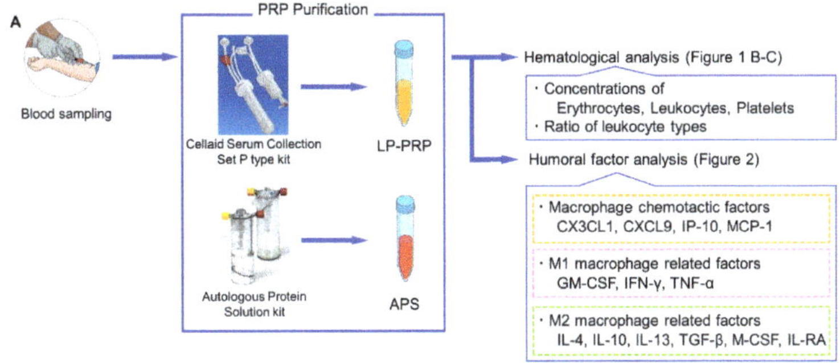

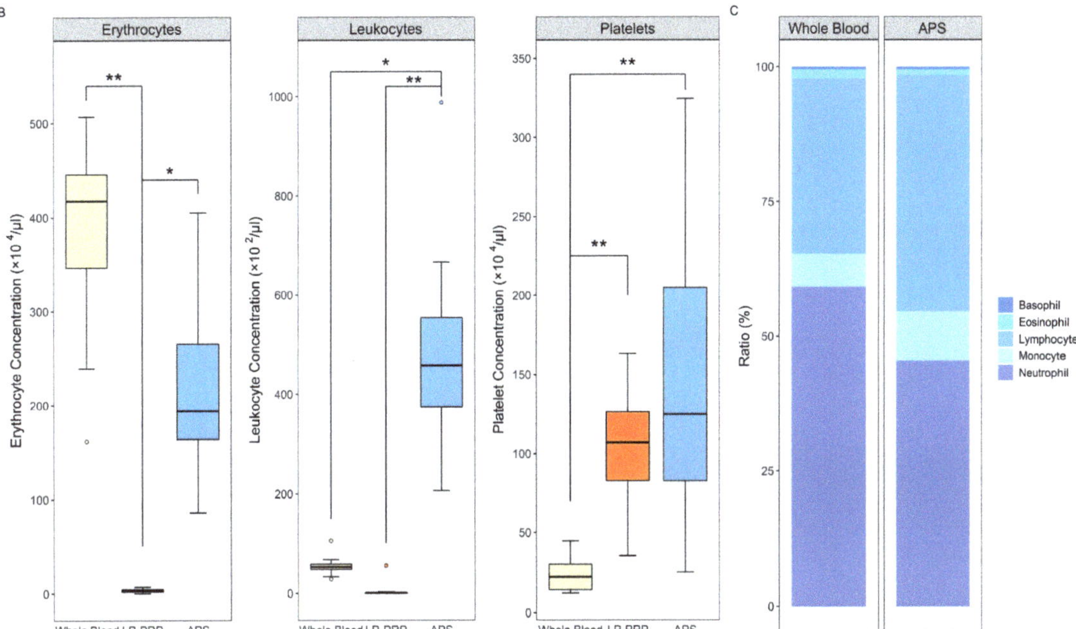

Figure 1. Purification and analyses of platelet-rich plasmas (PRPs) from two kits. (**A**) Flow of PRP purification and analyses. From each of the 12 healthy subjects, 120 mL of peripheral blood was collected and was added to 12 mL of anticoagulant citrate-dextrose solution. Leukocyte-poor PRP (LP-PRP) and Autologous Protein Solution (APS) were purified using their respective kits, and hematological analysis was performed immediately following purification. Humoral factor analysis was performed with frozen stored samples for various macrophage related factors. (**B**) Comparison of leukocyte, erythrocyte, and platelet concentrations in LP-PRP and APS. (**C**) Comparison of the ratios of leukocyte types in whole blood and APS. $n = 12$ PRP donors for each kit. * $p < 0.05$, ** $p < 0.01$.

2.2. Analysis of Humoral Factors

We investigated the effect of PRPs on macrophage polarization by analyzing the humoral factors in the respective PRPs. Macrophages respond to factors such as C-X-C motif chemokine ligand 9 (CXCL9), interferon gamma-induced protein 10 (IP-10), C-XXX-C motif chemokine ligand 1 (CX3CL1), and macrophage chemoattractant protein 1 (MCP-1), and are thereby recruited to areas of inflammation. M1 macrophages respond to and are polarized by granulocyte macrophage colony-stimulating factor (GM-CSF), TNF-α, IFN-γ, while M2 macrophages respond to and are polarized by macrophage colony-stimulating factor (M-CSF), IL-4, IL-10, IL-13, IL-1 receptor antagonist (IL-1RA), and transforming growth factor-beta (TGF-β) [14]. As a result of quantitative analysis, both LP-PRP and APS contained all of these factors. A comparison of the concentrations of these factors in LP-PRP and APS showed that the concentration of macrophage-recruitment chemokine IP-10 and CX3CL1 was significantly higher in APS (Figure 2A). Moreover, the concentrations of the M1 macrophage related factor TNF-α (Figure 2B) and the M2 macrophage related factors M-CSF, IL-10, IL-1RA, and TGF-β (Figure 2C) were also significantly higher in APS. These results showed that LP-PRP and APS differ in the concentrations of cytokines involved in macrophage polarization with higher concentrations of both M1 and M2 macrophage related factors in APS.

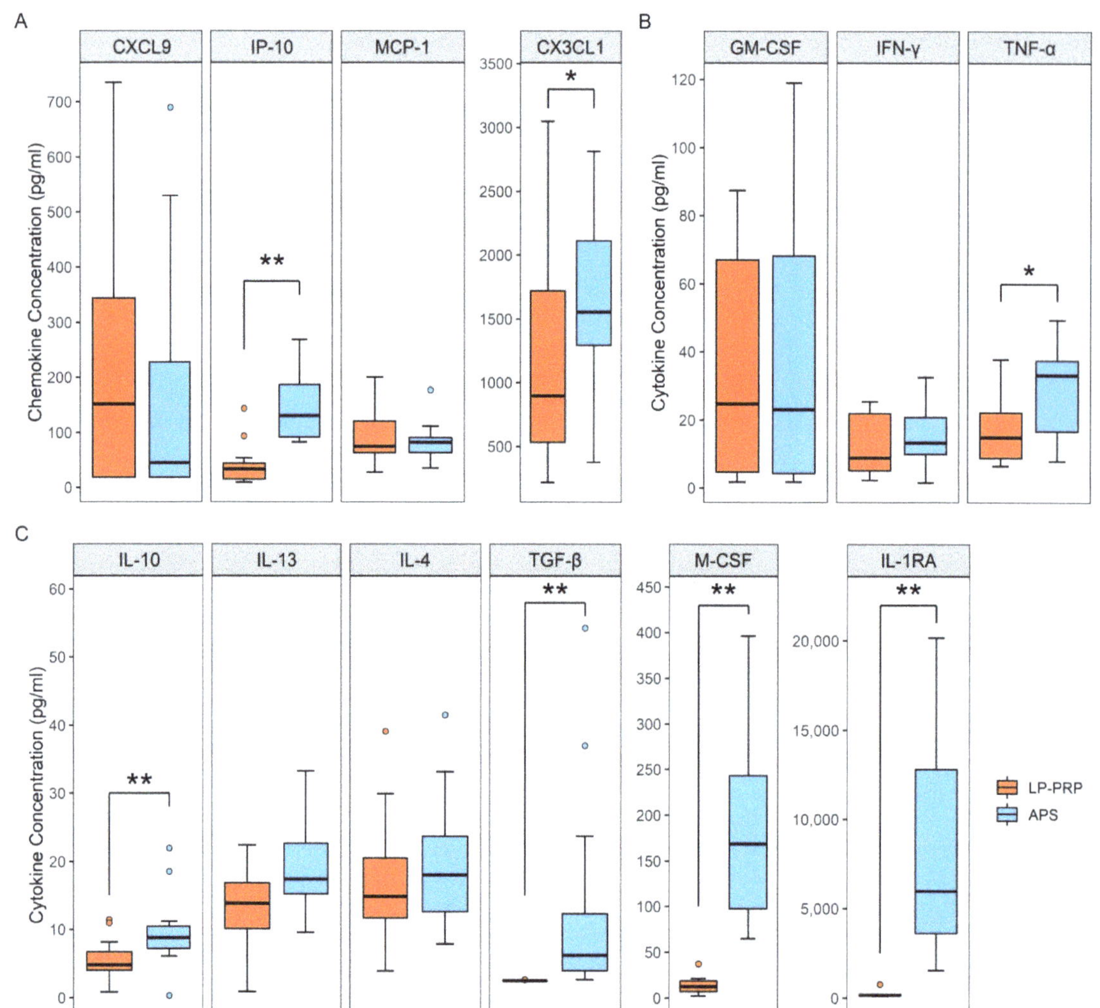

Figure 2. Analysis of humoral factors in PRPs. Comparison of concentrations of each factor in LP-PRP and APS. (**A**) Macrophage-recruitment factors (**B**) M1 macrophage related factors (**C**) M2 macrophage related factors. $n = 12$ PRP donors for each kit. * $p < 0.05$, ** $p < 0.01$.

2.3. Effect of the Addition of PRP Supernatants on Macrophage Phenotypes

Next, we investigated whether PRPs affect macrophage phenotypes and whether the effect differs depending on the purification kit. M1 macrophages express IL-1β, IL-6, and TNF-α, while M2 macrophages express MRC1, IL-10, and TGF-β [14]. Monocytes were isolated from human peripheral blood and cultured in M-CSF for six days, then supernatants from each PRP was added to the culture media, and the cells were cultured for another two days. Then, the effect of PRP supernatants on the expression of the M1/M2 macrophage marker genes was confirmed by quantitative reverse transcription-polymerase chain reaction (qRT-PCR) (Figure 3A). MDM control group served as negative control, and M1 and M2 polarized groups served as positive controls. The expression of IL-1β and TNF-α, which are M1 macrophage markers, was lower in the LP-PRP- and APS-added groups compared with the MDM control group, and the expression of IL-6 also tended to be lower. However, there was no significant difference in gene expression between the addition of PRP supernatants purified using the two kits (Figure 3B). The expression of MRC1, which is an M2 macrophage marker, was higher in the LP-PRP- and APS-added groups compared with the MDM control group. The expression of IL-10 was higher in the APS-added group compared with the MDM control group. The expression of TGF-β was higher in the LP-PRP-added group compared with the APS-added group, but no difference was detected between LP-PRP and APS-added groups compared with the MDM control group (Figure 3C).

Similar verification experiments were performed for cell surface markers using flow cytometry. M1 macrophages express CD80 and CD86, while M2 macrophages express CD163 and CD206 on the cell surface [14]. A typical histogram is shown in Figure 4A. The values were calculated and compared based on difference of the mean fluorescence intensities between the antibody and isotype control. The expression of CD80 and CD86, which are M1 macrophage markers, decreased in the LP-PRP- and APS-added groups compared with the MDM group, but no difference was observed between the purification kits (Figure 4B). On the other hand, there was no difference in the expression of the M2 macrophage markers CD163 and CD206 between the MDM control group and the LP-PRP- or APS-added groups (Figure 4C).

These results showed that the addition of PRP supernatants decreased the expression of M1 macrophage markers, but there was no difference between the purification kits. On the other hand, the expression of M2 macrophage surface markers tended to be maintained or increased by the addition of PRP supernatants while the gene expression of IL-10 increased in the APS-added group and TGF-β increased in the LP-PRP-added group.

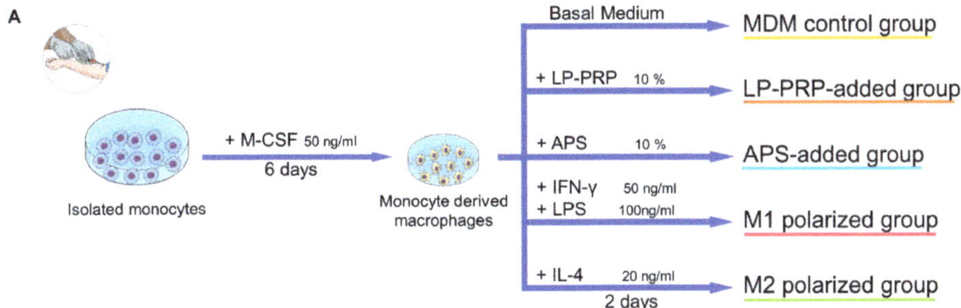

Figure 3. *Cont.*

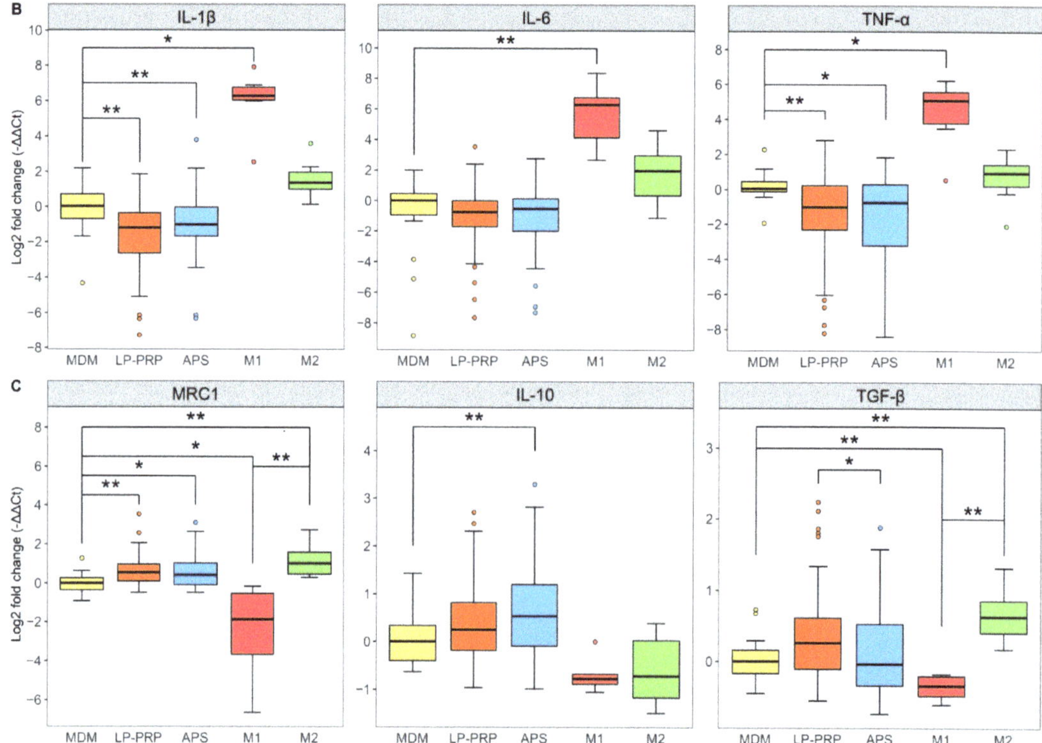

Figure 3. Effect of PRP supernatants on gene expression of M1/M2 macrophage markers. (**A**) Summary of this experiment. CD14 + monocytes were isolated from peripheral blood using density gradient centrifugation and magnetic beads. Monocytes were cultured in basal medium supplemented with 10% FBS containing M-CSF at 37 °C under 5% CO_2 for six days, then the medium was replaced by fresh basal medium supplemented with 10% FBS containing supernatants obtained from LP-PRP or APS, and the cells were cultured for another two days. (**B**) Gene expression of M1 macrophage markers (IL-1β, IL-6, TNF-α) and (**C**) M2 macrophage markers (MRC1, IL-10, TGF-β). Data were analyzed through qRT-PCR. -ΔΔCt values were calculated using GAPDH as an internal control. Monocyte-derived macrophages (MDM) control group served as negative control, and M1 and M2 polarized groups served as positive controls. MDM control group: 6 monocyte donors, 5 experiments, total $n = 30$; LP-PRP- and APS-added groups: 6 monocyte donors, 12 PRP donors, $n = 72$ group; M1 and M2 polarized groups: 6 monocyte donors, 2 or 3 experiments, $n = 14$ per group. * $p < 0.05$, ** $p < 0.01$.

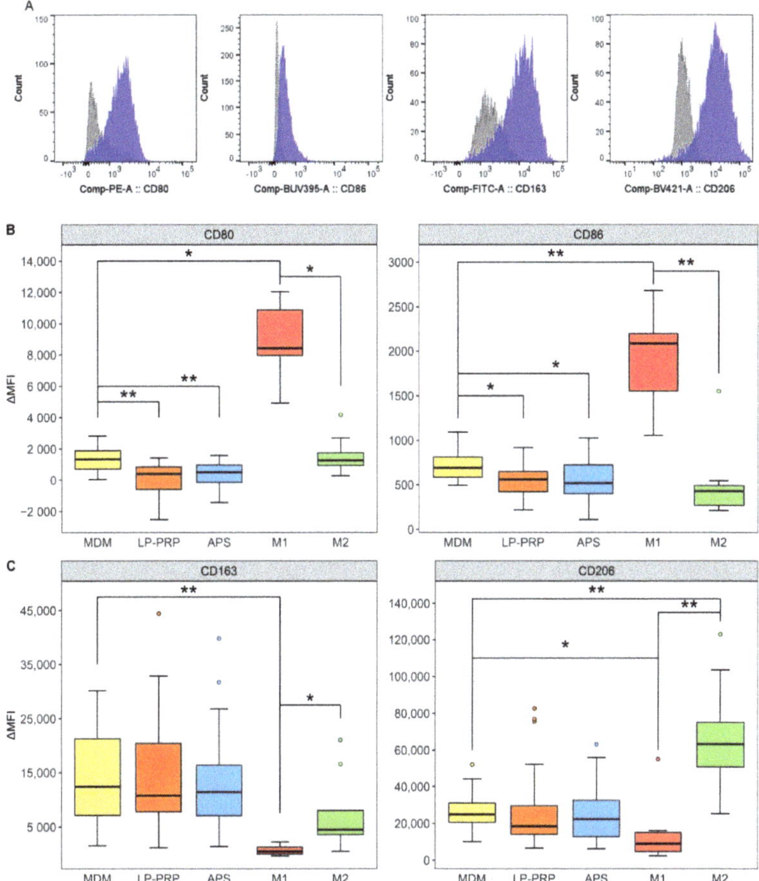

Figure 4. Flow cytometric analysis of M1/M2 macrophage cell surface markers. Cells were cultured under the same protocol as Figure 3. (**A**) A histogram representation of typical flow cytometry results. Gray and dashed lines: isotype control; blue and solid lines: signals for each antibody. Mean fluorescence intensity (MFI) values of each antibody were used to calculate ΔMFI values: ΔMFI = MFI Sample—MFI Isotype. (**B**) M1 macrophage markers. (**C**) M2 macrophage markers. MDM control group served as negative control, and M1 and M2 polarized groups served as positive controls. MDM control group: 6 monocyte donors, 5 experiments, $n = 30$; LP-PRP- and APS-added groups: 6 monocyte donors, 12 PRP donors, $n = 72$ per group; M1 and M2 polarized groups: 6 monocyte donors, 2 or 3 experiments, $n = 14$ per group. * $p < 0.05$, ** $p < 0.01$.

2.4. Effect of PRPs on M1 Macrophages

M1/M2 macrophages can change their phenotypes once polarized [24], and, thus, we investigated whether PRPs can induce the polarization of M1 macrophages to M2 macrophages. Monocytes were isolated from human peripheral blood and cultured in M-CSF for six days. MDMs were polarized to M1 macrophages and cultured for two days, then each PRP was added to the culture medium, and the cells were cultured for another two days (Figure 5A). MDM control and M1 polarized groups served as negative controls; M1-M2 and M2 polarized groups served as positive controls. The expressions of IL-1β and IL-6, which are M1 macrophage markers, decreased in the LP-PRP- and APS-added groups, and suppression was greater in the APS-added group than in the LP-PRP-added group (Figure 5B). Expression of the M2 macrophage marker MRC1 increased by the addition of PRPs and this increase was greater in the APS-added group than in the LP-PRP-added group. The expression of TGF-β was higher in the LP-PRP-added group than in the APS-added group (Figure 5C). These results showed that PRPs can repolarize M1 macrophages to M2 macrophages.

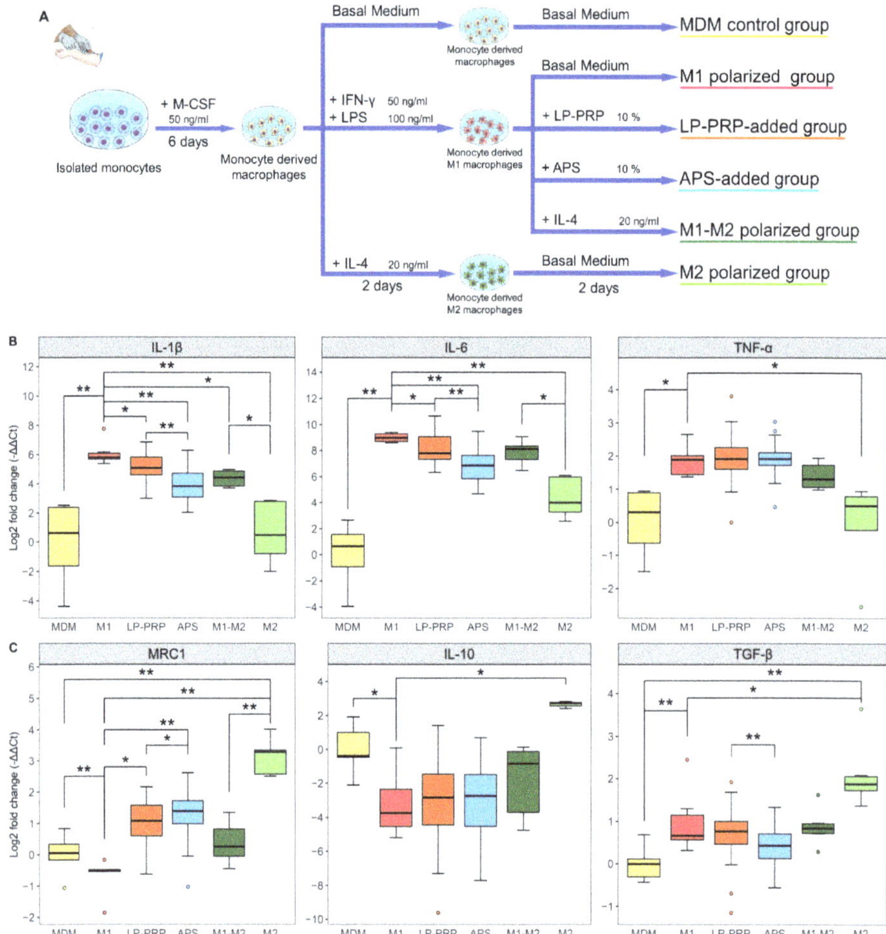

Figure 5. Effect of PRPs on M1 macrophages. (**A**) After CD14 + monocytes were isolated by the same method as described in Figure 3, they were cultured in a basal medium supplemented with 10% FBS containing M-CSF at 37 °C under 5% CO_2. After six days, the media was replaced by fresh basal medium supplemented with 10% FBS containing IFN-γ + LPS, and the cells were cultured for another two days to polarize them to M1 macrophages. The medium was removed, the basal medium supplemented with 10% FBS containing supernatants obtained from LP-PRP or APS was added, and the cells were cultured for another two days. (**B**) Expression of M1 macrophage markers (IL-1β, IL-6, TNF-α). (**C**) Expression of M2 macrophage markers (MRC1, IL-10, TGF-β). Data were analyzed through qRT-PCR. -ΔΔCt values were calculated using GAPDH as an internal control. MDM control and M1 polarized groups served as negative controls; M1-M2 and M2 polarized groups served as positive controls. MDM control, M1 polarized, M1-M2 polarized, M2 polarized groups: 6 monocyte donors, 1 experiment, *n* = 6 per group; LP-PRP- and APS-added groups: 6 monocyte donors, 12 PRP donors, *n* = 72 per group. * *p* <0.05, ** *p* <0.01.

3. Discussion

Platelets play an important role in the process of hemostasis and tissue repair by gathering at injury sites and secreting various cytokines and growth factors to initiate the repair process [25]. The use of PRPs is based on the hypothesis that they eliminate the imbalance of cytokines in joints following the injection of large amounts of platelets, which release anti-inflammatory cytokines and growth factors [26]. Many studies have been published regarding humoral factors in PRPs and their effects [27–29], and we have also previously reported that the concentrations of humoral factors differ between PRPs derived from healthy subjects and OAK patients [27]. However, only a few studies have reported on the effect of PRPs on macrophages and their polarizations.

Macrophages are present in all tissues, play an important role in innate immunity, and are essential for early tissue repair of damaged or inflamed areas [30,31]. Macrophages respond to IP-10, MCP-1, CXCL9, and CX3CL1, infiltrate damaged or inflamed areas, and polarize into two functionally different types (M1 and M2) depending on their microenvironment. M1 macrophages are stimulated by IFN-γ, TNF-α, and LPS and release pro-inflammatory factors such as IL-1β and TNF-α. M2 macrophages are stimulated by IL-4 and IL-13 and release anti-inflammatory factors such as IL-10 and growth factors such as TGF-β, which are known to suppress inflammation and induce tissue repair [32–34]. Inflammation occurs in most synovium of OAK patients, and pathology progresses with an imbalance of M1/M2 macrophages in the synovium and synovial fluid caused by an increase in M1 macrophages and a decrease in M2 macrophages [17,35,36].

Therefore, the purpose of this study was to compare in PRPs derived from two clinically relevant kits the concentrations of typical cytokines and growth factors associated with M1/M2 macrophages and verify the effect of PRP supernatants on macrophage polarization. We first investigated the blood components and humoral factors in LP-PRP and APS, which we have clinical experience administering to OAK patients [27]. We first confirmed that compared to whole blood, APS contained higher levels of leukocytes, while LP-PRP was leukocyte poor as expected.

Analysis of humoral factors in LP-PRP and APS showed that each contained different concentrations of humoral factors, among which both M1 and M2 macrophage related factors were found to be at a higher concentration in APS. APS is prepared with a dehydration process using polyacrylamide beads, which is reported to result in higher concentrations of humoral factors [37]. In fact, APS contained a higher concentration of M2 macrophage-related factors such as IL-10, an anti-inflammatory factor, and TGF-β, a growth factor. At the same time, M1 macrophage-related factors such as TNF-α, a pro-inflammatory factor, were contained at a higher concentration in APS than in LP-PRP. However, previous reports have shown that the effects of pro-inflammatory factors are negated by the higher concentrations of anti-inflammatory factors [38]. Moreover, an explanation for the fact that the concentrations of TGF-β and IL-1RA were negligible for LP-PRP is that platelets were not activated by external factors prior to quantitative analysis.

To investigate the effect of PRPs on M1/M2 macrophage polarization, PRP supernatants were added to the culture media of monocyte-derived macrophages. Specifically, blood components were removed through centrifugation to prevent addition of monocytes derived from PRPs and to remove the effect of leukocytes contained in PRPs. The addition of PRP supernatants to macrophages resulted in the reduction of the expression of M1 macrophage markers. Furthermore, when PRP supernatants were added to the culture media of macrophages, they suppressed the M1 polarization of monocyte-derived macrophages (Figure 3B, Figure 4B) and promoted the polarization of monocyte–derived M1 macrophages to M2 macrophages (Figure 5). The effect of PRP supernatants on the macrophage phenotype observed in this study (Figure 4B,C) suggest that PRPs have little effect on M1 macrophage related factors that may polarize macrophages to M1 macrophages. These results indicate that PRPs may improve symptoms in OAK patients by polarizing M1 macrophages in joints to M2 macrophages.

In addition, reports have suggested that M2 macrophages can be categorized into three subsets: M2a macrophages induced by IL-4 and IL-13, expressing MRC1 and IL-10; M2b induced by signals from the immune complex, expressing IL-10 and major histocompatibility complex class II; and M2c induced by IL-10 and glucocorticoids, expressing MRC1, IL-10, and TGF-β [32–34]. Our results suggest that LP-PRP promotes the polarization to M2c macrophages and that APS specifically promotes the polarization to M2a macrophages. M2a macrophages are mainly related to anti-inflammatory activity while M2c macrophages to tissue repair [14,39]. As such, the mechanism of action of LP-PRP and APS may differ in the elimination of the imbalance of M1/M2 macrophages in the joint.

PRPs have also been shown to enhance macrophage infiltration into tissues in tendon repair [20]. Furthermore, it is possible that monocytes contained in APS itself may differentiate and polarize to M2 macrophages once administered to the joint. Taken together, administration of PRPs to the knee joint of OAK patients may help eliminate the imbalance of M1/M2 macrophages in several ways: polarizing M1 macrophages already present in the joints to M2 macrophages; populating the joint with macrophages from the surrounding area and polarizing them to M2 macrophages; and populating the joint with monocytes contained PRPs and polarizing them to M2 macrophages.

This study has some limitations. First, because of the difficulty in recruitment, PRPs and monocytes used in this study were derived from peripheral blood of healthy subjects and not OAK patients. Second, we removed cellular components through centrifugation in experiments involving the addition of PRP supernatants. In addition to cytokines and growth factors, APS contains high concentrations of leukocytes, which may affect macrophage polarization, and their effect when administered to the joint must be further elucidated. For example, TNF-R released from leukocytes has been reported to inhibit the polarization to M1 macrophages [37]. Third, 60 mL of peripheral blood provide about 6 mL of LP-PRP or 2.5 mL of APS. Thus, the clinically relevant amount of PRPs used for intra-articular injection would result in different doses whereas equal amounts of LP-PRP or APS supernatants were added to the media in the macrophage experiments. Fourth, in standard treatment, PRPs are used immediately after purification whereas here they were frozen once.

This study is the first report on the effect of clinically relevant PRPs on macrophages and macrophage phenotypes. Our findings suggest that PRPs may help improve symptoms via modulation of macrophages in the joint, and warrant a further investigation of their effect on macrophages as a possible mechanism of action by which PRPs promote the tissue repair process.

4. Materials and Methods

4.1. Ethics Statement

This study was reviewed and approved by the Institutional Review Board of the Tokai University School of Medicine (18R-134) and was conducted in compliance with relevant guidelines. Written informed consent was obtained from all participants.

4.2. PRP Purification

To purify PRPs, 120 mL of peripheral blood collected from each of the 12 healthy subjects (M = 5, F = 7, Age = 38.6 ± 11.0 years) was added to 12 mL of anticoagulant citrate-dextrose solution A (ACD-A; TERUMO, Tokyo, Japan) using two 60 mL syringes. Excess peripheral blood was centrifuged at 2200× g for 10 min and the top plasma layer was collected and stored at −80 °C until use.

LP-PRP was purified using a Cellaid Serum Collection Set P type kit (JMS, Hiroshima, Japan). This kit consists of primary and secondary containers connected at the top by multiple tubes. Blood (20 mL) containing ACD-A was injected into the primary container and centrifuged at 200× g for 15 min. The plasma layer containing platelets was transferred to the secondary container via a tube at the top and centrifuged at 1200× g for 15 min. Excess plasma was returned to the primary container, the pelletized platelets were disrupted by tapping, and then 2 mL of LP-PRP was collected. Approximately 6 mL of LP-PRP was collected from 60 mL of peripheral blood using 3 kits. One hundred micro liters was used immediately after for hematological analysis.

APS was purified using an Autologous Protein Solution (APS) kit (Zimmer Biomet, Warsaw, IN, USA). This kit consists of two independent tubes (a GPS3 III system and an APS Separator). Blood (60 mL) containing ACD-A was injected into the cell separation tube (GPS III system) and centrifuged at $745\times g$ for 15 min using a dedicated centrifuge (Zimmer Biomet) and 6 mL of the upper layer (PRP layer) was collected. This 6 mL was added to the APS Separator and centrifuged at $219\times g$ for 2 min in the same centrifuge and approximately 2.5 mL of APS was collected. One hundred micro liters was used immediately after for hematological analysis. The collected PRPs were stored at $-80\,^{\circ}\text{C}$ until use.

4.3. Hematological Analysis

The leukocyte, erythrocyte, and platelet concentrations of whole-blood, LP-PRP, and APS samples and leukocyte compositions of whole-blood and APS samples were determined using an automated hematology analyzer (XT-1800i; Sysmex, Kobe, Japan) immediately after preparation.

4.4. Analysis of Humoral Factors

The concentrations of humoral factors in plasma, LP-PRP, and APS were measured using a flow cytometry bead-based immunoassay (LEGENDplex™ Custom Human 13-plex panel, BioLegend, San Diego, CA, USA) according to the manufacturer's protocol. Plasma, LP-PRP, and APS were centrifuged at $16139\times g$ at $4\,^{\circ}\text{C}$ for 5 min to remove cellular components and the supernatants were analyzed without any external activation. GM-CSF, M-CSF, IFN-γ, IL-4, IL-10, IL-13, IL-1RA, free active TGF-β1, TNF-α, MCP-1, IP-10, CXCL9, CXCL10, and CX3CL1 were measured simultaneously. Data were acquired using a FACS Verse™ Flow Cytometer (BD Bioscience, San Diego, CA, USA) and analyzed using BioLegend's cloud-based LEGENDplex™ Data Analysis Software.

4.5. Isolation of Monocytes

Peripheral blood mononuclear cells (PBMCs) were separated from the buffy-coat of six healthy donors (M = 3, F = 3, Age = 32.0 ± 1.7 years) using a density gradient (Histopaque 1077, Sigma-Aldrich, St. Louis, MO, USA). PBMCs were washed with wash buffer consisting of Dulbecco's phosphate-buffered saline (DPBS; Gibco, Waltham, MA, USA) and 1% bovine serum albumin (BSA; Sigma-Aldrich) and centrifuged at $538\times g$ at $4\,^{\circ}\text{C}$ for 5 min. Contaminating red blood cells were hemolyzed with red blood cell lysing buffer (Sigma-Aldrich) for 10 min at $37\,^{\circ}\text{C}$. Wash buffer was added and the cell suspension was centrifuged at $538\times g$ at $4\,^{\circ}\text{C}$ for 5 min. The cell pellet was disrupted, FcR blocking reagent (Miltenyi Biotech, Bergisch, Gladbach, Germany) was added, and the mixture was kept at room temperature for 15 min to inhibit Fc receptor-mediated nonspecific antibody binding. Cells were stained with mouse anti-human CD14-allophycocyanin (APC) (BD Bioscience) at $4\,^{\circ}\text{C}$ for 30 min. After washing with wash buffer, the anti-APC microbeads (Miltenyi Biotech) were bound to cells at $4\,^{\circ}\text{C}$ for 20 min. The cells were washed with wash buffer and the CD14 + monocytes were isolated using an autoMACS Pro Separator (Miltenyi Biotech). Collected cells were counted with a particle counter (Sysmex).

4.6. Cell Culture

A modified version of a previously reported protocol was used [40–42].

The cells were cultured in Roswell Park Memorial Institute (RPMI) 1640 media containing GlutaMAX™ supplement (Gibco) supplemented with heat-inactivated 10% fetal bovine serum (AusGeneX, Molendinar, Australia) and 1% penicillin-streptomycin (FujiFilm, Tokyo, Japan) (hereinafter called basal medium).

4.6.1. Preparation of Monocyte-Derived Macrophages and Addition of PRP Supernatants

A summary of the process is shown in Figure 3A. Isolated monocytes were seeded at a density of 1×10^5 cells/cm^2 on an Upcell Multi 24 well plate (CellSeed, Tokyo, Japan) containing basal medium supplemented with 20 ng/mL M-CSF (Peprotech, Rocky Hill, NJ, USA), and incubated at 37 °C, 5% CO$_2$. After six days, the medium was replaced with medium with each of the following supplements: monocyte-derived macrophages (MDM control group)—basal medium; M1 polarized group—basal medium + 50 ng/mL IFN-γ (Peprotech) + 100 ng/mL lipopolysaccharides from *Escherichia coli* (LPS, Sigma-Aldrich); M2 polarized group—basal medium + 20 ng/mL IL-4 (Peprotech); LP-PRP-added group—basal medium + 10% LP-PRP; APS-added group—basal medium + 10% APS. PRPs were centrifuged at 16,139× *g* at 4 °C for 5 min to remove cellular components and the supernatants were used. The cells were cultured for another two days and then used for analysis. MDM control group: 6 monocyte donors, 5 experiments, total *n* = 30; LP-PRP- and APS-added groups: 6 monocyte donors, 12 PRP donors, *n* = 72 group; M1 and M2 polarized groups: 6 monocyte donors, 2 or 3 experiments, *n* = 14 per group.

4.6.2. Preparation of M1 Macrophages and Addition of PRP Supernatants

A summary of the process is shown in Figure 5A. Isolated monocytes were seeded at a density of 1×10^5 cells/cm^2 on a 96 well plate (Thermo Fisher Scientific, Tokyo, Japan) with basal medium supplemented with 20 ng/mL M-CSF and incubated at 37 °C, 5% CO$_2$.

After six days, the medium was removed. Basal medium was added to MDM control group, and basal medium + 20 ng/mL IL-4 was added to M2 polarized group. Basal medium + 50 ng/mL IFN-γ + 100 ng/mL LPS was added to the other groups. All cells were cultured for two days, and then the medium was replaced by the following: MDM control, M2 polarized, and M1polarized groups—basal medium; M1-M2 polarized group—basal medium + 20 ng/mL IL-4; LP-PRP-added group—basal medium + 10% LP-PRP; and APS-added group—basal medium + 10% APS. PRPs were centrifuged at 16,139× *g* at 4 °C for 5 min to remove cellular components and the supernatants were used. After two days, the medium was removed, the cells were washed with DPBS, and the total RNA was collected using Isogen II reagent (Nippon Gene, Tokyo, Japan). MDM control, M1 polarized, M1-M2 polarized, M2 polarized groups: 6 monocyte donors, 1 experiment, *n* = 6 per group; LP-PRP- and APS-added groups: 6 monocyte donors, 12 PRP donors, *n* = 72 per group.

4.7. Flow Cytometry

To collect cells, FACS buffer (DPBS + 1% BSA) was added to the Upcell Multi 24 well plate (method 4.5.1). The plates were kept at room temperature for 30 min to promote detachment of the cells and the cells were collected by pipetting the contents of each well. FcR blocking reagent was added to the cell suspension and the cell suspension was kept at room temperature for 15 min to inhibit Fc receptor-mediated nonspecific antibody binding. Each cell suspension was divided into two tubes. In one tube, the cells were mixed with the following six mouse monoclonal anti-human antibodies and the cell suspension was kept at 4 °C for 30 min for multiple staining: CD80-PE (Clone: L307.4), CD86-BUV395 (Clone: 2331), CD163-FITC (Clone: GHI/61), CD206-BV421 (Clone: 19.2), CD14-APC (Clone: M5E2), CD45-BV605 (Clone: HI30) (BD Bioscience). Cells in the other tube were mixed with nonspecific fluorescent mouse IgGs as a negative control. The cells were reacted at 4 °C for 30 min and then washed with FACS buffer. Data on the stained cells were acquired using a BD LSR Fortessa ™ Flow Cytometer (BD Bioscience) and analyzed using FlowJo (Tree Star, Ashland, OR, USA).

4.8. Gene Expression Analysis

Cells were lysed using Isogen II reagent for RNA extraction and stored at −80 °C until use. The lysate was then thawed at room temperature, and 40% (final volume) deionized water was added. The sample was vortexed and centrifuged at 16,139× *g* and 4 °C for 15 min, 75% of the supernatant was transferred to a new tube, and an equal amount of 70% ethanol was added and then vortexed. Total RNA was extracted using an RNeasy Micro kit (QIAGEN, Valencia, CA, USA) according to the manufacturer's protocol. The total RNA quantity and quality were determined using a NanoDrop Lite (Thermo Fisher Scientific). cDNA was synthesized from RNA (40 ng) using a QuantiTect reverse transcription kit (QIAGEN) and a Thermal Cycler GeneAmp PCR System 9700 (Thermo Fisher Scientific). cDNA (5 ng) was pre-amplified using TaqMan PreAmp Master Mix (Applied Biosystems, Waltham, MA, USA). For pre-amplification, the reaction volume was adjusted to 10 μL according to the manufacturer's instructions. Using the Thermal Cycler GeneAmp PCR System 9700, the DNA was denatured at 95 °C for 10 min, and then at 95 °C for 15 s and 60 °C for 4 min for 14 cycles. The amplified product was diluted 20-fold with TE buffer (1 ×) and used for qRT-PCR. qRT-PCR was performed using TaqMan fast advanced master mix (Applied Biosystems) with a QuantStudio 3 Real-Time PCR System (Applied Biosystems) at 50 °C for 2 min, 95 °C for 2 min, and then 95 °C for 1 s and 60 °C for 20 s for 40 cycles. The probes used for pre-amplification and qRT-PCR were GAPDH (Hs_02758991_g1), IL-1β (Hs_01555410_m1), IL-6 (Hs_00985639_m1), TNF-α (Hs_00174128_m1), MRC1 (Hs_00267207_m1), IL-10 (Hs_00961622_m1), and TGF-β (Hs_00998133_m1). The value of each gene expression ($-\Delta\Delta$Ct value) was obtained against the Ct of the internal control GAPDH and normalized to the control sample.

4.9. Statistical Analysis

Numerical results were statistically analyzed using SPSS® Statistics Software 26 (IBM, Armonk, NY, USA). The data obtained was tested for normality of distribution through the Kolomogorov–Smirnov test and was rejected. Thus, the Wilcoxon signed-rank test was used for two-group comparisons, and Friedman's test was used to compare three or more groups. The significance level was set at $p < 0.05$.

5. Conclusions

PRPs purified using two clinically relevant kits contained different concentrations of humoral factors, among which both M1 and M2 macrophage-related factors were contained at a higher concentration in APS. The addition of PRP supernatants to the culture media of monocytes and M1 macrophages promoted their polarization to M2 macrophages, possibly in different ways. These results warrant a further investigation of the possibility that PRPs may act on M1 macrophages in the joint and synovium to improve OAK symptoms.

Author Contributions: Conceptualization, M.S., E.T., and R.U.; data curation R.U. and M.M.; formal analysis, R.U., investigation, R.U., M.M. and S.W.; resources, S.W., H.O. and M.W.; writing—original draft preparation, R.U.; writing—review and editing, M.S.; supervision, M.S. and M.W. All authors have read and agreed to the published version of the manuscript.

Funding: This research received no external funding.

Institutional Review Board Statement: This study was reviewed and approved by the Institutional Review Board of the Tokai University School of Medicine (18R-134) and was conducted in compliance with relevant guidelines.

Informed Consent Statement: Informed consent was obtained from all subjects involved in the study.

Data Availability Statement: The data presented in this study are available on request from the corresponding author. The data are not publicly available because of confidentiality issues.

Acknowledgments: We would like to express our deep gratitude to the Support Center for Medical Research and Education for their cooperation in advancing this research.

Conflicts of Interest: The authors declare no conflict of interest.

Abbreviations

APS	Autologous protein solution
CX3CL1	C-XXX-C motif chemokine ligand 1
CXCL9	C-X-C motif chemokine ligand 9
GAPDH	Glyceraldehyde 3-phosphate dehydrogenase
GM-CSF	Granulocyte macrophage colony-stimulating factor
IFN-γ	Interferon-gamma
IL-10	Interleukin-10
IL-13	Interleukin-13
IL-1β	Interleukin-1 beta
IL-1RA	Interleukin-1 receptor antagonist
IL-4	Interleukin-4
IL-6	Interleukin-6
IP-10	Interferon gamma-induced protein-10
LP-PRP	Leukocyte-poor platelet rich plasma
LR-PRP	Leukocyte-rich platelet rich plasma
MCP-1	Macrophage chemoattractant protein-1
M-CSF	Macrophage colony-stimulating factor
MDM	Monocyte derived macrophage
MRC1	Mannose receptor 1
OAK	Osteoarthritis of the knee
PRP	Platelet rich plasma
TGF-β	Transforming growth factor beta
TNF-α	Tumor necrosis factor-alpha

References

1. Hügle, T.; Geurts, J. What drives osteoarthritis?—Synovialversussubchondral bone pathology. *Rheumatology* **2016**, *56*, 1461–1471. [CrossRef]
2. Loeser, R.F.; Goldring, S.R.; Scanzello, C.R.; Goldring, M.B. Osteoarthritis: A disease of the joint as an organ. *Arthritis Rheum.* **2012**, *64*, 1697–1707. [CrossRef] [PubMed]
3. Yoshimura, N.; Muraki, S.; Oka, H.; Mabuchi, A.; En-Yo, Y.; Yoshida, M.; Saika, A.; Yoshida, H.; Suzuki, T.; Yamamoto, S.; et al. Prevalence of knee osteoarthritis, lumbar spondylosis, and osteoporosis in Japanese men and women: The research on osteoarthritis/osteoporosis against disability study. *J. Bone Miner. Metab.* **2009**, *27*, 620–628. [CrossRef] [PubMed]
4. McAlindon, T.E.; Bannuru, R.R.; Sullivan, M.C.; Arden, N.K.; Berenbaum, F.; Bierma-Zeinstra, S.M.; Hawker, G.A.; Henrotin, Y.; Hunter, D.J.; Kawaguchi, H.; et al. Response to Letter to the Editor entitled "Comments on 'OARSI guidelines for the non-surgical management of knee osteoarthritis'". *Osteoarthr. Cartil.* **2014**, *22*, 890. [CrossRef] [PubMed]
5. Zhang, W.; Moskowitz, R.; Nuki, G.; Abramson, S.; Altman, R.; Arden, N.; Bierma-Zeinstra, S.; Brandt, K.; Croft, P.; Doherty, M.; et al. OARSI recommendations for the management of hip and knee osteoarthritis, Part II: OARSI evidence-based, expert consensus guidelines. *Osteoarthr. Cartil.* **2008**, *16*, 137–162. [CrossRef]
6. Woodell-May, J.; Matuska, A.; Oyster, M.; Welch, Z.; O'Shaughnessey, K.; Hoeppner, J. Autologous protein solution inhibits MMP-13 production by IL-1β and TNFα-stimulated human articular chondrocytes. *J. Orthop. Res.* **2011**, *29*, 1320–1326. [CrossRef]
7. El-Sharkawy, H.; Kantarci, A.; Deady, J.; Hasturk, H.; Liu, H.; Alshahat, M.; Van Dyke, T.E. Platelet-Rich Plasma: Growth Factors and Pro- and Anti-Inflammatory Properties. *J. Periodontol.* **2007**, *78*, 661–669. [CrossRef]
8. Kon, E.; Engebretsen, L.; Verdonk, P.; Nehrer, S.; Filardo, G. Clinical Outcomes of Knee Osteoarthritis Treated with an Autologous Protein Solution Injection: A 1-Year Pilot Double-Blinded Randomized Controlled Trial. *Am. J. Sports Med.* **2018**, *46*, 171–180. [CrossRef] [PubMed]
9. Mourão, C.F.D.A.B.; De Mello-Machado, R.C.; Javid, K.; Moraschini, V. The use of leukocyte- and platelet-rich fibrin in the management of soft tissue healing and pain in post-extraction sockets: A randomized clinical trial. *J. Cranio-Maxillofac. Surg.* **2020**, *48*, 452–457. [CrossRef] [PubMed]
10. Gordon, S. Alternative activation of macrophages. *Nat. Rev. Immunol.* **2003**, *3*, 23–35. [CrossRef] [PubMed]
11. Arnold, C.E.; Gordon, P.; Barker, R.N.; Wilson, H.M. The activation status of human macrophages presenting antigen determines the efficiency of Th17 responses. *Immunobiology* **2015**, *220*, 10–19. [CrossRef] [PubMed]
12. Utomo, L.; Bastiaansen-Jenniskens, Y.; Verhaar, J.; Van Osch, G. Cartilage inflammation and degeneration is enhanced by pro-inflammatory (M1) macrophages in vitro, but not inhibited directly by anti-inflammatory (M2) macrophages. *Osteoarthr. Cartil.* **2016**, *24*, 2162–2170. [CrossRef] [PubMed]
13. Lee, H.-W.; Choi, H.-J.; Ha, S.-J.; Lee, K.-T.; Kwon, Y.-G. Recruitment of monocytes/macrophages in different tumor microenvironments. *Biochim. et Biophys. Acta (BBA) Bioenerg.* **2013**, *1835*, 170–179. [CrossRef]

14. Mantovani, A.; Sica, A.; Sozzani, S.; Allavena, P.; Vecchi, A.; Locati, M. The chemokine system in diverse forms of macrophage activation and polarization. *Trends Immunol.* **2004**, *25*, 677–686. [CrossRef] [PubMed]
15. Zhang, S.; Chuah, S.J.; Lai, R.C.; Hui, J.H.P.; Lim, S.K.; Toh, W. MSC exosomes mediate cartilage repair by enhancing proliferation, attenuating apoptosis and modulating immune reactivity. *Biomaterials* **2018**, *156*, 16–27. [CrossRef]
16. Liu, B.; Zhang, M.; Zhao, J.; Zheng, M.; Yang, H. Imbalance of M1/M2 macrophages is linked to severity level of knee osteoarthritis. *Exp. Ther. Med.* **2018**, *16*, 5009–5014. [CrossRef] [PubMed]
17. Sun, A.R.; Panchal, S.K.; Friis, T.; Sekar, S.; Crawford, R.; Brown, L.; Xiao, Y.; Prasadam, I. Obesity-associated metabolic syndrome spontaneously induces infiltration of pro-inflammatory macrophage in synovium and promotes osteoarthritis. *PLoS ONE* **2017**, *12*, e0183693. [CrossRef]
18. Manferdini, C.; Paolella, F.; Gabusi, E.; Gambari, L.; Piacentini, A.; Filardo, G.; Fleury-Cappellesso, S.; Barbero, A.; Murphy, M.; Lisignoli, G. Adipose stromal cells mediated switching of the pro-inflammatory profile of M1-like macrophages is facilitated by PGE2: In vitro evaluation. *Osteoarthr. Cartil.* **2017**, *25*, 1161–1171. [CrossRef]
19. Woodell-May, J.E.W.; Sommerfeld, S.D. Role of Inflammation and the Immune System in the Progression of Osteoarthritis. *J. Orthop. Res.* **2020**, *38*, 253–257. [CrossRef] [PubMed]
20. Nishio, H.; Saita, Y.; Kobayashi, Y.; Takaku, T.; Fukusato, S.; Uchino, S.; Wakayama, T.; Ikeda, H.; Kaneko, K. Platelet-rich plasma promotes recruitment of macrophages in the process of tendon healing. *Regen. Ther.* **2020**, *14*, 262–270. [CrossRef] [PubMed]
21. Omar, N.N.; Rashed, R.R.; El-Hazek, R.M.; El-Sabbagh, W.A.; Rashed, E.R.; El-Ghazaly, M.A. Platelet-rich plasma-induced feedback inhibition of activin A/follistatin signaling: A mechanism for tumor-low risk skin rejuvenation in irradiated rats. *J. Photochem. Photobiol. B: Biol.* **2018**, *180*, 17–24. [CrossRef]
22. Escobar, G.; Escobar, A.; Ascui, G.; Tempio, F.I.; Ortiz, M.C.; Pérez, C.A.; López, M.N. Pure platelet-rich plasma and supernatant of calcium-activated P-PRP induce different phenotypes of human macrophages. *Regen. Med.* **2018**, *13*, 427–441. [CrossRef]
23. Kon, E.; Di Matteo, B.; Delgado, D.; Cole, B.J.; Dorotei, A.; Dragoo, J.L.; Filardo, G.; Fortier, L.A.; Giuffrida, A.; Jo, C.H.; et al. Platelet-rich plasma for the treatment of knee osteoarthritis: An expert opinion and proposal for a novel classification and coding system. *Expert Opin. Biol. Ther.* **2020**, *20*, 1447–1460. [CrossRef] [PubMed]
24. Stout, R.D.; Jiang, C.; Matta, B.; Tietzel, I.; Watkins, S.K.; Suttles, J. Macrophages Sequentially Change Their Functional Phenotype in Response to Changes in Microenvironmental Influences. *J. Immunol.* **2005**, *175*, 342–349. [CrossRef]
25. Gawaz, M.P.; Vogel, S. Platelets in tissue repair: Control of apoptosis and interactions with regenerative cells. *Blood* **2013**, *122*, 2550–2554. [CrossRef] [PubMed]
26. Knop, E.; De Paula, L.E.; Fuller, R. Platelet-rich plasma for osteoarthritis treatment. *Rev. Bras. de Reum. (Engl. Ed.)* **2016**, *56*, 152–164. [CrossRef]
27. Wasai, S.; Sato, M.; Maehara, M.; Toyoda, E.; Uchiyama, R.; Takahashi, T.; Okada, E.; Iwasaki, Y.; Suzuki, S.; Watanabe, M. Characteristics of autologous protein solution and leucocyte-poor platelet-rich plasma for the treatment of osteoarthritis of the knee. *Sci. Rep.* **2020**, *10*, 1–11. [CrossRef] [PubMed]
28. Mussano, F.; Genova, T.; Munaron, L.; Petrillo, S.; Erovigni, F.; Carossa, S. Cytokine, chemokine, and growth factor profile of platelet-rich plasma. *Platelets* **2016**, *27*, 467–471. [CrossRef]
29. Castillo, T.N.; Pouliot, M.A.; Kim, H.J.; Dragoo, J.L. Comparison of Growth Factor and Platelet Concentration From Commercial Platelet-Rich Plasma Separation Systems. *Am. J. Sports Med.* **2010**, *39*, 266–271. [CrossRef]
30. Julier, Z.; Park, A.J.; Briquez, P.S.; Martino, M.M. Promoting tissue regeneration by modulating the immune system. *Acta Biomater.* **2017**, *53*, 13–28. [CrossRef]
31. Novak, M.L.; Koh, T.J. Macrophage phenotypes during tissue repair. *J. Leukoc. Biol.* **2013**, *93*, 875–881. [CrossRef] [PubMed]
32. Benoit, M.; Desnues, B.; Mege, J.-L. Macrophage Polarization in Bacterial Infections. *J. Immunol.* **2008**, *181*, 3733–3739. [CrossRef]
33. Gordon, S.; Taylor, P.R. Monocyte and macrophage heterogeneity. *Nat. Rev. Immunol.* **2005**, *5*, 953–964. [CrossRef] [PubMed]
34. Mantovani, A.; Biswas, S.K.; Galdiero, M.R.; Sica, A.; Locati, M. Macrophage plasticity and polarization in tissue repair and remodelling. *J. Pathol.* **2013**, *229*, 176–185. [CrossRef] [PubMed]
35. Pessler, F.; Dai, L.; Diaz-Torne, C.; Gomez-Vaquero, C.; Paessler, M.E.; Zheng, D.-H.; Einhorn, E.; Range, U.; Scanzello, C.; Schumacher, H.R. The synovitis of "non-inflammatory" orthopaedic arthropathies: A quantitative histological and immunohistochemical analysis. *Ann. Rheum. Dis.* **2008**, *67*, 1184–1187. [CrossRef] [PubMed]
36. Zhang, H.; Lin, C.; Zeng, C.; Wang, Z.; Wang, H.; Lu, J.; Liu, X.; Shao, Y.; Zhao, C.; Pan, J.; et al. Synovial macrophage M1 polarisation exacerbates experimental osteoarthritis partially through R-spondin-2. *Ann. Rheum. Dis.* **2018**, *77*, 1524–1534. [CrossRef] [PubMed]
37. O'Shaughnessey, K.; Matuska, A.; Hoeppner, J.; Farr, J.; Klaassen, M.; Kaeding, C.; Lattermann, C.; King, W.; Woodell-May, J. Autologous protein solution prepared from the blood of osteoarthritic patients contains an enhanced profile of anti-inflammatory cytokines and anabolic growth factors. *J. Orthop. Res.* **2014**, *32*, 1349–1355. [CrossRef]
38. Wojdasiewicz, P.; Poniatowski, Ł.A.; Szukiewicz, D. The Role of Inflammatory and Anti-Inflammatory Cytokines in the Pathogenesis of Osteoarthritis. *Mediat. Inflamm.* **2014**, *2014*, 1–19. [CrossRef]
39. Rőszer, T. Understanding the Mysterious M2 Macrophage through Activation Markers and Effector Mechanisms. *Mediat. Inflamm.* **2015**, *2015*, 1–16. [CrossRef]
40. Vogel, D.Y.; Glim, J.E.; Stavenuiter, A.W.; Breur, M.; Heijnen, P.; Amor, S.; Dijkstra, C.D.; Beelen, R.H. Human macrophage polarization in vitro: Maturation and activation methods compared. *Immunobiology* **2014**, *219*, 695–703. [CrossRef]

41. Bertani, F.R.; Mozetic, P.; Fioramonti, M.; Iuliani, M.; Ribelli, G.; Pantano, F.; Santini, D.; Tonini, G.; Trombetta, M.; Businaro, L.; et al. Classification of M1/M2-polarized human macrophages by label-free hyperspectral reflectance confocal microscopy and multivariate analysis. *Sci. Rep.* **2017**, *7*, 1–9. [CrossRef] [PubMed]
42. Tjiu, J.-W.; Chen, J.-S.; Shun, C.-T.; Lin, S.-J.; Liao, Y.-H.; Chu, C.-Y.; Tsai, T.-F.; Chiu, H.-C.; Dai, Y.-S.; Inoue, H.; et al. Tumor-Associated Macrophage-Induced Invasion and Angiogenesis of Human Basal Cell Carcinoma Cells by Cyclooxygenase-2 Induction. *J. Investig. Dermatol.* **2009**, *129*, 1016–1025. [CrossRef] [PubMed]

International Journal of
Molecular Sciences

Commentary

Characteristics of MSCs in Synovial Fluid and Mode of Action of Intra-Articular Injections of Synovial MSCs in Knee Osteoarthritis

Ichiro Sekiya *, Hisako Katano and Nobutake Ozeki

Center for Stem Cells and Regenerative Medicine, Tokyo Medical and Dental University, 1-5-45 Yushima, Bunkyo-ku, Tokyo 113-8510, Japan; katano.arm@tmd.ac.jp (H.K.); Ozeki.arm@tmd.ac.jp (N.O.)
* Correspondence: sekiya.arm@tmd.ac.jp

Abstract: We have been studying mesenchymal stem cells (MSCs) in synovial fluid and the intra-articular injection of synovial MSCs in osteoarthritis (OA) knees. Here, mainly based on our own findings, we overview the characteristics of endogenous MSCs in the synovial fluid of OA knees and their mode of action when injected exogenously into OA knees. Many MSCs similar to synovial MSCs were detected in the synovial fluid of human OA knees, and their number correlated with the radiological OA grade. Our suspended synovium culture model demonstrated the release of MSCs from the synovium through a medium into a non-contacting culture dish. In OA knees, endogenous MSCs possibly mobilize in a similar manner from the synovium through the synovial fluid and act protectively. However, the number of mobilized MSCs is limited; therefore, OA progresses in its natural course. Synovial MSC injections inhibited the progression of cartilage degeneration in a rat OA model. Injected synovial MSCs migrated into the synovium, maintained their MSC properties, and increased the gene expressions of TSG-6, PRG-4, and BMP-2. Exogenous synovial MSCs can promote anti-inflammation, lubrication, and cartilage matrix synthesis in OA knees. Based on our findings, we have initiated a human clinical study of synovial MSC injections in OA knees.

Keywords: MSC; synovial fluid; mode of action; injection; synovium; knee; osteoarthritis; TSG-6; PRG-4; BMP-2

Citation: Sekiya, I.; Katano, H.; Ozeki, N. Characteristics of MSCs in Synovial Fluid and Mode of Action of Intra-Articular Injections of Synovial MSCs in Knee Osteoarthritis. *Int. J. Mol. Sci.* **2021**, *22*, 2838. https://doi.org/10.3390/ijms22062838

Academic Editor: Masato Sato

Received: 28 January 2021
Accepted: 1 March 2021
Published: 11 March 2021

Publisher's Note: MDPI stays neutral with regard to jurisdictional claims in published maps and institutional affiliations.

1. Introduction

Mesenchymal stem cells (MSCs) are derived from mesenchymal tissue and have the functional capacity to self-renew and generate a number of differentiated progeny [1]. These cells participate in tissue homoeostasis, remodeling, and repair by ensuring the replacement of mature cells that are lost during the course of physiological turnover, senescence, injury, or disease [2]. MSCs can be isolated from bone marrow as well as from various adult mesenchymal tissues, including the synovium [3,4]. Reports on the intra-articular injection of MSCs for the treatment of knee osteoarthritis (OA), the most prevalent degenerative joint disease, have increased in recent years. The incidence of OA is rising due to the aging of populations [5], and OA currently has no effective disease-modifying drugs due to its complicated chronic pathology. However, a systematic review has shown that MSC injections can improve OA pain in many cases and increase cartilage volume in some cases [6–8]. We have been actively studying MSCs obtained from synovial fluid, as well as the intra-articular injection of these MSCs in OA knees. This paper provides an overview, based largely on our own findings, of the role of endogenous MSCs in the synovial fluid of OA knees and their mode of action when injected exogenously into the synovium as a therapy for OA knees.

2. Main

2.1. MSCs in Synovial Fluid of OA Knees

We aspirated synovial fluid from OA knees, plated the cell components on a 10 cm-diameter dish, and cultured the cells for 14 days. Staining with crystal violet revealed a large number of cell colonies (Figure 1a). These colony-forming cells could differentiate into chondrocytes and adipocytes and undergo calcification, indicating a multi-differentiation potential [9]. Thus, the colony-forming cells from synovial fluid of OA knees had the characteristics of MSCs, and the number of colonies reflected the number of MSCs in the synovial fluid. MSCs in synovial fluid were seldom detected in normal volunteer knees but were found in much greater numbers in OA knees, and their number correlated with the radiological OA grade (Figure 1b) [9].

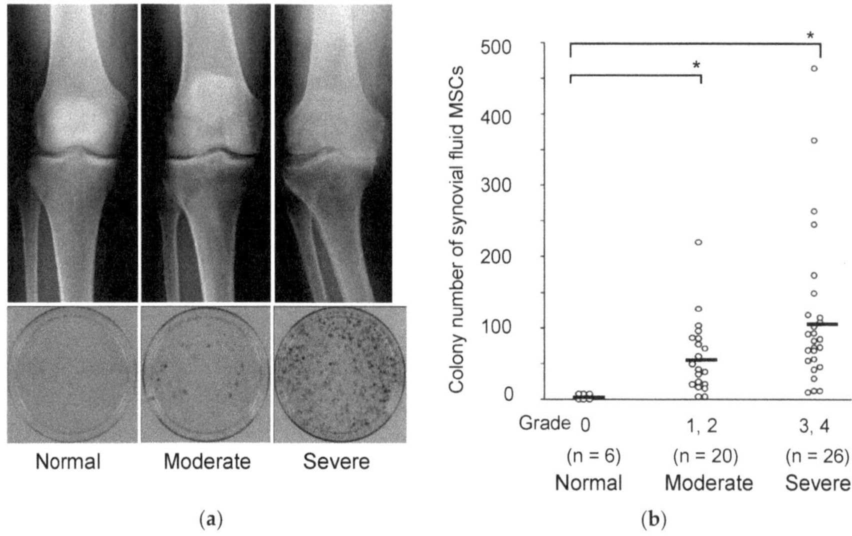

Figure 1. MSCs in synovial fluid derived from osteoarthritis patients by radiographic grade. (**a**) Radiographic images of the knees and representative dishes showing colonies of synovial fluid MSCs. (**b**) Relationship between the Kellgren–Lawrence grading and the colony number of synovial fluid MSCs per synovial fluid volume (mL). Average values are shown as bars (* $p < 0.05$) (reproduced from [9]).

2.2. MSCs from Synovial Fluid Resemble Synovial MSCs

We collected the synovial fluid, bone marrow, and synovium from the same OA knee during knee arthroplasty and prepared MSCs from those tissues. Morphologically, the synovial fluid MSCs appeared to be closer in character to synovial MSCs than to bone marrow MSCs, as the synovial MSCs and synovial fluid MSCs were narrower than bone marrow MSCs and their nuclei were more obvious (Figure 2a). We also prepared mRNA derived from each MSC type from three knees and conducted a comprehensive gene expression analysis using microarrays. Hierarchical clustering analysis demonstrated that the gene profiles in the synovial fluid MSCs were more similar to those of synovial MSCs than those of bone marrow MSCs (Figure 2b) [9]. These results indicate that the MSCs in synovial fluid are similar to the MSCs of the synovium.

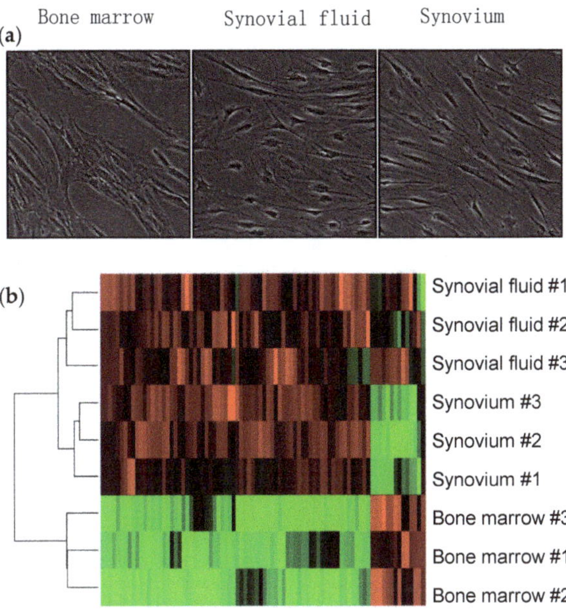

Figure 2. Comparison of synovium MSCs, synovial fluid MSCs, and bone marrow MSCs from osteoarthritis patients. (**a**) Representative morphologies. (**b**) Comparison of the gene expression profiles by hierarchical clustering analysis (reproduced from [9]).

2.3. Migration of MSCs from the Synovium via the Synovial Fluid in a Tissue Culture System

Synovial fluid from OA knee contains MSCs. One of the possible reservoirs of the MSCs found in synovial fluid is the synovium itself, and synovial fluid may induce the mobilization of MSCs into synovial fluid in OA patients. We investigated whether synovial fluid could expand synovial MSCs in a tissue culture system. The synovial fluid in OA knees contains a high concentration of transforming growth factor β (TGFβ), so the effect of TGFβ was also examined. Autologous synovial fluid promoted a greater expansion of synovial MSCs than was observed with treatment with α minimum essential medium (αMEM) + fetal bovine serum (FBS), and the addition of TGFβ to αMEM+FBS increased this expansion to a similar level in samples from all 11 OA donors (Figure 3). The synovial cells expanded in synovial fluid retained their multipotentiality and showed surface markers characteristic of MSCs. The addition of an anti-TGFβ neutralizing antibody to the synovial fluid partially inhibited synovial cell expansion. These experiments demonstrated that autologous synovial fluid enhanced the expansion of MSCs in tissue cultures of synovium from OA patients by promoting cell migration, and this effect was partially induced by TGFβ [10].

2.4. MSCs Released from Suspended Synovium

Although the origin of the MSCs found in synovial fluid could be the synovium, no direct evidence had been demonstrated to confirm that MSCs were mobilized from the synovium into the synovial fluid. We developed a novel in vitro model, the "suspended synovium culture model", to provide this evidence. The synovium was harvested during total knee arthroplasty, cut into approximately 1 g specimens, and washed thoroughly to remove blood. Each synovium specimen was then sutured with 4-0 nylon thread and suspended in a 100 mL bottle. A 35 mm-diameter culture dish containing 40 mL αMEM+FBS had previously been placed at the bottom of the bottle. The specimens were cultured for 7 days (Figure 4a) and then stained with crystal violet staining. Cell colonies were confirmed from the specimens from all 28 donors (Figure 4b). The colony-forming cells consisted of spindle cells (Figure 4c), and they formed a cartilage pellet that was positively stained with safranin-o, differentiated into adipocytes, and calcified when cultured in the appropriate differentiation medium (Figure 4d). This suspended synovium culture model confirmed the release of MSCs and their passage through the medium in a bottle into a non-contacting culture dish [11].

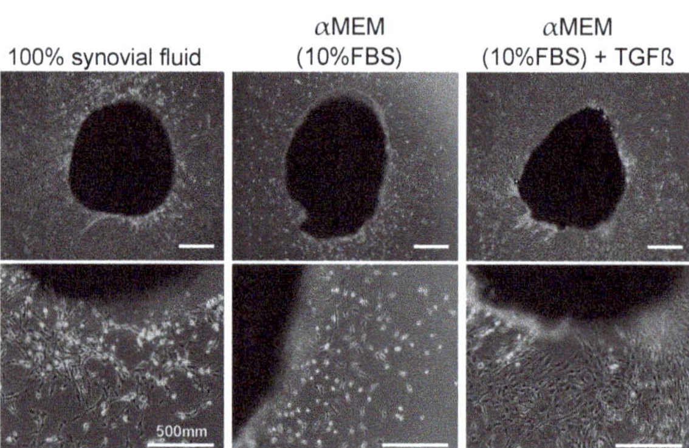

Figure 3. Morphology of an explant culture of synovium. Synovial tissues were cultured for 8 days in autologous synovial fluid, αMEM (10%FBS), or αMEM (10%FBS) with 10 ng/mL of TGFβ (reproduced from [10]).

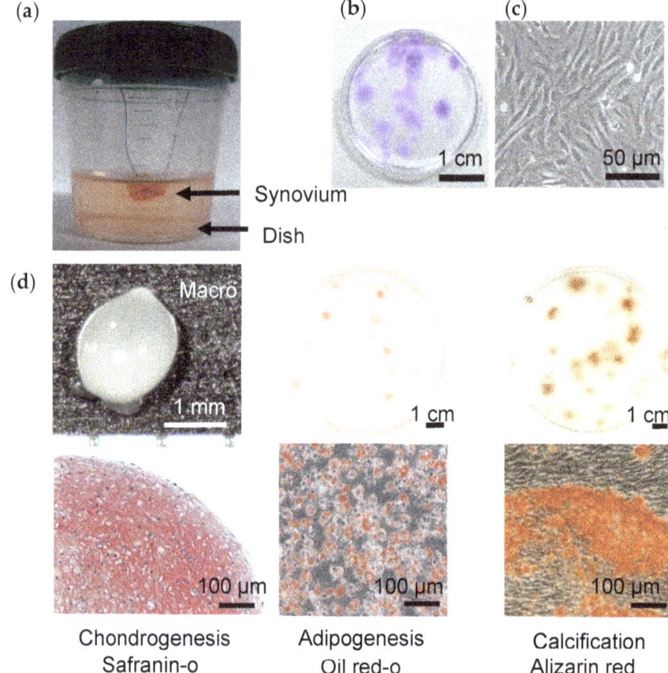

Figure 4. MSCs released from suspended synovium. (**a**) Suspended synovium culture model. Approximately 1 g of synovium was suspended in a bottle containing a culture dish at the bottom. (**b**) Cell morphology. After culturing for 7 days, the dish was observed. (**c**) Culture dishes stained with crystal violet. (**d**) Chondrogenesis, adipogenesis, and calcification (reproduced from [11]).

2.5. MSCs in Synovial Fluid in Intra-Articular Tissue Injuries

We previously demonstrated that the MSCs in synovial fluid increased in number after anterior cruciate ligament (ACL) injury in humans [12]. Cluster analysis of gene profiles demonstrated that the synovial fluid MSCs from patients with ACL injury were more similar to synovial MSCs than to bone marrow MSCs, as previously observed in patients with OA [12]. We also attempted injection of synovial MSCs into the knee joint in a rabbit model of partial ACL. A greater number of injected MSCs was observed in the injured area than in the uninjured area of the ligament [12]. Kanaya et al. injected MSCs into the knees of rats with a partially resected ACL and demonstrated adhesion of the cells to the injured site and a contribution of MSCs to the repair of the resection [13].

We also examined the correlation between the number of MSCs in synovial fluid and the cartilage degeneration score evaluated arthroscopically in 22 patients with ACL injury. The MSC number in the synovial fluid was correlated with the cartilage degeneration score [9]. Use of a rabbit cartilage defect model also confirmed adhesion of injected synovial MSCs to the cartilage lesion and the promotion of cartilage regeneration [14].

MSCs in synovial fluid were also greater in number in knees with a meniscus injury than in normal knees. The number of MSCs in synovial fluid was positively correlated with the duration after meniscus injury [15]. The synovial MSCs injected into the knee joint adhered to the meniscus lesion and promoted meniscal regeneration in partial meniscectomy models in rats [16,17], rabbits [18], and pigs [19].

These findings suggest that synovium serves as a reservoir of MSCs that are mobilized following intra-articular tissue injuries and that migrate to the injury site to participate in the repair response. However, the number of endogenous MSCs in the synovial fluid is probably too small to contribute to the repair of intra-articular tissue injuries by natural processes.

2.6. Possible Roles of Endogenous MSCs in the Synovial Fluid of OA Knee

In the adult cartilage, enzymatic activities and the mechanical stress imposed on the joints inevitably lead to cartilage damage. Under normal circumstances, this damage is overcome by the turnover of matrix components synthesized by chondrocytes. Thus, in the normal adult articular cartilage, a balance exists between cartilage anabolism and catabolism. In OA, the catabolism becomes stronger than the anabolic capacities of chondrocytes; therefore, the cartilage matrix degenerates and the joint cartilage is damaged [20]. MSCs are rare in the synovial fluid in normal knees but are much more prevalent in OA knees (Figure 5). We speculate that the endogenous MSCs mobilize from synovium through synovial fluid and act protectively. However, the number of MSCs that can be mobilized is limited, so OA undergoes its natural progression. We therefore investigated whether the intra-articular injection of exogenous synovial MSCs could inhibit the progression of osteoarthritis.

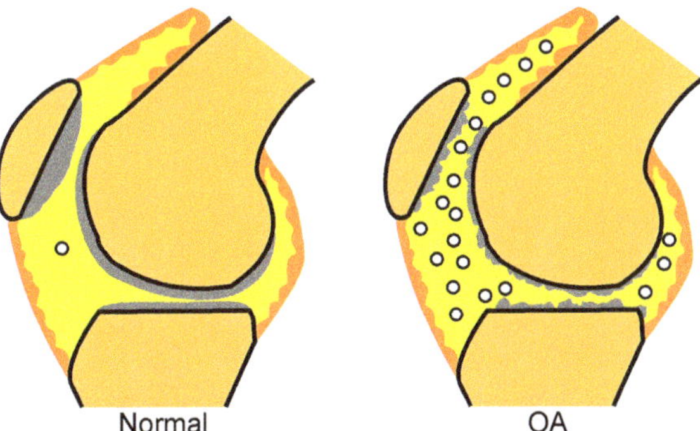

Figure 5. MSCs in synovial fluid. MSCs in synovial fluid are seldom detected in normal knees but are found in greater numbers in OA knees.

2.7. Synovial MSC Injections in a Rat OA Model

We completely transected the ACL of the rats and allowed them to walk freely in the cage. The rats were given intra-articular injections of PBS alone (control group) or 1×10^6 synovial MSCs once (single group) or weekly (weekly group), beginning one week after the surgery. In the control group, the cartilage fissured at 8 weeks and extensively disappeared by 12 weeks (Figure 6). In the single group, the cartilage disappeared over time but was better preserved than in the control group. In the weekly group, the cartilage was essentially preserved at 12 weeks [21]. The MSC injections inhibited the progression of cartilage degeneration in a frequency-dependent manner.

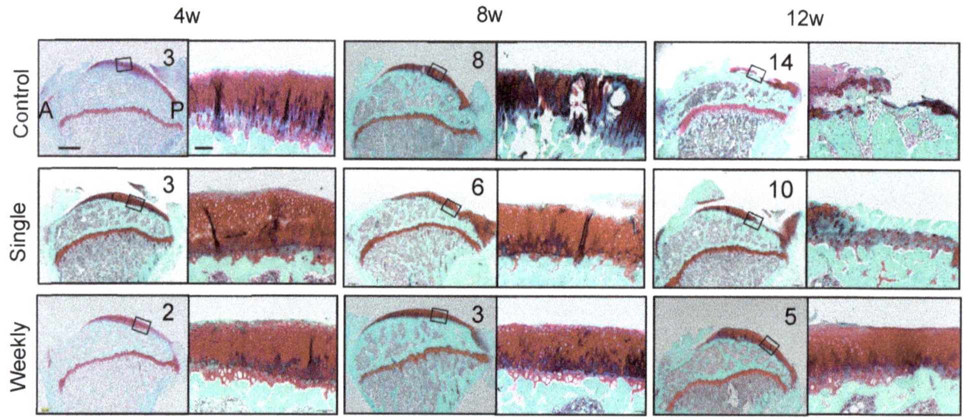

Figure 6. The effect of intra-articular injections of synovial MSCs in ACL-transected rats. Synovial MSCs were injected once or weekly. Histological sections of tibial cartilage stained with safranin-o and the Osteoarthritis Research Society International (OARSI) scores are shown. A, anterior; P, posterior (reproduced from [21]).

2.8. Distribution of Synovial MSCs after Injection

We injected rat synovial MSCs expressing LacZ into the knee to determine the distribution of the cells at 1 day. The MSCs with blue staining for X-gal were widely observed in the synovium (Figure 7a) but not in the cartilage or meniscus (Figure 7b). We also injected luciferase-expressing rat synovial MSCs into the knee and analyzed them with an in vivo imaging system to investigate the activity of the cells and their migration out of the joint. The fluorescence was more prominent in the ACL-transected knee injected once than in the intact knee injected once, but the fluorescence was no longer detectable in either knee after 14 days. By contrast, a sustained fluorescence was observed after 14 days in the group that received weekly injections. The injected synovial MSCs were engrafted into the synovium, and weekly injections maintained a high cellular activity. The injected MSCs did not migrate outside the knee joint [21].

2.9. Surface Epitopes and Differentiation Potential of MSCs after Migration

We injected green fluorescent protein (GFP)-expressing MSCs (GFP+ MSCs) into the knee, selected GFP+ cells migrating into the synovium using flow cytometry, and examined the MSC properties of the migrated cells (Figure 7c). The percentage of GFP+ cells relative to the total viable cells decreased gradually over time (Figure 7d), but the percentage of CD90+ cells relative to the total GFP+ cells remained at approximately 90% after 28 days (Figure 7e). The GFP+ cells that were sorted after migrating into the synovium retained their multi-differentiation ability for chondrogenesis, adipogenesis, and calcification (Figure 7f) [21]. Therefore, the synovial MSCs injected into the knee joint and migrated to the synovium maintained the properties of MSCs.

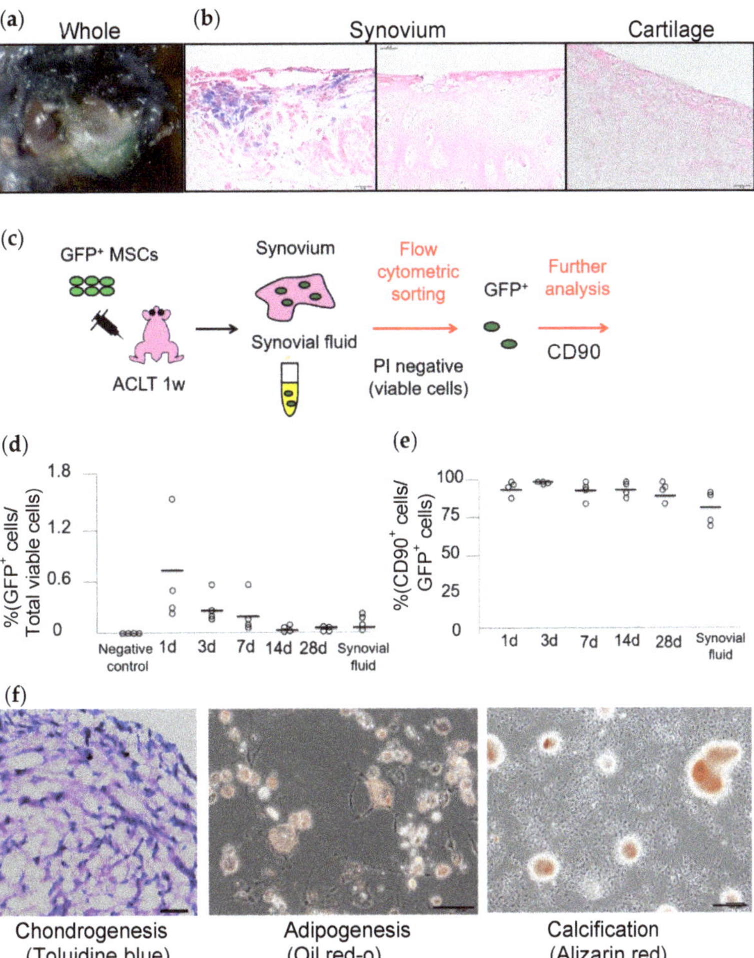

Figure 7. Distribution and properties of synovial MSCs injected into the knee. (**a**) The whole knee joint one day after injection of synovial MSCs expressing the LacZ gene into a rat knee. The ACL had been transected 7 days before. (**b**) Histological sections stained with X-gal. (**c**) Schema for the flow-cytometry assay. GFP-expressing MSCs were injected into the knees of rats 7 days after ACL transection and the synovium was harvested 1 week later. After enzymatic digestion, GFP+ cells were sorted for further analysis. PI, propidium iodide. (**d**) Ratio of GFP+ cells relative to total viable cells. (**e**) Ratio of CD90+ cells relative to total GFP+ cells. (**f**) Multi-differentiation potential of the sorted GFP+ cells 28 days after injection (reproduced from [21]).

2.10. Gene Expression of Synovial MSCs after Migration to Synovium

We used species-specific gene expression to analyze the gene expression changes of human synovial MSCs that migrated to rat synovium. Approximately 1% of the 1×10^6 human MSC injected into the rat synovium of knees with ACL transection had migrated one day after the intra-articular injection. The human transcriptomes were then compared between the MSCs in the synovium one day after injection and a control consisting of un-injected rat synovium mixed with 1×10^4 human MSCs (Figure 8). An analysis of human mRNA using microarrays showed that 5 genes had undergone a more than 100-fold increase, 21 genes had a greater than 50-fold increase, and 255 genes had a greater than 10-fold increase in the human MSCs that had migrated to the rat synovium. Human proteoglycan-4 (PRG-4) and human bone morphogenetic protein-2 (BMP-2) were among the 10 most highly expressed human transcripts (Table 1). Further expression analysis of human mRNA by reverse transcription- polymerase chain reaction (RT-PCR) showed significant increases in the expression of human PRG-4, human BMP-2, human bone morphogenetic protein-6 (BMP-6), and human TNF-stimulated gene-6 (TSG-6) [21]. PRG-4, also known as lubricin, is produced by synovial cells or by superficial zone chondrocytes and has a key role in the homeostasis and maintenance of cartilage [22]. BMP-2 and BMP-6 have important biological effects on chondrocyte differentiation, cartilage matrix synthesis, and cartilage protection [23]. TSG-6 is secreted by transplanted MSCs and suppresses inflammation [24].

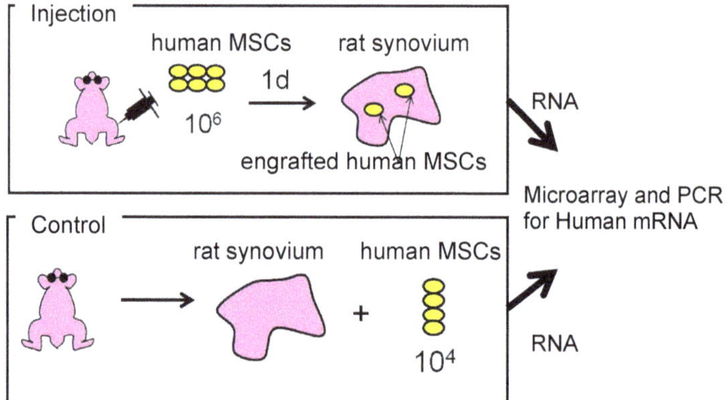

Figure 8. Scheme for species-specific gene expression analysis. We injected 1×10^6 human synovial MSCs into the knees of rats 7 days after ACL transection. One day later, we harvested rat synovium and prepared total RNA for microarrays and RT-PCR. We also prepared total RNA from a mixture of harvested, un-injected control rat synovium, and 1×10^4 human synovial MSCs as a control (reproduced from [21]).

Table 1. The top 10 human transcripts upregulated in the human MSCs that migrated within the synovium.

Gene Symbol	Gene Title	Fold Change
TFPI2	Tissue factor pathway inhibitor 2	252.6
PRG4	Proteoglycan 4	162.3
PTHLH	Parathyroid hormone-like hormone	130.5
T.L	Transcribed locus	107.5
LOC285359///PDCL3	Phosducin-like 3 pseudogene///phosducin-like 3	102.6
PLA2G4A	Phospholipase A2, group IVA (cytosolic, calcium-dependent)	92.5
BMP2	Bone morphogenetic protein 2	87.4
RPS4Y1	Ribosomal protein S4, Y-linked 1	75.5
CACNA1D	Calcium channel, voltage-dependent, L type, alpha 1D subunit	63.6
COL15A1	Collagen, type XV, alpha 1	63.3

2.11. Mode of Action in Synovial MSC Injection Therapy for OA

Synovial MSCs injected into the knee joint mostly migrated to the synovium, and the cells maintained their MSC properties without differentiating into other lineages. Synovial MSCs act as anti-inflammatory agents through TSG-6 expression, as lubrication agents by PRG-4 expression, and in cartilage matrix synthesis by BMP expression (Figure 9). Consequently, exogenous MSCs can inhibit the progression of OA.

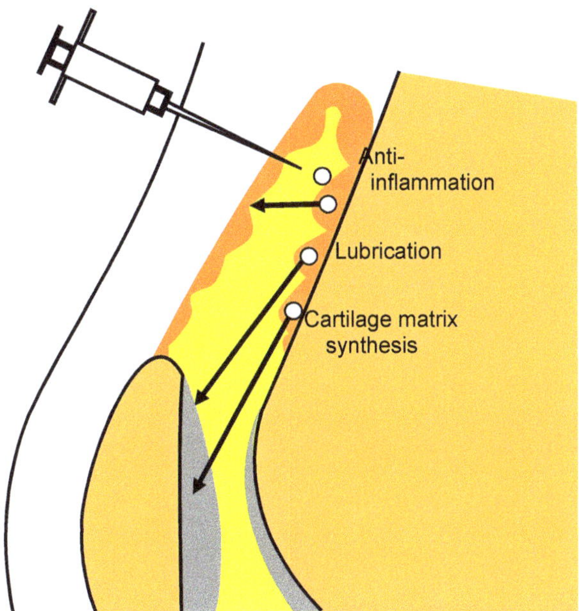

Figure 9. Mode of action in synovial MSC injection therapy for OA. Synovial MSCs injected into the knee joint migrate to the synovium and maintain the properties of MSCs. Synovial MSCs promote anti-inflammation, lubrication, and cartilage matrix synthesis. Consequently, synovial MSC injections inhibit the progression of OA.

2.12. Clinical Study of Synovial MSC Injections into OA Knees

We have previously reported that synovial MSC transplantation into cartilage defects in human patients improved MRI findings and clinical scores [25], and that synovial MSC transplantation into a repaired meniscus improved clinical outcomes [26]. Based on the results of these clinical and preclinical studies, we started a novel clinical study, "Intra-articular injections of synovial stem cells for osteoarthritis of the knee (UMIN 000026732)". to follow the effects of synovial MSC injections into osteoarthritic knees in 14 human patients. We harvested synovial tissue arthroscopically from each individual under local anesthesia, cultured the synovial MSCs with autologous serum, and injected two million cells into the knee twice at 15-week intervals (Figure 10a). In this clinical study, the last patient's treatment has already been completed and we are currently in the process of analyzing the MRI and clinical outcomes. The preliminary results from fully automatic 3D MRI analysis indicate that some participants, who had shown decreased cartilage thickness in the posteromedial region of the femoral cartilage before injection, now show increased thickness after injection [27,28] (Figure 10b).

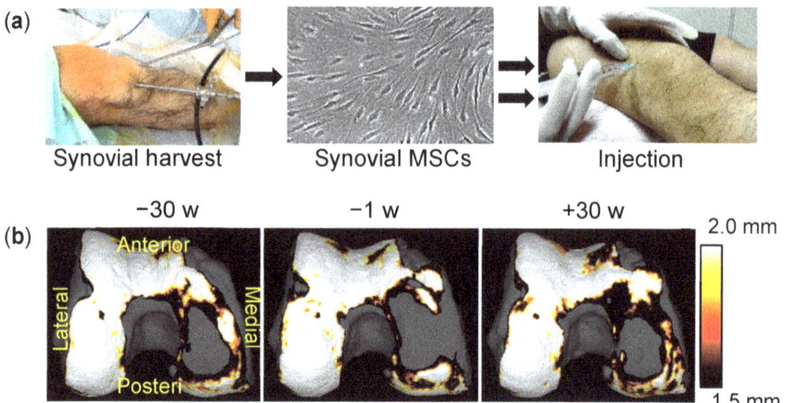

Figure 10. Clinical study of synovial MSC injections into OA knees. (**a**) Scheme of a clinical study of autologous synovial MSC injections into the OA knees of human patients. The synovium was harvested and enzymatically digested. The resulting synovial MSCs were expanded with autologous human serum and then injected into the knee twice at a 15-week interval. (**b**) Cartilage thickness mappings in a representative patient who underwent synovial MSC injections into a knee at time 0 and 15 weeks. MRI examinations were performed at −30, −1, and +30 weeks. Using the software we developed, the cartilage area was automatically extracted and visualized in three dimensions. Femoral cartilage at the posteromedial region at −30 weeks had decreased at −1 week but increased again at +30 weeks.

3. Concluding Remarks

In OA knees, MSCs may possibly mobilize from the synovium through synovial fluid and act protectively. However, the number of mobilized MSCs is limited, so OA progresses in its natural course. By contrast, synovial MSCs injected into the knee joint of a rat OA model migrate to the synovium, where they act as anti-inflammation, lubrication, and cartilage matrix synthesis agents, and inhibit the progression of OA. We have started a clinical study in which synovial MSCs are injected twice into the OA knees of human patients.

Author Contributions: Conceptualization, I.S.; methodology, N.O.; data curation, N.O.; writing—original draft preparation, I.S.; writing—review and editing, I.S., H.K., and N.O.; visualization, H.K. and I.S.; funding acquisition, I.S. and H.K. All authors have read and agreed to the published version of the manuscript.

Funding: This research was funded by the Japan Agency for Medical Research and Development, grant number JP20bk0104103 to I.S.

Institutional Review Board Statement: Not applicable.

Informed Consent Statement: Not applicable.

Data Availability Statement: No new data were created or analyzed in this study.

Acknowledgments: We thank Ellen Roider for English editing.

Conflicts of Interest: The authors declare no conflict of interest.

References

1. Horwitz, E.M.; Le Blanc, K.; Dominici, M.; Mueller, I.; Slaper-Cortenbach, I.; Marini, F.C.; Deans, R.J.; Krause, D.S.; Keating, A. Clarification of the nomenclature for MSC: The International Society for Cellular Therapy position statement. *Cytotherapy* **2005**, *7*, 393–395. [CrossRef] [PubMed]
2. De Bari, C.; Kurth, T.B.; Augello, A. Mesenchymal stem cells from development to postnatal joint homeostasis, aging, and disease. *Birth Defects Res. Part C Embryo Today Rev.* **2010**, *90*, 257–271. [CrossRef] [PubMed]

3.   De Bari, C.; Dell'Accio, F.; Tylzanowski, P.; Luyten, F.P. Multipotent mesenchymal stem cells from adult human synovial mem-brane. *Arthritis Rheum.* **2001**, *44*, 1942. [CrossRef]

4.   Sakaguchi, Y.; Sekiya, I.; Yagishita, K.; Muneta, T. Comparison of human stem cells derived from various mesenchymal tissues: Superiority of synovium as a cell source. *Arthritis Rheum.* **2005**, *52*, 2521–2529. [CrossRef]

5.   Bijlsma, J.W.; Berenbaum, F.; Lafeber, F.P. Osteoarthritis: An update with relevance for clinical practice. *Lancet* **2011**, *377*, 2115–2126. [CrossRef]

6.   Doyle, E.C.; Wragg, N.M.; Wilson, S.L. Intraarticular injection of bone marrow-derived mesenchymal stem cells enhances regeneration in knee osteoarthritis. *Knee Surgery Sports Traumatol. Arthrosc.* **2020**, *28*, 3827–3842. [CrossRef]

7.   Ha, C.W.; Park, Y.B.; Kim, S.H.; Lee, H.J. Intra-articular Mesenchymal Stem Cells in Osteoarthritis of the Knee: A Systematic Re-view of Clinical Outcomes and Evidence of Cartilage Repair. *Arthroscopy* **2019**, *35*, 277–288.e2. [CrossRef]

8.   McIntyre, J.A.; Jones, I.A.; Han, B.; Vangsness, J.C.T. Intra-articular Mesenchymal Stem Cell Therapy for the Human Joint: A Systematic Review. *Am. J. Sports Med.* **2018**, *46*, 3550–3563. [CrossRef]

9.   Sekiya, I.; Ojima, M.; Suzuki, S.; Yamaga, M.; Horie, M.; Koga, H.; Tsuji, K.; Miyaguchi, K.; Ogishima, S.; Tanaka, H.; et al. Human mesenchymal stem cells in synovial fluid increase in the knee with degenerated cartilage and osteoarthritis. *J. Orthop. Res.* **2011**, *30*, 943–949. [CrossRef]

10.   Zhang, S.; Muneta, T.; Morito, T.; Mochizuki, T.; Sekiya, I. Autologous synovial fluid enhances migration of mesenchymal stem cells from synovium of osteoarthritis patients in tissue culture system. *J. Orthop. Res.* **2008**, *26*, 1413–1418. [CrossRef]

11.   Katagiri, K.; Matsukura, Y.; Muneta, T.; Ozeki, N.; Mizuno, M.; Katano, H.; Sekiya, I. Fibrous Synovium Releases Higher Numbers of Mesenchymal Stem Cells Than Adipose Synovium in a Suspended Synovium Culture Model. *Arthrosc. J. Arthrosc. Relat. Surg.* **2017**, *33*, 800–810. [CrossRef]

12.   Morito, T.; Muneta, T.; Hara, K.; Ju, Y.-J.; Mochizuki, T.; Makino, H.; Umezawa, A.; Sekiya, I. Synovial fluid-derived mesenchymal stem cells increase after intra-articular ligament injury in humans. *Rheumatology* **2008**, *47*, 1137–1143. [CrossRef]

13.   Kanaya, A.; Deie, M.; Adachi, N.; Nishimori, M.; Yanada, S.; Ochi, M. Intra-articular Injection of Mesenchymal Stromal Cells in Partially Torn Anterior Cruciate Ligaments in a Rat Model. *Arthrosc. J. Arthrosc. Relat. Surg.* **2007**, *23*, 610–617. [CrossRef] [PubMed]

14.   Koga, H.; Shimaya, M.; Muneta, T.; Nimura, A.; Morito, T.; Hayashi, M.; Suzuki, S.; Ju, Y.-J.; Mochizuki, T.; Sekiya, I. Local adherent technique for transplanting mesenchymal stem cells as a potential treatment of cartilage defect. *Arthritis Res. Ther.* **2008**, *10*, R84. [CrossRef] [PubMed]

15.   Matsukura, Y.; Muneta, T.; Tsuji, K.; Koga, H.; Sekiya, I. Mesenchymal Stem Cells in Synovial Fluid Increase After Meniscus Injury. *Clin. Orthop. Relat. Res.* **2014**, *472*, 1357–1364. [CrossRef] [PubMed]

16.   Horie, M.; Sekiya, I.; Muneta, T.; Ichinose, S.; Matsumoto, K.; Saito, H.; Murakami, T.; Kobayashi, E. Intra-articular Injected Synovial Stem Cells Differentiate into Meniscal Cells Directly and Promote Meniscal Regeneration Without Mobilization to Distant Organs in Rat Massive Meniscal Defect. *Stem Cells* **2009**, *27*, 878–887. [CrossRef] [PubMed]

17.   Mizuno, K.; Muneta, T.; Morito, T.; Ichinose, S.; Koga, H.; Nimura, A.; Mochizuki, T.; Sekiya, I. Exogenous synovial stem cells adhere to defect of me-niscus and differentiate into cartilage cells. *J. Med. Dent. Sci.* **2008**, *55*, 101–111.

18.   Hatsushika, D.; Muneta, T.; Horie, M.; Koga, H.; Tsuji, K.; Sekiya, I. Intraarticular injection of synovial stem cells promotes meniscal regeneration in a rabbit massive meniscal defect model. *J. Orthop. Res.* **2013**, *31*, 1354–1359. [CrossRef]

19.   Hatsushika, D.; Muneta, T.; Nakamura, T.; Horie, M.; Koga, H.; Nakagawa, Y.; Tsuji, K.; Hishikawa, S.; Kobayashi, E.; Sekiya, I. Repetitive allogeneic intraarticular injections of synovial mesenchymal stem cells promote meniscus regeneration in a porcine massive meniscus defect model. *Osteoarthr. Cartil.* **2014**, *22*, 941–950. [CrossRef]

20.   Aigner, T.; Soeder, S.; Haag, J. IL-1ß and BMPs-Interactive players of cartilage matrix degradation and regeneration. *Eur. Cells Mater.* **2006**, *12*, 49–56. [CrossRef]

21.   Ozeki, N.; Muneta, T.; Koga, H.; Nakagawa, Y.; Mizuno, M.; Tsuji, K.; Mabuchi, Y.; Akazawa, C.; Kobayashi, E.; Matsumoto, K.; et al. Not single but periodic injections of synovial mesenchymal stem cells maintain viable cells in knees and inhibit osteoarthritis progression in rats. *Osteoarthr. Cartil.* **2016**, *24*, 1061–1070. [CrossRef] [PubMed]

22.   Rhee, D.K.; Marcelino, J.; Baker, M.; Gong, Y.; Smits, P.; Lefebvre, V.; Jay, G.D.; Stewart, M.; Wang, H.; Warman, M.L.; et al. The secreted glycoprotein lubricin protects cartilage sur-faces and inhibits synovial cell overgrowth. *J. Clin. Investig.* **2005**, *115*, 622–631. [CrossRef] [PubMed]

23.   Sekiya, I.; Larson, B.L.; Vuoristo, J.T.; Reger, R.L.; Prockop, D.J. Comparison of effect of BMP-2, -4, and -6 on in vitro cartilage formation of human adult stem cells from bone marrow stroma. *Cell Tissue Res.* **2005**, *320*, 269–276. [CrossRef]

24.   Lee, R.H.; Pulin, A.A.; Seo, M.J.; Kota, D.J.; Ylostalo, J.; Larson, B.L.; Semprun-Prieto, L.; Delafontaine, P.; Prockop, D.J. Intravenous hMSCs Improve Myocardial Infarction in Mice because Cells Embolized in Lung Are Activated to Secrete the Anti-inflammatory Protein TSG-6. *Cell Stem Cell* **2009**, *5*, 54–63. [CrossRef]

25.   Sekiya, I.; Muneta, T.; Horie, M.; Koga, H. Arthroscopic Transplantation of Synovial Stem Cells Improves Clinical Outcomes in Knees with Cartilage Defects. *Clin. Orthop. Relat. Res.* **2015**, *473*, 2316–2326. [CrossRef]

26.   Sekiya, I.; Koga, H.; Otabe, K.; Nakagawa, Y.; Katano, H.; Ozeki, N.; Mizuno, M.; Horie, M.; Kohno, Y.; Katagiri, K.; et al. Additional Use of Synovial Mesenchymal Stem Cell Transplantation Following Surgical Repair of a Complex Degenerative Tear of the Medial Meniscus of the Knee: A Case Report. *Cell Transplant.* **2019**, *28*, 1445–1454. [CrossRef]

27. Aoki, H.; Ozeki, N.; Katano, H.; Hyodo, A.; Miura, Y.; Matsuda, J.; Takanashi, K.; Suzuki, K.; Masumoto, J.; Okanouchi, N.; et al. Relationship between medial meniscus extrusion and cartilage measurements in the knee by fully automatic three-dimensional MRI analysis. *BMC Musculoskelet. Disord.* **2020**, *21*, 3. [CrossRef]
28. Hyodo, A.; Ozeki, N.; Kohno, Y.; Suzuki, S.; Mizuno, M.; Otabe, K.; Katano, H.; Tomita, M.; Nakagawa, Y.; Koga, H.; et al. Projected Cartilage Area Ratio Determined by 3-Dimensional MRI Analysis: Validation of a Novel Technique to Evaluate Articular Cartilage. *JB JS Open Access* **2019**, *4*, e0010. [CrossRef]

International Journal of
Molecular Sciences

Article

Development of Injectable Polydactyly-Derived Chondrocyte Sheets

Shiho Wasai [1,2], Eriko Toyoda [1,2], Takumi Takahashi [1,2], Miki Maehara [1,2], Eri Okada [1,2], Ryoka Uchiyama [1,2], Tadashi Akamatsu [3], Masahiko Watanabe [1,2] and Masato Sato [1,2,*]

1   Department of Orthopaedic Surgery, Surgical Science, Tokai University School of Medicine, 143 Shimokasuya, Isehara, Kanagawa 259-1193, Japan; sw12345@tsc.u-tokai.ac.jp (S.W.); etoyoda@tokai-u.jp (E.T.); takumi.takahashi22@gmail.com (T.T.); m-maehara@tsc.u-tokai.ac.jp (M.M.); okada-eri@tokai-u.jp (E.O.); 9bmrm002@mail.u-tokai.ac.jp (R.U.); masahiko@is.icc.u-tokai.ac.jp (M.W.)
2   Center for Musculoskeletal Innovative Research and Advancement (C-MiRA), Graduate School of Medicine, Tokai University, 143 Shimokasuya, Isehara, Kanagawa 259-1193, Japan
3   Department of Plastic Surgery, Surgical Science, Tokai University School of Medicine, 143 Shimokasuya, Isehara, Kanagawa 259-1193, Japan; akamatu@is.icc.u-tokai.ac.jp
*   Correspondence: sato-m@is.icc.u-tokai.ac.jp; Tel.: +81-46-393-1121; Fax: +81-46-396-4404

Abstract: We are conducting a clinical study of the use of allogeneic polydactyly-derived chondrocyte sheets (PD sheets) for the repair of articular cartilage damage caused by osteoarthritis. However, the transplantation of PD sheets requires highly invasive surgery. To establish a less invasive treatment, we are currently developing injectable fragments of PD sheets (PD sheets-mini). Polydactyly-derived chondrocytes were seeded in RepCell™ or conventional temperature-responsive inserts and cultured. Cell counts and viability, histology, enzyme-linked immunosorbent assay (ELISA), quantitative real-time polymerase chain reaction (qPCR), and flow cytometry were used to characterize PD sheets-mini and PD sheets collected from each culture. To examine the effects of injection on cell viability, PD sheets-mini were tested in four experimental conditions: non-injection control, 18 gauge (G) needle, 23G needle, and syringe only. PD sheets-mini produced similar amounts of humoral factors as PD sheets. No histological differences were observed between PD sheets and PD sheets-mini. Except for *COL2A1*, expression of cartilage-related genes did not differ between the two types of PD sheet. No significant differences were observed between injection conditions. PD sheets-mini have characteristics that resemble PD sheets. The cell viability of PD sheets-mini was not significantly affected by needle gauge size. Intra-articular injection may be a feasible, less invasive method to transplant PD sheets-mini.

Keywords: cartilage regeneration; cell sheet; osteoarthritis; minimally invasive treatment

Citation: Wasai, S.; Toyoda, E.; Takahashi, T.; Maehara, M.; Okada, E.; Uchiyama, R.; Akamatsu, T.; Watanabe, M.; Sato, M. Development of Injectable Polydactyly-Derived Chondrocyte Sheets. *Int. J. Mol. Sci.* **2021**, *22*, 3198. https://doi.org/10.3390/ijms22063198

Academic Editor: Bernd Rolauffs

Received: 18 February 2021
Accepted: 17 March 2021
Published: 21 March 2021

Publisher's Note: MDPI stays neutral with regard to jurisdictional claims in published maps and institutional affiliations.

1. Introduction

Articular cartilage is a unique tissue that comprises mainly extracellular matrix (ECM), in the form of collagens and proteoglycans, and has a hypocellular and avascular structure [1]. Given its tissue properties, articular cartilage can tolerate intensive and repetitive physical stress but, once damaged, is difficult to repair [2]. Diffuse damage to articular cartilage gradually promotes joint degeneration, which eventually leads to osteoarthritis (OA), especially in the knee joint.

OA of the knee (OAK) is the most common joint disease, involves cartilage defects of varied depth and breadth, and is the most frequent cause of knee pain and dysfunction. OA progresses slowly with age, reduces the quality of life and work productivity, and ultimately imposes socioeconomic costs worldwide [3]. The Framingham osteoarthritis study found that the age-standardized prevalence of OAK reached 19.2% (18.6% for men and 19.3% for women) in people older than 45 years. In Japan, where the aging population is increasing, the number of OAK patients is expected to continue increasing in the future [4–6].

Current methods of treatment for OAK include patient education, exercise, oral treatments, conservative therapy by intra-articular injection of hyaluronic acid or steroids, and surgical treatments, including arthroscopic debridement of the knee joint, osteotomy around the knee such as a high tibial osteotomy, unicompartmental knee arthroplasty, and total knee arthroplasty [7,8].

Autologous chondrocyte implantation (ACI) is performed as a regenerative treatment for cartilage defects. However, most of the indications for ACI are cartilage defects caused by trauma and osteochondritis dissecans, and ACI does not have a sufficient therapeutic effect on OAK. Moreover, because OAK is characterized by heterogeneous cartilage defects, no fully effective regenerative treatment has been developed to stop the progression of OAK or to induce cartilage regeneration [9–11].

Cell sheet engineering is a technique that enables the collection of cells in sheet form simply by lowering the temperature from 37 °C to below 32 °C without using enzymes such as trypsin and growing the cells in a temperature-responsive culture dish with the temperature-responsive polymer poly (N-isopropyl acrylamide) [12]. Cell sheets have been applied as a regenerative treatment of various tissues including the cornea [13], esophagus [14], myocardium [15], and periodontium [16]. We have been engaged in research on cartilage regenerative medicine using this cell sheet engineering since 2004. At present, autologous chondrocyte sheet transplantation is available as an advanced medical care B for the treatment of OAK in Japan. However, two surgical procedures are required to fabricate and transplant an autologous chondrocyte sheet. In addition, the proliferative capacity of adult chondrocytes varies greatly among individuals. To overcome these issues, we explored the possibility of using allogeneic cell sources [17].

By focusing on the immunotolerant properties of cartilage that has already been reported [18], particulated juvenile cartilage implants (De Novo ® NT Graft; Zimmer, Warsaw, IN, USA) are used clinically in the United States [19]. However, the use of this product in Japan is difficult, due to the difficulty in ensuring traceability. Therefore, we focused on surgical remains obtained from patients with polydactyly as a source of allogeneic chondrocytes.

We are currently conducting a clinical study of the use of allogeneic polydactyly-derived chondrocyte sheets (PD sheets) for articular cartilage repair [20]. However, the transplantation of PD sheets requires open-knee surgery and is currently performed in conjunction with open-wedge high tibial osteotomy, which is highly invasive.

Injection is a promising approach to deliver substances for less-invasive types of regenerative medicine. To establish a less-invasive treatment for OAK, we are currently developing injectable PD sheets fabricated in RepCell™ plates (CellSeed Inc., Tokyo, Japan), a temperature-responsive culture dish with 3 mm × 3 mm grid walls etched into the culture surface. Our aim is to develop sheet transplantation treatments that can be delivered through needle injection into the knee joint (Figure 1).

The purposes of this study were to examine the potential clinical applications of injectable fragments of PD sheets (PD sheets-mini) by comparing their sheet characteristics with those of conventional PD sheets and to examine the effects of PD sheets-mini injected using needles of different gauge size on cell viability.

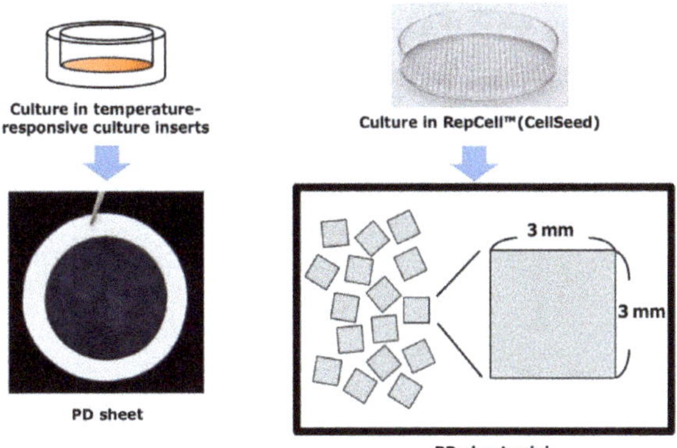

Figure 1. Fabrication of polydactyly-derived chondrocyte (PD) sheets and PD sheets-mini. PD sheets and PD sheets-mini were fabricated in temperature-responsive culture inserts called UpCell®and temperature-responsive culture dishes called RepCell™ (CellSeed Inc.), respectively. RepCell™ is a unique culture dish with grid walls etched onto the culture surface creating 3 mm × 3 mm grids.

2. Results

2.1. Comparison of Sheet Characteristics of Polydactyly-Derived Chondrocyte (PD) Sheets and PD Sheets-Mini

2.1.1. Cell Count and Viability

After culturing for 2 weeks, both PD sheets-mini and PD sheets were collected, and the cell count corrected by the area of the culture surface was calculated. The cell count and viability did not differ significantly between PD sheets-mini and PD sheets (Figure 2a,b).

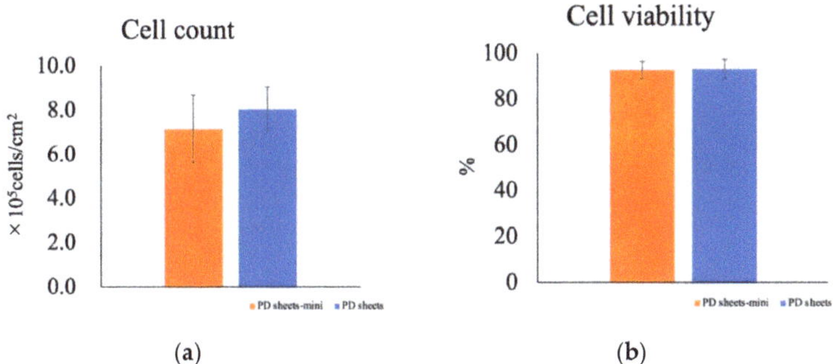

Figure 2. Comparison of cell count and cell viability for PD sheets and PD sheets-mini. Values are expressed as mean ± standard deviation. (**a**) Cell count of PD sheets and PD sheets-mini. (**b**) Cell viability of PD sheets and PD sheets-mini. The cell counts and cell viability did not differ between the two types of cell sheets.

2.1.2. Flow Cytometric Analysis

PD sheets-mini and PD sheets showed similar surface markers (Figure 3). Both sheets were negative for CD31, a vascular endothelium marker, and CD45, a blood cell marker: PD sheets-mini CD31 0.7%; CD45, 0.8%; PD sheets CD31, 0.8%; CD45, 1.4%. Both types of sheet were positive for the mesenchymal stem cell (MSC) markers CD29, CD44, CD73, CD81, CD90, and CD105: PD sheets-mini CD29, 98.4%; CD44, 99.9%; CD73, 99.9%; CD81, 100%; CD90, 99.9%; CD105, 88.1%; PD sheets CD29, 98.9%; CD44, 99.9%; CD73, 100%; CD81, 100%; CD90, 99.9%; CD105, 90.0%.

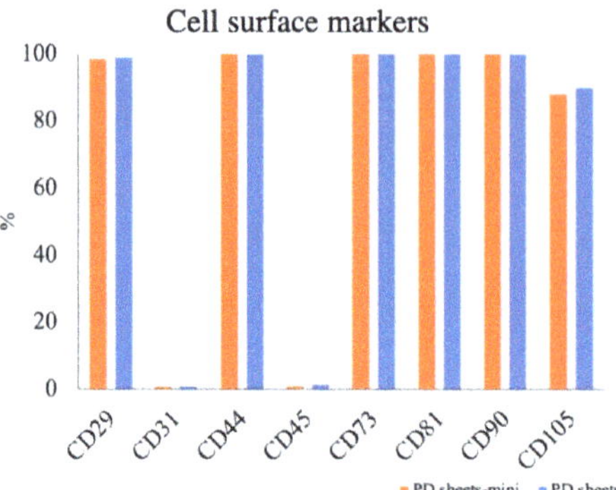

Figure 3. Flow cytometric analysis comparing PD sheets and PD sheets-mini. Both sheets were positive for mesenchymal stem cell markers such as CD29, CD44, CD73, CD81, CD90, and CD105, and negative for vascular endothelium and blood cell markers such as CD31 and CD45.

2.1.3. Measurement of the Amounts of Humoral Factors

The amounts of humoral factors were corrected by the culture surface area (Figure 4). There was no significant difference between PD sheets-mini and PD sheets in the amounts of transforming growth factor beta-1 (TGF-β1): PD sheets-mini 0.87 ± 0.09 ng/cm^2 and PD sheets 0.68 ± 0.36 ng/cm^2. PD sheets-mini produced similar amounts of melanoma inhibitory activity (MIA) as PD sheets: PD sheets-mini 5.52 ± 3.69 ng/cm^2 and PD sheets 5.93 ± 5.07 ng/cm^2. Similarly, the amounts of humoral factors produced did not differ between PD sheets and PD sheets-mini for the following markers: endothelial cell-specific marker 1 (ESM-1), PD sheets-mini, 0.31 ± 0.11 ng/cm^2 and PD sheets, 0.42 ± 0.24 ng/cm^2; Dickkopf-1 protein (DKK-1) PD sheets-mini, 1.19 ± 0.30 ng/cm^2 and PD sheets 1.06 ± 0.52 ng/cm^2; and monocyte chemoattractant protein 1 (MCP-1), PD sheets-mini, 0.40 ± 0.09 ng/cm^2 and PD sheets 0.34 ± 0.06 ng/cm^2. The amounts of enzymes that decompose the extracellular matrix, matrix metalloproteinase (MMP)-3 and MMP-13, also did not differ between PD sheets and PD sheets-mini. The respective values were PD sheets-mini, 3.79 ± 2.53 ng/cm^2 and PD sheets 2.73 ± 1.91 ng/cm^2 and PD sheets-mini, 0.56 ± 0.12 ng/cm^2 and PD sheets 0.48 ± 0.16 ng/cm^2.

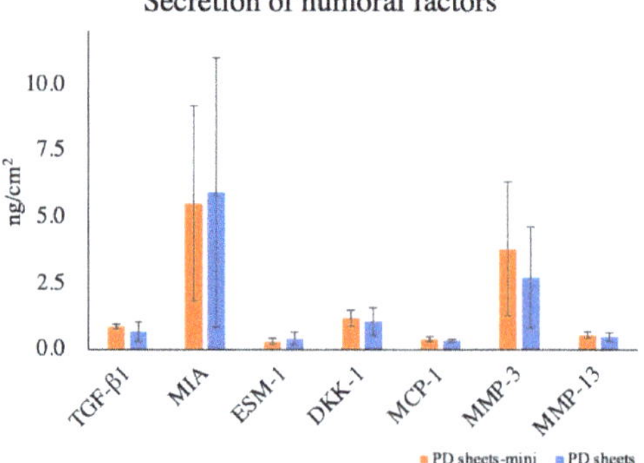

Figure 4. Amounts of secreted humoral factors. The amounts of transforming growth factor beta-1 (TGF-β1), melanoma inhibitory activity (MIA), endothelial cell-specific marker 1 (ESM-1), Dickkopf-1 protein (DKK-1), monocyte chemoattractant protein 1 (MCP-1), matrix metalloproteinase (MMP)-3, and MMP-13, humoral factors related to cartilage, did not differ significantly between PD sheets-mini and PD sheets.

2.1.4. Gene Expression Analysis

The relative gene expression of PD sheets was set at 1.0. The relative ratios of gene expression of PD sheets-mini were 0.68 times for *COL1A1*, 0.19 times for *COL2A1*, 0.73 times for *SOX9*, 0.63 times for *ACAN*, 1.23 times for *RUNX2*, and 1.30 times for *MMP3*. Except for *COL2A1*, the expression of cartilage-related genes did not differ between PD sheets and PD sheets-mini. The expression of *COL10A1* was not detected (Figure 5).

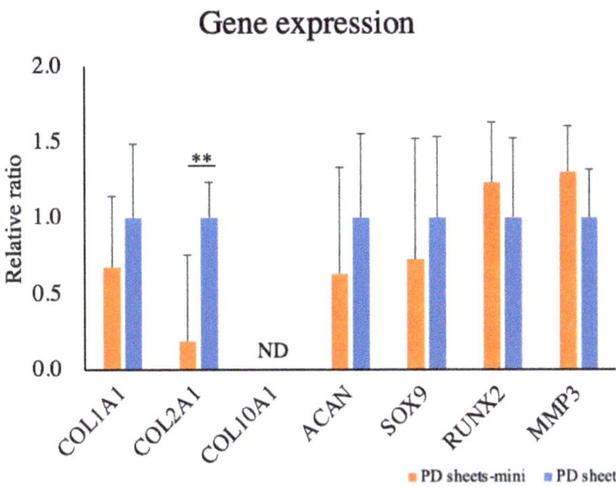

Figure 5. Cartilage-related gene expression analysis in PD sheets and PD sheets-mini. The relative gene expression of PD sheets was set at 1.0. The values are expressed as mean ± standard deviation. ** $p < 0.01$; ND, not detected. Only the expression of *COL2A1* differed between PD sheets and PD sheets-mini.

2.1.5. Histological Analysis

Sections of both PD sheets and PD sheets-mini showed similar multiple stratification of chondrocytes derived from polydactyly tissues. Both PD sheets-mini and PD sheets were stained weakly or not at all with Safranin O and toluidine blue. Histological analysis revealed no differences between PD sheets and PD sheets-mini (Figure 6).

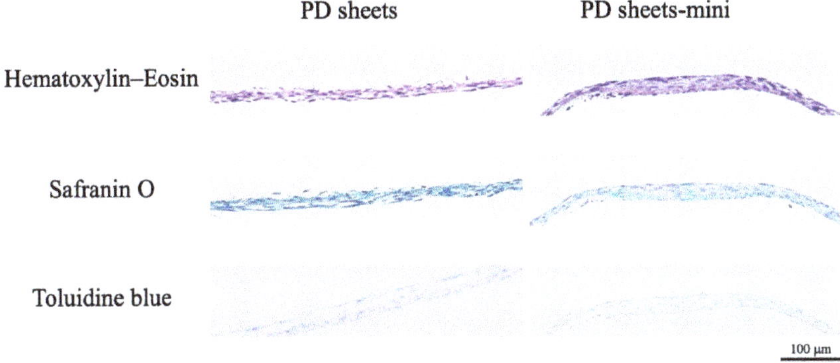

Figure 6. Histological analysis of PD sheets and PD sheets-mini. Sections of both PD sheets and PD sheets-mini were stained with hematoxylin–eosin, safranin O, and toluidine blue (×10, scale bar = 100 µm).

2.2. Effects of Injection of PD Sheets-Mini on Cell Viability

Cell viability decreased with time after injection but did not differ significantly between the four injection conditions immediately after ($p = 0.748$), 4 h ($p = 0.987$), and 24 h ($p = 0.994$) after injection. Cell viability of the conventional PD sheets after detachment did not differ between the four conditions (Figure 7a,b).

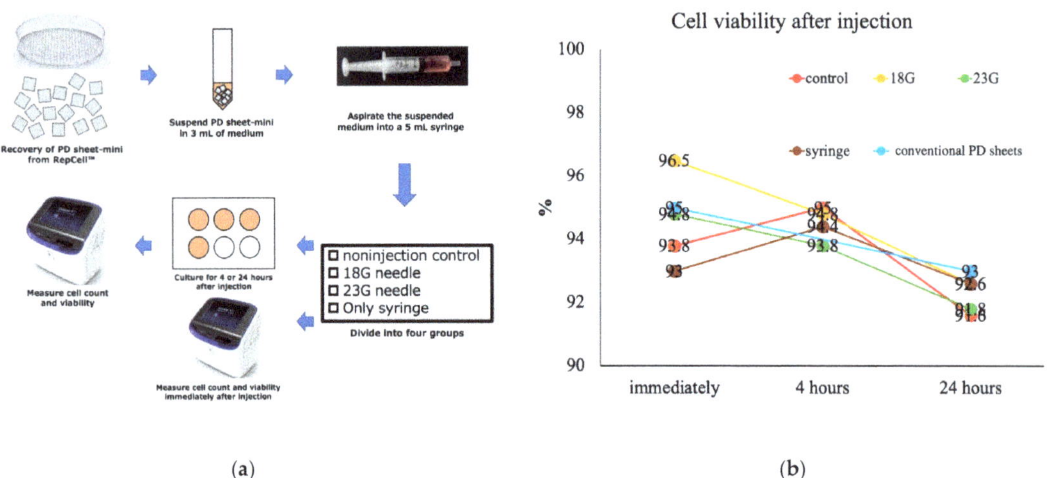

(**a**)   (**b**)

Figure 7. Evaluation of the effect of injection on cell viability of PD sheets-mini. (**a**) To examine the effect of injection on cell viability, the PD sheets-mini suspension was examined under four conditions: non-injection control, 18 gauge (G) needle injection, 23G needle injection, and syringe injection, and cell viability was measured at 0, 4, and 24 h after injection. At 0 and 4 h after injection, cell viability of conventional PD sheets was compared. (**b**) Comparison of cell viability of PD sheets-mini between the four conditions immediately after, 4 h, and 24 h after injection. Cell viability decreased with time but did not differ significantly between the four conditions at any time.

3. Discussion

Cartilage regeneration is difficult when induced only with concentrated humoral factors without cells, and the presence of cells producing cartilage-related humoral factors is considered to be important. Cell transplantation therapy by injection is a promising approach for less-invasive regenerative medicine, but the retention of transplanted cells at the injected sites is an important issue [21].

In cell sheet engineering, cells can be collected in sheets without the need for enzymes such as trypsin. Cell sheets of most types of cells can be transplanted while retaining the extracellular matrix and cell adhesion factors.

Wang et al. reported that the ability for cell attachment and proliferation was maintained even after MSC sheets were fragmented and injected through a needle, and this method may be superior to that using dissociated MSCs [21]. We believe that transplanted cells engraft at the cartilage defect sites or in the vicinity, even temporarily, and that humoral factors secreted from the cells are important for cartilage regeneration. The results of gene expression and histological evaluation suggest that dedifferentiation of chondrocytes occurs in both PD sheet and PD sheet-mini. We consider that cartilage repair is promoted by the recipient's chondrocytes, which are influenced by the humoral factors secreted by both sheets, rather than replacing both of the transplanted sheets with cartilage tissue.

In conducting clinical studies on the transplantation of autologous chondrocyte sheets [20] and polydactyly-derived chondrocyte sheets for articular cartilage regeneration, we have reported the characteristics and therapeutic effects of both types of sheet in animal experiments using mini-pigs, rabbits, and rats [22–25]. Our previous flow cytometric analysis studies have shown that PD sheets express MSC markers and our histological analyses have shown that PD sheets are multiply stratified without stacking, as in autologous chondrocyte sheets [17]. Similar results to these studies were obtained for both PD sheets and PD sheets-mini in this study.

MIA and TGF-β1 secreted from autologous chondrocyte sheets and PD sheets play an important role in cartilage differentiation and repair. MIA, which is a cartilage differentiation marker and anabolic factor, modulates chondrogenic differentiation via MIA-regulated extracellular signal-regulated kinase (ERK)-signaling [26]. TGF-b is known to be very beneficial for cartilage as it stimulates chondrocytes to induce elevation of proteoglycan and collagen type II production and also counteracts the main catabolic factors in OA [27]. Importantly, more TGF-β1 is secreted from PD sheets than from autologous chondrocyte sheets [17,28]. We also measured the production of ESM-1 and DKK-1, which we previously reported as humoral factors whose production correlates positively with the effectiveness of PD sheets for articular cartilage repair. ESM-1 is a proteoglycan involved in angiogenesis and is expressed in the vascular endothelium and adipocytes. DKK-1 is a protein that has an inhibitory effect on the Wnt signaling pathway and is involved in bone metabolism and differentiation [29]. In this study, the secretion of these humoral factors did not differ significantly between PD sheets and PD sheets-mini and, therefore, we consider that both sheets would have an equivalent effect of cartilage regeneration.

MCP-1, a chemokine with monocyte/macrophage migration activity, is also called C–C chemokine ligand 2 (CCL2). CCL2 provokes an inflammatory response through the C–C chemokine receptor 2 (CCR2), and the CCL2–CCR2 signal plays a role in cartilage degeneration. However, Jablonski et al. reported that, in the absence of CCR2, CCL2 may be required for cartilage regeneration [30]. Based on the report by Jablonski et al., we measured MCP-1 production in this study, but found no significant difference in MCP-1 production between PD sheet and PD sheet-mini.

MMP-3 and MMP-13 are known as catabolic factors that lead to cartilage degeneration by degrading ECM [31,32]. We have reported that PD sheets produce less MMP-3 than adult autologous chondrocytes, suggesting that polydactyly-derived chondrocytes have an advantage as a cell source [17]. In this study, there was no significant difference in the production of MMP-3 and MMP-13 by PD sheets-mini compared with the conventional PD sheets.

For all humoral factors measured in this study, we found no significant differences in the amounts secreted between both types of cell sheet. The new injection form of the chondrocyte sheets used in this study (PD sheets-mini) had similar characteristics as the PD sheets. However, the expression of *COL2A1* was lower in the PD sheets-mini, and this may reflect the culture of sheets in a flat culture dish instead of an insert, unlike in conventional PD sheets. From the results of clinical studies and accompanying animal experiments that we have conducted so far, we consider that the humoral factors secreted by the transplanted chondrocyte sheets play a more important role in the regeneration of articular cartilage than the genes expressed by the sheets [20,28,29].

Injection of MSCs has been reported to have no effects on the trilineage differentiation of osteogenic, adipogenic and chondrogenic cell lines, or on cell viability [33], although it has been suggested that apoptosis may occur early after injection through a 21 gauge (G) or 23G needle [34]. However, those studies examined this issue at the cellular level, and no reports have examined the effects of injection of mini-tissues such as cell sheets.

Our findings that the injection condition had almost no effect on the cell viability of PD sheets-mini suggest that injection of PD sheets-mini is feasible. Even if cell apoptosis occurs after injection through a 21G or 23G needle, the size of the needle usually used for arthrocentesis of knee is 18G, which should have less effect on cells.

PD sheets-mini have sheet characteristics that closely resemble those of PD sheets. Our results suggest that the cell viability of PD sheets-mini is not significantly affected by the clinically relevant needle gauge size. Intra-articular injection may be a feasible method for the administration of PD sheets-mini as a less-invasive method. In the future, it is necessary to examine the effect of cartilage treatment of PD sheets-mini in vivo.

4. Materials and Methods

4.1. Collection of Polydactyly-Derived Chondrocytes

Cartilage tissue was obtained from six polydactyly patients (average age, 15.5 months; range, 12–23 months; three boys and three girls) who underwent surgery at Tokai University Hospital.

Cartilage tissue was finely minced with scissors and treated with the following procedure to isolate chondrocytes from cartilage tissue. The minced cartilage tissue was incubated in Dulbecco's modified Eagle's medium/F12 (DMEM/F12; Gibco, Grand Island, NY, USA) supplemented with 20% fetal bovine serum (FBS; SAFC Biosciences, Lenexa, KS, USA), 1% antibiotic–antimycotic solution (AB; Gibco), and 5 mg/mL collagenase type 1 (CLS1; Worthington, Mannheim, Germany) for 1.5 h at 37 °C in a humidified atmosphere of 5% CO_2 and 95% air. The collected cell suspension was washed and passed through a 100 μm strainer (BD Falcon, Franklin Lakes, NJ, USA).

The isolated chondrocytes were seeded at a density of 1×10^4 cells/cm^2 in six-well culture plates (Corning, Corning, NY, USA) in DMEM/F12 supplemented with 20% FBS and 1% AB, and incubated at 37 °C. After 4 days, 100 μg/mL ascorbic acid (Nissin Pharmaceutical, Yamagata, Japan) was added to the medium, and the medium was replaced every 3 or 4 days thereafter. When the chondrocytes reached confluency, they were passaged once or twice, and passaged chondrocytes were cryopreserved at -180 °C.

4.2. Fabrication of PD Sheets and PD Sheets-Mini

PD sheets and PD sheets-mini were fabricated using a temperature-responsive culture insert and RepCell™ plate (CellSeed Inc.). The latter is a unique temperature-responsive culture plate with 3 mm × 3 mm grid walls etched into the culture surface. The areas of the culture dishes were 4.2 cm^2 and 8.8 cm^2, respectively (Figure 1).

To fabricate PD sheets, cryopreserved chondrocytes isolated from polydactyly patients were thawed, passaged once, and then seeded in temperature-responsive culture inserts (CellSeed Inc.) at 1×10^4 cells/cm^2. After culturing for 2 weeks with the medium changed every 3 or 4 days, the culture plates were allowed to stand at 25 °C for 30 min to detach PD sheets from the inserts, and each sheet was collected by hooking it onto a paper ring. PD sheets were checked visually to confirm the strength and to identify any tearing.

PD sheets-mini were fabricated using a method similar to that for PD sheets. The lysed chondrocytes were passaged and then seeded at 1×10^4 cells/cm^2 in RepCell™ plates. After culturing for 2 weeks with the medium changed every other day, each RepCell™ plate was kept at 25 °C for 30 min, and multiple 3 mm × 3 mm-sized fragments of PD sheets were collected (Figure 1).

4.3. Comparison of Sheet Characteristics between PD Sheets and PD Sheets-Mini

4.3.1. Cell Count and Viability

PD sheets and PD sheets-mini were washed in Dulbecco's phosphate-buffered saline (DPBS; Gibco). Each type of sheet was then incubated in TripLE Express® (Gibco) at 37 °C for 15 min and centrifuged at 1500 rpm for 5 min. The sheets were resuspended in 0.25 mg/mL Collagenase P (Roche, Basel, Switzerland) at 37 °C for up to 30 min and then centrifuged at 1500 rpm for 5 min. Cells isolated from both types of sheets were resuspended in DMEM/F12, and cell count and viability were determined using trypan blue exclusion.

4.3.2. Flow Cytometric Analysis

After the cell count and viability were measured, isolated cells were washed with DPBS containing 0.2% bovine serum albumin (BSA; Sigma-Aldrich, St. Louis, MO, USA) and 1 mM ethylenediaminetetraacetic acid (EDTA; Gibco). About 1.5×10^5 cells were mixed in each tube with the following antibodies: hCD31–fluorescein isothiocyanate (FITC) (clone: 5.6E; Beckman Coulter, Brea, CA, USA), hCD44–FITC (clone: G44-26; BD Biosciences, Franklin, NJ, USA), hCD45–FITC (clone: J33; Beckman Coulter), hCD73–FITC (clone: AD2;

BD Biosciences), hCD81–FITC (clone: JS-81; BD Biosciences), hCD90–allophycocyanin (clone: 5E10; BD Biosciences), and CD105–phycoerythrin (clone: 266; BD Biosciences).

The cells were incubated for 90 min at 4 °C and then washed with DPBS containing 0.2% BSA and 1 mM EDTA. Fluoroprobe-labeled mouse IgG1 antibody (Beckman Coulter) and rat IgG2b antibody (BioLegend, San Diego, CA, USA) were used as negative controls. Stained cells were analyzed using a FACS Verse TM cell sorter (BD Biosciences).

4.3.3. Measurement of Humoral Factors

PD sheets and PD sheets-mini fabricated as described above were used. PD sheets collected by hooking on paper rings were cultured for 24 h in 3 mL of DMEM/F12 supplemented with 5% FBS and 1% AB in an adherent culture dish. PD sheets-mini were harvested from RepCell™ plates after the plate was left at room temperature for more than 30 min. PD sheets-mini were suspended in 3 mL of DMEM/F12 supplemented with 5% FBS and 1% AB, and cultured for 24 h in a culture dish. Each supernatant was collected and centrifuged at 15,000× g for 10 min to remove cell debris. The concentrations of TGF-β1, MIA, ESM-1, DKK-1, MCP-1, MMP-3, and MMP-13 were measured using the enzyme-linked immunosorbent assays (ELISAs) described below.

TGF-β1, MIA, and ESM-1 concentrations were measured using their respective single-sample ELISA kits: Human TGF-β1 Quantikine ELISA Kit (R&D Systems, Minneapolis, MN, USA), MIA ELISA (Roche), and Human ESM-1 ELISA Kit (Abcam, Cambridge, UK). The other humoral factors were measured using Quantibody Multiplex ELISA Array (RayBiotech Life, Inc., Norcross, GA, USA).

The signal detected for blank medium containing 5% FBS was subtracted to adjust for proteins contained in FBS. Measurements were repeated at least twice for each donor, and averages were used. The amount of humoral factor per area of the culture surface was calculated.

4.3.4. Gene Expression Analysis

Each of the PD sheets and PD sheets-mini fabricated by culturing for 2 weeks was disrupted in TRIzol Reagent (Life Technologies, Carlsbad, CA, USA) and crushed at 1500 rpm for 3 min using a SHAKE Master Neo instrument (Bio Medical Science, Tokyo, Japan). Total RNA was extracted using an RNeasy Mini Kit (Qiagen, Hilden, Germany). Next, first-strand cDNA was reverse transcribed from 1 mg of total RNA and treated with a QuantiTect Reverse Transcription Kit (Qiagen) in a SimpliAmp Thermal Cycler (Thermo Fisher Scientific, Waltham, MA, USA) at 42 °C for 15 min and 95 °C for 3 min. Quantitative real-time polymerase chain reaction (qPCR) was performed using an Applied Biosystems 7300 Real-Time PCR system (Applied Biosystems, Waltham, MA, USA). The SYBR™ Green primers used are shown in Table 1. Glyceraldehyde-3-phosphate dehydrogenase (GAPDH) primers were used to normalize the samples.

Table 1. Genes and their primers used for quantitative real-time polymerase chain reaction (qPCR).

Genes	Forward Primer Sequence	Reverse Primer Sequence
ACAN	AGGAGACAGAGGGACACGTC	TCCACTGGTAGTCTTGGGCAT
COL1A1	GTCGAGGGCCAAGACGAAG	CAGATCACGTCATCGCACAAC
COL2A1	GTGGAGCAGCAAGAGCAA	TGTTGGGAGCCAGATTGT
COL10A1	ATGCTGCCACAAATACCCTTT	GGTAGTGGGCCTTTTATGCCT
GAPDH	AGAAGGCTGGGGCTCATTTG	AGGGGCCATCCACAGTCTTC
MMP3	ATGATGAACAATGGACAAAGGA	GAGTGAAAGAGACCCAGGGA
RUNX2	ACCATGGTGGAGATCATCG	CGCCATGACAGTAACCACAG
SOX9	AACGCCGAGCTCAGCAAGA	CCGCGGCTGGTACTTGTAATC

4.3.5. Histological Analysis

The harvested PD sheets and PD sheets-mini were fixed with 20% formalin (Wako Pure Chemical, Osaka, Japan) and embedded in paraffin wax. The embedded tissue was cut into 20-μm-thick sections, which were deparaffinized and stained with hematoxylin–eosin, Safranin O, and toluidine blue according to standard methods.

4.4. Analysis of the Effect of Injection of PD Sheets-Mini on Cell Viability

PD sheets-mini harvested from one RepCell™ plate were suspended in 3 mL of medium. To analyze the effect of injection on cell viability, suspensions of PD sheets-mini were tested for the following four injection conditions: non-injection control, 18G needle, 23G needle, and syringe only. For the injection condition, a 3 mL suspension was aspirated slowly into a 5 mL syringe and then injected through a needle or the syringe itself.

Cell viability was examined using the trypan blue exclusion assay at 0, 4, and 24 h after injection. Cell viability was compared among the four injection conditions, and also between PD sheets-mini and after detachment of cells from PD sheets at 0 and 24 h (Figure 7a).

4.5. Statistical Analysis

Significant differences between groups were identified using the Mann–Whitney U test for non-normally distributed quantitative data. The Kruskal–Wallis test was used to detect significant differences between three groups or more. All statistical analyses were performed using IBM SPSS Statistics (v. 25.0; IBM, Armonk, NY, USA). Probability values of <0.05 were considered to be significant.

5. Conclusions

PD sheets-mini have sheet characteristics that closely resemble those of PD sheets. Our results suggest that the cell viability of PD sheets-mini is not significantly affected by injection using a clinically relevant needle gauge size. Intra-articular injection may be a feasible way to administer PD sheets-mini in a less-invasive method as a point-of-care treatment for OAK.

6. Patents

The work reported in this study is patent pending; the application number is 2020-148200 in Japan, and the application date was 3 September 2020.

Author Contributions: Conceptualization, M.S. and S.W.; formal analysis, S.W.; funding acquisition, M.S.; investigation, S.W., T.T., M.M., E.O., E.T. and R.U.; resources, S.W., T.A. and M.S.; data curation, S.W.; writing—original draft preparation, S.W.; writing—review and editing, M.S.; supervision, M.S. and M.W. All authors have read and agreed to the published version of the manuscript.

Funding: This research received no external funding.

Institutional Review Board Statement: This study was reviewed and approved by the Provision Plan of Class 1 Regenerative Medicine (jRCTa030190242) in accordance with the Act on the Safety of Regenerative Medicine and conducted in compliance with relevant guidelines.

Informed Consent Statement: Written informed consent was obtained from all subjects involved in the study.

Data Availability Statement: The data presented in this study are available on request from the corresponding author. The data are not publicly available because of confidentiality issues.

Acknowledgments: We would like to thank Saori Shirasuna, Ayako Watanabe (Tokai University) and Yuka Kawaguchi (CellSeed Inc.) for their clinical and technical contributions to this study. We also thank all members of the Department of Plastic Surgery, Surgical Science, Tokai University School of Medicine, for assistance in the collection of chondrocytes. Finally, we are grateful to the Support Center for Medical Research and Education at Tokai University for technical support in this study.

Conflicts of Interest: M.S. receives research funding from CellSeed Inc.

References

1. Buckwalter, J.A.; Mankin, H.J. Articular Cartilage: Tissue Design and Chondrocyte-Matrix Interactions. *Instr. Course Lect.* **1998**, *47*, 477–486.
2. Pearle, A.D.; Warren, R.F.; Rodeo, S.A. Basic Science of Articular Cartilage and Osteoarthritis. *Clin. Sports Med.* **2005**, *24*, 1–12. [CrossRef] [PubMed]
3. World Health Organization (WHO). The World Health Report 2002–Reducing Risks, Promoting Healthy Life, Annex Table 3 Burden of disease in DALYs by Cause, Sex and Mortality Stratum in WHO Regions, Estimates for 2001, 196–197. 2002. Available online: https://www.who.int/whr/2002/en/ (accessed on 24 January 2021).
4. Yoshimura, N.; Muraki, S.; Oka, H.; Mabuchi, A.; En-Yo, Y.; Yoshida, M.; Saika, A.; Yoshida, H.; Suzuki, T.; Yamamoto, S.; et al. Prevalence of Knee Osteoarthritis, Lumbar Spondylosis, and Osteoporosis in Japanese Men and Women: The Research on Osteoarthritis/Osteoporosis against Disability Study. *J. Bone Min. Metab.* **2009**, *27*, 620–628. [CrossRef] [PubMed]
5. Felson, D.T.; Naimark, A.; Anderson, J.; Kazis, L.; Castelli, W.; Meenan, R.F. The Prevalence of Knee Osteoarthritis in the Elderly. The Framingham Osteoarthritis Study. *Arthritis Rheum.* **1987**, *30*, 914–918. [CrossRef]
6. Lawrence, R.C.; Felson, D.T.; Helmick, C.G.; Arnold, L.M.; Choi, H.; Deyo, R.A.; Gabriel, S.; Hirsch, R.; Hochberg, M.C.; Hunder, G.G.; et al. Estimates of the Prevalence of Arthritis and Other Rheumatic Conditions in the United States: Part II. *Arthritis Rheum.* **2008**, *58*, 26–35. [CrossRef] [PubMed]
7. McAlindon, T.E.; Bannuru, R.R.; Sullivan, M.C.; Arden, N.K.; Berenbaum, F.; Bierma-Zeinstra, S.M.; Hawker, G.A.; Henrotin, Y.; Hunter, D.J.; Kawaguchi, H.; et al. OARSI Guidelines for the Non-Surgical Management of Knee Osteoarthritis. *Osteoarthr. Cartil.* **2014**, *22*, 363–388. [CrossRef] [PubMed]
8. Zhang, W.; Moskowitz, R.W.; Nuki, G.; Abramson, S.; Altman, R.D.; Arden, N.; Bierma-Zeinstra, S.; Brandt, K.D.; Croft, P.; Doherty, M.; et al. OARSI Recommendations for the Management of Hip and Knee Osteoarthritis, Part II: OARSI Evidence-Based, Expert Consensus Guidelines. *Osteoarthr. Cartil.* **2008**, *16*, 137–162. [CrossRef]
9. Harris, J.D.; Siston, R.A.; Pan, X.; Flanigan, D.C. Autologous Chondrocyte Implantation: A Systematic Review. *J. Bone Jt. Surg. Am. Vol.* **2010**, *92*, 2220–2233. [CrossRef]
10. Nam, J.; Perera, P.; Liu, J.; Rath, B.; Deschner, J.; Gassner, R.; Butterfield, T.A.; Agarwal, S. Sequential Alterations in Catabolic and Anabolic Gene Expression Parallel Pathological Changes during Progression of Monoiodoacetate-Induced Arthritis. *PLoS ONE* **2011**, *6*, e24320. [CrossRef]
11. Maldonado, M.; Nam, J. The Role of Changes in Extracellular Matrix of Cartilage in the Presence of Inflammation on the Pathology of Osteoarthritis. *Biomed. Res. Int.* **2013**, *2013*, 1–10. [CrossRef] [PubMed]
12. Yamato, M.; Okano, T. Cell Sheet Engineering. *Mater. Today* **2004**, *7*, 42–47. [CrossRef]
13. Nishida, K.; Yamato, M.; Hayashida, Y.; Watanabe, K.; Yamamoto, K.; Adachi, E.; Nagai, S.; Kikuchi, A.; Maeda, N.; Watanabe, H.; et al. Corneal Reconstruction with Tissue-Engineered Cell Sheets Composed of Autologous Oral Mucosal Epithelium. *N. Engl. J. Med.* **2004**, *351*, 1187–1196. [CrossRef] [PubMed]
14. Ohki, T.; Yamato, M.; Murakami, D.; Takagi, R.; Yang, J.; Namiki, H.; Okano, T.; Takasaki, K. Treatment of Oesophageal Ulcerations Using Endoscopic Transplantation of Tissue-Engineered Autologous Oral Mucosal Epithelial Cell Sheets in a Canine Model. *Gut* **2006**, *55*, 1704–1710. [CrossRef] [PubMed]
15. Sawa, Y.; Miyagawa, S.; Sakaguchi, T.; Fujita, T.; Matsuyama, A.; Saito, A.; Shimizu, T.; Okano, T. Tissue Engineered Myoblast Sheets Improved Cardiac Function Sufficiently to Discontinue LVAS in a Patient with DCM: Report of a Case. *Surg. Today* **2012**, *42*, 181–184. [CrossRef] [PubMed]
16. Iwata, T.; Yamato, M.; Tsuchioka, H.; Takagi, R.; Mukobata, S.; Washio, K.; Okano, T.; Ishikawa, I. Periodontal Regeneration with Multi-Layered Periodontal Ligament-Derived Cell Sheets in a Canine Model. *Biomaterials* **2009**, *30*, 2716–2723. [CrossRef]
17. Maehara, M.; Sato, M.; Toyoda, E.; Takahashi, T.; Okada, E.; Kotoku, T.; Watanabe, M. Characterization of Polydactyly-Derived Chondrocyte Sheets versus Adult Chondrocyte Sheets for Articular Cartilage Repair. *Inflamm. Regen.* **2017**, *37*, 22. [CrossRef]
18. Abe, S.; Nochi, H.; Ito, H. Alloreactivity and Immunosuppressive Properties of Articular Chondrocytes from Osteoarthritic Cartilage. *J. Orthop. Surg. (Hong Kong)* **2016**, *24*, 232–239. [CrossRef]
19. Arthroscopic de Novo NT®Juvenile Allograft Cartilage Implantation in the Talus: A Case Presentation Elsevier Enhanced Reader. Available online: https://reader.elsevier.com/reader/sd/pii/S1067251611005928?token=622A192DC76E2E202E1C689C6650F8F3 6EDC0DBE5020429169FCA045999F9E590932C7478071DE98C741DD881B327855 (accessed on 16 March 2021).
20. Sato, M.; Yamato, M.; Mitani, G.; Takagaki, T.; Hamahashi, K.; Nakamura, Y.; Ishihara, M.; Matoba, R.; Kobayashi, H.; Okano, T.; et al. Combined Surgery and Chondrocyte Cell-Sheet Transplantation Improves Clinical and Structural Outcomes in Knee Osteoarthritis. *NPJ Regen. Med.* **2019**, *4*, 4. [CrossRef] [PubMed]
21. Wang, C.-C.; Chen, C.-H.; Lin, W.-W.; Hwang, S.-M.; Hsieh, P.C.H.; Lai, P.-H.; Yeh, Y.-C.; Chang, Y.; Sung, H.-W. Direct Intramyocardial Injection of Mesenchymal Stem Cell Sheet Fragments Improves Cardiac Functions after Infarction. *Cardiovasc. Res.* **2008**, *77*, 515–524. [CrossRef] [PubMed]
22. Ebihara, G.; Sato, M.; Yamato, M.; Mitani, G.; Kutsuna, T.; Nagai, T.; Ito, S.; Ukai, T.; Kobayashi, M.; Kokubo, M.; et al. Cartilage Repair in Transplanted Scaffold-Free Chondrocyte Sheets Using a Minipig Model. *Biomaterials* **2012**, *33*, 3846–3851. [CrossRef]
23. Takahashi, T.; Sato, M.; Toyoda, E.; Maehara, M.; Takizawa, D.; Maruki, H.; Tominaga, A.; Okada, E.; Okazaki, K.; Watanabe, M. Rabbit Xenogeneic Transplantation Model for Evaluating Human Chondrocyte Sheets Used in Articular Cartilage Repair. *J Tissue Eng. Regen. Med.* **2018**. [CrossRef]

24. Takatori, N.; Sato, M.; Toyoda, E.; Takahashi, T.; Okada, E.; Maehara, M.; Watanabe, M. Cartilage Repair and Inhibition of the Progression of Cartilage Degeneration after Transplantation of Allogeneic Chondrocyte Sheets in a Nontraumatic Early Arthritis Model. *Regen. Ther.* **2018**, *9*, 24–31. [CrossRef]
25. Takizawa, D.; Sato, M.; Okada, E.; Takahashi, T.; Maehara, M.; Tominaga, A.; Sogo, Y.; Toyoda, E.; Watanabe, M. Regenerative Effects of Human Chondrocyte Sheets in a Xenogeneic Transplantation Model Using Immune-deficient Rats. *J. Tissue Eng. Regen. Med.* **2020**. [CrossRef]
26. Schubert, T.; Schlegel, J.; Schmid, R.; Opolka, A.; Grässel, S.; Humphries, M.; Bosserhoff, A.-K. Modulation of Cartilage Differentiation by Melanoma Inhibiting Activity/Cartilage-Derived Retinoic Acid-Sensitive Protein (MIA/CD-RAP). *Exp. Mol. Med.* **2010**, *42*, 166. [CrossRef] [PubMed]
27. Blaney Davidson, E.N.; van der Kraan, P.M.; van den Berg, W.B. TGF-β and Osteoarthritis. *Osteoarthr. Cartil.* **2007**, *15*, 597–604. [CrossRef]
28. Hamahashi, K.; Sato, M.; Yamato, M.; Kokubo, M.; Mitani, G.; Ito, S.; Nagai, T.; Ebihara, G.; Kutsuna, T.; Okano, T.; et al. Studies of the Humoral Factors Produced by Layered Chondrocyte Sheets: Humoral Factors Produced by Layered Chondrocyte Sheets. *J. Tissue Eng. Regen. Med.* **2015**, *9*, 24–30. [CrossRef] [PubMed]
29. Toyoda, E.; Sato, M.; Takahashi, T.; Maehara, M.; Okada, E.; Wasai, S.; Iijima, H.; Nonaka, K.; Kawaguchi, Y.; Watanabe, M. Transcriptomic and Proteomic Analyses Reveal the Potential Mode of Action of Chondrocyte Sheets in Hyaline Cartilage Regeneration. *IJMS* **2019**, *21*, 149. [CrossRef]
30. Jablonski, C.L.; Leonard, C.; Salo, P.; Krawetz, R.J. CCL2 But Not CCR2 Is Required for Spontaneous Articular Cartilage Regeneration Post-Injury. *J. Orthop. Res.* **2019**, *37*, 2561–2574. [CrossRef] [PubMed]
31. Goldring, M.B. Osteoarthritis and Cartilage: The Role of Cytokines. *Curr. Rheumatol. Rep.* **2000**, *2*, 459–465. [CrossRef] [PubMed]
32. Rigoglou, S.; Papavassiliou, A.G. The NF-KB Signalling Pathway in Osteoarthritis. *Int. J. Biochem. Cell Biol.* **2013**, *45*, 2580–2584. [CrossRef]
33. Walker, P.A.; Jimenez, F.; Gerber, M.H.; Aroom, K.R.; Shah, S.K.; Harting, M.T.; Gill, B.S.; Savitz, S.I.; Cox, C.S. Effect of Needle Diameter and Flow Rate on Rat and Human Mesenchymal Stromal Cell Characterization and Viability. *Tissue Eng. Part C Methods* **2010**, *16*, 989–997. [CrossRef] [PubMed]
34. Garvican, E.R.; Cree, S.; Bull, L.; Smith, R.K.; Dudhia, J. Viability of Equine Mesenchymal Stem Cells during Transport and Implantation. *Stem Cell Res. Ther.* **2014**, *5*, 1. [CrossRef] [PubMed]

International Journal of
Molecular Sciences

Review

Recent Updates of Diagnosis, Pathophysiology, and Treatment on Osteoarthritis of the Knee

Sunhee Jang [1,2,†], Kijun Lee [1,2,†] and Ji Hyeon Ju [1,2,*]

1 Catholic iPSC Research Center, College of Medicine, The Catholic University of Korea, Seoul 06591, Korea;
   sunshinerosa@naver.com (S.J.); bluebusker@gmail.com (K.L.)
2 Division of Rheumatology, Department of Internal Medicine, Seoul St. Mary's Hospital, College of Medicine,
   The Catholic University of Korea, Seoul 06591, Korea
* Correspondence: juji@catholic.ac.kr; Tel.: +82-2-2258-6893; Fax: +82-2-3476-2274
† These authors contributed equally to this work.

Abstract: Osteoarthritis (OA) is a degenerative and chronic joint disease characterized by clinical symptoms and distortion of joint tissues. It primarily damages joint cartilage, causing pain, swelling, and stiffness around the joint. It is the major cause of disability and pain. The prevalence of OA is expected to increase gradually with the aging population and increasing prevalence of obesity. Many potential therapeutic advances have been made in recent years due to the improved understanding of the underlying mechanisms, diagnosis, and management of OA. Embryonic stem cells and induced pluripotent stem cells differentiate into chondrocytes or mesenchymal stem cells (MSCs) and can be used as a source of injectable treatments in the OA joint cavity. MSCs are known to be the most studied cell therapy products in cell-based OA therapy owing to their ability to differentiate into chondrocytes and their immunomodulatory properties. They have the potential to improve cartilage recovery and ultimately restore healthy joints. However, despite currently available therapies and advances in research, unfulfilled medical needs persist for OA treatment. In this review, we focused on the contents of non-cellular and cellular therapies for OA, and briefly summarized the results of clinical trials for cell-based OA therapy to lay a solid application basis for clinical research.

Keywords: osteoarthritis; diagnosis; management; surgery; cell therapy; embryonic stem cells; induced pluripotent stem cells; mesenchymal stem cells

Citation: Jang, S.; Lee, K.; Ju, J.H. Recent Updates of Diagnosis, Pathophysiology, and Treatment on Osteoarthritis of the Knee. *Int. J. Mol. Sci.* **2021**, 22, 2619. https://doi.org/10.3390/ijms22052619

Academic Editor: Masato Sato

Received: 30 January 2021
Accepted: 26 February 2021
Published: 5 March 2021

Publisher's Note: MDPI stays neutral with regard to jurisdictional claims in published maps and institutional affiliations.

1. Introduction

Osteoarthritis (OA) is the most common chronic articular disease and remains one of the few chronic aging disorders with few effective treatments, none of which have been proven to delay disease progression. It can affect small, medium, and large joints, although in terms of a painful disease, the knee is most frequently affected in up to 10% of men and 13% of women aged above 60 years, with evidence of symptomatic OA of the knee in the United States [1].

OA can be defined pathologically, radiographically, and clinically. The most common method for radiographic definition is the Kellgren–Lawrence (KL) radiographic grading system and atlas, which has been used for more than 40 years. This overall joint scoring system grades OA into five levels from 0 to 4, defining OA by the presence of a definite osteophyte (Grade $\geq$ 2), and more severe grades by the presumed successive appearance of joint space narrowing, sclerosis, cysts, and deformity [1,2]. However, all patients with radiographic OA do not have a clinical condition, and all patients with joint symptoms do not demonstrate radiographic OA [1]. Therefore, OA must be diagnosed using a variety of pathological, clinical, and radiological methods [3].

2. Osteoarthritis of the Knee

The knee is the largest synovial joint in humans and consists of bone structures (distal femur, proximal tibia, and patella); cartilage (meniscus and free cartilage); ligaments; infrapatellar fat pad; and synovium. The synovium is responsible for the production of synovial fluid that lubricates and nourishes the vascular cartilage. However, considering the frequent use and high stress on this joint, it is a frequent site of painful conditions, particularly OA [4,5]. Disease evaluation of OA is generally slow and can take years. Successively, the disease can also go through stages or show gradual evolution over time, making the severity and symptoms of the disease worse [6].

3. Mechanisms/Pathophysiology

The diarthrodial joint connects two adjacent bones, covered with a special articular cartilage layer, and wrapped in a synovial bursa [7]. The articular cartilage is composed of water (>70%) and organic extracellular matrix components, mainly type II collagen, aggrecan, or other proteoglycans [6]. Chondrocytes detect mechanical stress and changes in the pericellular matrix primarily through receptors on the components of the extracellular matrix. The change in response to mechanical or inflammatory stimulation results in the upregulation of aggrecanase and collagenase. Moreover, receptors on resting chondrocytes are protected from interacting with certain matrix components by the unique composition of the pericellular matrix. Type II collagen-containing networks in the interregional regions are generally not degraded as they are coated with proteoglycans [8].

The main cartilage matrix-degrading enzymes are zinc-dependent metalloproteinases (MMPs) belonging to the MMP and A Disintegrin and Metalloproteinase with Thrombospondin motifs (ADAMTS) families. MMPs include the collagenases MMP-1 and MMP-13 (highly efficient against type II collagen as a substrate), MMP-3 (a potent aggrecanase), and MMP activator [9].

The importance of proteoglycan depletion in cartilage erosion has been demonstrated in ADAMT5 (the primary aggrecanase) knockout mice protected from progression using a surgical OA model [10]. However, aggrecan depletion, by itself, does not drive OA progression, as suggested by studies wherein MMP13 knockout mice exhibited inhibition of cartilage erosion; however, it does not prevent aggrecan depletion [11]. When the collagen network begins to break down, irreversible cartilage breakdown proceeds [8]. Partly, overloading and inflammation leading to cartilage degradation can cause OA. Prostaglandin E2 is one of the major catabolic factors associated with OA, where MMP is crucial for cartilage degeneration. Thus, the mechanosensitive microsomal prostaglandin E synthase type 1 enzyme represents a potential therapeutic target in OA [12].

In OA, there is a gradual disappearance of cartilage associated with chondrocyte loss and phenotypic transformation, including cluster formation and activation of catabolic phenotypic and hypertrophic differentiation. Remodeling of subchondral bone occurs with the development of blood vessels located in structures (vascular channels) that contain osteoblasts and sensory nerves. Vascular channels should facilitate biochemical communication between the bone and cartilage. In response to multiple stimulations, chondrocytes modify the phenotype and express a subset of factors (such as cytokines, chemokines, alarmins, damage-associated molecular pattern, and adipokines). All these mediators act as paracrine factors, begin a vicious cycle of cartilage breakdown, reach the synovial fluid, and trigger an inflammatory process with the production of synovial macrophages and fibroblasts of the factor [8]. Vascular channels have sensory nerve terminations, and the associated innervation of articular cartilage may contribute to tibiofemoral pain in OA across a wide range of structural disease severities [13]. Figure 1 illustrates the associated mechanism [6,8].

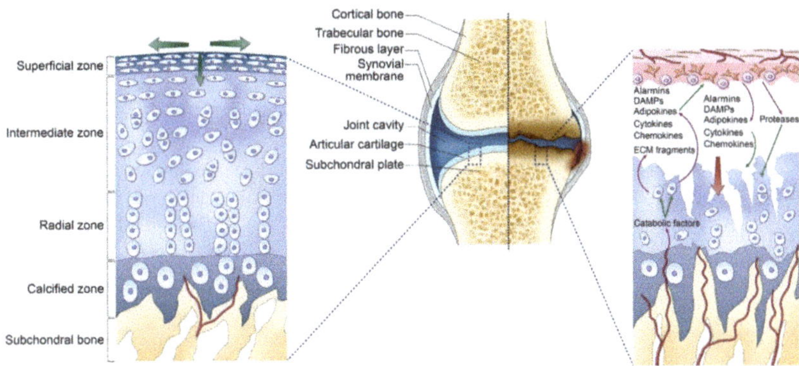

Figure 1. Mechanisms for the osteoarthritis of the knee. Healthy articular cartilage (Left)—Because of absence of vessels within cartilage, chondrocytes can live in a hypoxic environment. Hypoxia is important for chondrocyte function and survival. The main function of cartilage is the absorption and the removal of mechanical load, which is necessary to maintain cartilage homeostasis. Osteoarthritis articular cartilage (Right)—Development of vessels (called vascular channels) are supposed to facilitate biochemical communication between the bone and the cartilage (such as cytokines, chemokines, alarmins). It initiates a vicious cycle of cartilage degradation.

The infrapatellar fat pad (IFP) is in close contact with the synovial membrane, and due to the metabolic properties of adipose tissue, IFP may affect the functioning of the synovial membrane [14]. In a recent study, it was found that IFP in OA patients of the knee was more inflamed and vascularized compared to the IFP in anterior cruciate ligament reconstruction patients [15]. It has been recognized that IFP secretes adipocytokines and has more inflammation and fibrotic changes than control group [15]. All of these studies could support the new idea that the IFP and the synovial membrane could be considered a morpho-functional unit [14].

Cartilage is a highly specialized connective tissue, and its damage is the main feature of OA. The primary determinants of OA are namely, aging, genetic predisposition, metabolic syndrome, or trauma, and activation of the inflammatory pathways occurs in cartilage [16]. Chondrocytes form a vicious cycle leading to the progression of OA by producing inflammatory mediators that can lead to cartilage damage and changes in adjacent joint tissue. Therefore, the components of inflammatory pathways to discover disease-modifying OA drugs should be known [8].

4. Diagnosis

Although OA is an extremely common illness, its diagnosis may be difficult. Diagnostic criteria were developed for OA of the knee. The primary goal of the diagnostic criteria is to differentiate OA from other arthritis, such as rheumatoid arthritis and ankylosing spondylitis [6].

The American College of Rheumatology (ACR) classification criteria for OA of the knee were used widely [17]. One study demonstrated that crepitus is specific for patellofemoral joint OA rather than tibiofemoral joint OA, as suggested by the magnetic resonance imaging definition [18]. Another arthroscopy-based study reported an association between crepitus and cartilage pathology in both compartments of the knee [19].

Cartilage degeneration and other skeletal changes can be examined radiographically and quantified using the semi-quantitative grading scale known as the KL scale [20], Ahlbäck classification [21], and knee osteoarthritis grading system (KOGS) [22]. The original definitions of the KL scale, Ahlbäck classification, and KOGS are shown in Table 1. A KL grade picture is shown in Figure 2 [21].

Table 1. Comparison of the original definitions of the Kellgren–Lawrence (KL) scale, Ahlbäck clasScheme 20.

Grade	KL Scale	Ahlbäck Classification	KOGS
Grade 0	No pathological features of osteoarthritis (OA)		
Grade 1	Suspicious narrowing of the joint space and possible osseous lip	Joint space narrowing, with or without subchondral sclerosis. Joint space narrowing is defined by this system as a joint space <3 mm, or less than half of the space in the other compartment, or less than half of the space of the homologous compartment of the other knee	An isolated medial, lateral tibiofemoral, or patella-femoral joint OA with ligament stability and two functionally intact compartments
Grade 2	Clear bone tissue and possible stenosis of the joint space	Obliteration of the joint space	Deteriorating isolated lesion with ligament stability and a correctible coronal subluxation
Grade 3	Moderate multiple bone tissue, clear narrowing of the joint space, slight sclerosis, and possible deformity of the ends of the bones	Bone defect/loss < 5 mm	Includes an isolated medial or lateral tibiofemoral OA and concomitant pathologies such as anterior cruciate ligament deficiency (3A) or grooving of patella-femoral joint or patellectomy (3B)
Grade 4	Large bone tissue, marked narrowing of the joint space, severe sclerosis, and clear deformities of the ends of the bones	Bone defect/loss between 5 mm and 10 mm	Includes cases of bi-compartmental tibiofemoral OA without concomitant ligament instability (4A) and with ligament instability (4B)
Grade 5		Bone defect/loss >10 mm, often with subluxation and arthritis of the other compartment	

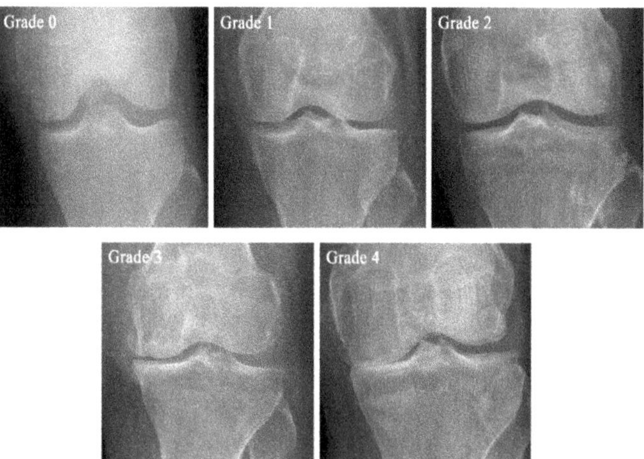

Figure 2. The example of the Kellgren–Lawrence (KL) scale. KL classification is the most widely used radiographic scale. The radiograph was recorded at St. Mary's Hospital in Seoul.

5. Risk Factor

Previous knee trauma increases the risk of osteoarthritis of the knee by 3.86 times [23]. Old age, female, overweight and obesity, repetitive use of joints, bone density, muscle weakness, and joint relaxation all play an important role in the development of knee OA [23]. Also, frequent squatting is a risk factor for knee OA [23].

6. Current Point-of-Care Treatment

Conventional Management

Currently, various guidelines have been developed to standardize and recommend available treatments by academic and professional societies. Table 2 shows the available treatment options from the Osteoarthritis Research Society International (OARSI), ACR, and the American Academy of Orthopedic Surgeons (AAOS) publications [24–26].

Table 2. Osteoarthritis of the knee management recommendations from the three societies.

Treatment	OARSI	ACR	AAOS
Exercise (Land-based)	Appropriate	Strong recommendation	Strong recommendation
Exercise (Water-based)	Appropriate	Strong recommendation	Strong recommendation
Transcutaneous electrical nerve stimulation	Uncertain	Strong recommendation against use	Inconclusive
Cane (Walking stick)	Appropriate	Strong recommendation	
Weight control	Appropriate	Strong recommendation	Moderate recommendation
Chondroitin or Glucosamine	Not appropriate for disease modification, Uncertain (Sx relief)	Strong recommendation against use	Recommendation against use
Acetaminophen	Without comorbidities: appropriate	Conditional recommendation	Inconclusive
Duloxetine	Appropriate	Conditional recommendation	No recommendation
Oral NSAIDs	Without comorbidities: appropriate; With comorbidities: Uncertain	Strong recommendation	Strong recommendation
Topical NSAIDs	Appropriate	Conditional recommendation against use	Strong recommendation
Opioids	Uncertain	No recommendation	Recommended (only tramadol)
Intra-articular corticosteroids	Appropriate	Strong recommendation	Inconclusive
Intra-articular viscosupplementation	Uncertain	Conditional recommendation against use	Recommendation against use

OARSI: Osteoarthritis Research Society International; ACR: American College of Rheumatology; AAOS: American Academy of Orthopedic Surgeons; NSAIDs: non-steroidal anti-inflammatory drugs.

7. Interventional Management

Multiple substances that are delivered through intra-articular (IA) injections have been explored. The idea is that local treatments (IA injection) will have less systemic side effects and placing the drug inside the joint will have a more direct effect. Studies have shown that IA therapy is more effective than oral non-steroidal anti-inflammatory drugs and other systemic pharmacological treatments; however, it has also revealed that some of its benefits may be secondary to the IA placebo effect [27,28].

Also, injectable drug delivery alone, including new treatments, may not provide significant benefits for OA treatment. Because IA injections cannot target the complexity of the pathological mechanisms [29]. It means that relatively new concept is the multimodal approach to the IA injections, which is needed to significant effect on the entire knee.

7.1. Intra-Articular Corticosteroid Injection

In the knee joint of OA, IA corticosteroid injections are usually conditionally recommended over other forms of IA injection. There are few one-to-one comparisons; however, the evidence for the efficacy of glucocorticoid injections is significantly higher than that of other drugs [25]. Corticoids act directly on nuclear receptors, disrupting the inflammatory cascade at several levels, causing immunosuppressive and anti-inflammatory effects. They are part of the pain relief mechanism and reduce the action and production of interleukin-1, prostaglandins, leukotriene, MMP9, and MMP-11, which are believed to increase joint mobility in the knee OA [4]. The clinical anti-inflammatory effects of these actions include decrease in erythema, heat, swelling, and tenderness of the inflammatory joints, and increases in relative viscosity with an increase in hyaluronic acid (HA) concentration [28]. Therefore, IA corticosteroid injections reduce acute pain episodes and increase joint mobility, particularly when there is evidence of inflammation and joint effusion during OA redness [30].

The current Food and Drug Administration approved immediate release corticosteroids for IA usages namely, methylprednisolone acetate, triamcinolone acetate, triamcinolone hexacetonide, betamethasone acetate, betamethasone sodium phosphate, and dexamethasone [28].

In summary, research evidence shows that IA corticosteroid injections provide a short-term reduction in OA pain and act as an adjunct to key therapy for moderate to severe pain relief in patients with OA [31].

7.2. Intra-Articular Hyaluronic Acid Injection (Viscosupplementation)

HA, a viscoelastic mucopolysaccharide component of synovial fluid, is produced from harvested rooster combs or through in vitro bacterial fermentation [32]. HA is a high-molecular-weight glycosaminoglycan that consists of a repeating sequence of disaccharide units composed of N-acetyl glucosamine and glucuronic acid [33]. Viscous supplementation through an IA injection of HA is aimed at restoring the beneficial environment present in non-arthritic joints. Additionally, the safety profile of such injections for painful knee OA is well established [32].

Previous studies demonstrated obvious benefits of intra-articular HA injection; however, according to the 2019 ACR/European League Against Rheumatism study, the benefit was restricted to studies with a higher risk of bias compared to saline injection. Therefore, in recent years, HA injection has been conditionally recommended to control joint symptoms when glucocorticoid injection or other interventions fail [25].

7.3. Intra-Articular Platelet-Rich Plasma Injection

The IA platelet-rich plasma (PRP) injection has emerged as a good treatment for knee OA. Several randomized controlled trials have been shown that PRP is a safe and effective treatment. At this time, IA PRP is not a standard treatment of the knee OA, but it is similar in efficacy to HA, and appears to be more effective than HA in young, active patients with low-grade OA [34].

8. Surgery

The goals of surgery for patients with OA are to reduce pain, minimize disability, and improve quality of life. Treatment should be individualized according to the functional condition of the patients, severity of the disease, and nature of the underlying disease. Surgical intervention for patients with OA is generally performed when a less invasive treatment is unsuccessful.

8.1. Total Knee Replacement Surgery (Total Knee Arthroplasty)

Total knee replacement (TKR) surgery involves excising the damaged ends of the tibia and femur and capping both using a prosthesis. Both prostheses comprise durable plastic.

These new surfaces move smoothly with each other. Partial recovery takes 6 weeks and complete recovery takes up to 1 year [35].

A randomized, controlled trial of TKR demonstrated that non-surgical treatment after TKR is superior to non-surgical treatment alone in providing pain relief in patients with moderate to severe OA of the knee and improving function and quality of life after 1 year. However, clinically relevant improvements were seen in both groups, and patients who received TKR had more severe side effects [36].

In addition, IFP resection during TKR is the subject of an ongoing debate without clear consensus [37].

8.2. Partial Knee Replacement Surgery (Unicompartmental Knee Arthroplasty)

Unicompartmental knee arthroplasty is an alternative to TKR for patients whose disease is limited to a single area of the knee, particularly the isolated tibiofemoral compartment (medial or lateral). As partial knee replacement is performed using smaller incisions, patients can generally be discharged earlier than those who undergo TKR and can return sooner to normal activities, including work and sports [38].

8.3. Knee Osteotomy (High Tibial Osteotomy or Femoral Osteotomy)

High tibial osteotomy is a surgery to realign the knee joint. It is more important for the treatment of cartilage damage or OA of the medial compartment with varus deformity. High tibial osteotomy creates a postoperative valgus limb alignment by lateral movement of the load-bearing axis of the lower limb [39].

8.4. Knee Arthroscopy

Knee arthroscopy is most commonly performed to treat OA or meniscus problems. An arthroscopy requires a small incision in the skin with the insertion of a camera on a stick. Another incision is needed to insert other instruments and treat the disease [40].

8.5. Knee Cartilage Repair and Cartilage Restoration

Many surgical techniques have been developed to address focal cartilage defects. Cartilage surgery strategies include palliative (chondroplasty and debris removal); repair (perforation and microfracture); or restoration (auto chondral cell transplant, osteochondral autograft, and bone cartilage allograft) [41].

9. Cellular & Experimental Therapy

9.1. Cellular Therapy

Several cell therapeutic attempts have been made to regenerate damaged joint cartilage. Autologous chondrocyte implantation (ACI) has been proposed as a surgical technique for partial cartilage lesions [42]. ACI is known as the most traditional cell-based therapy that has evolved with a high success rate. However, since it is limited to the damaged cartilage area, it is difficult to use in general OA treatment [43–45]. Mechanical, biological, and chemical scaffold-based approaches have also been developed to allow autologous chondrocytes to fill the cartilage lesions. With the scaffold, chondrocytes are less prone to dedifferentiation, and more favorable cartilage can be produced [46]. However, the limited number of primary chondrocytes has shown therapeutic limitations. Consequently, stem cell-based therapies have been developed to compensate for these shortcomings (Figure 3). Stem cells are undifferentiated cells capable of differentiating into various specialized cells such as bone cells, chondrocytes, and adipocytes [47]. Additionally, stem cells are characterized by the ability to release cytokine secretions that can downregulate several important inflammations [48]. Three other types of stem cells, embryonic stem cells (ESCs), induced pluripotent stem cells (iPSCs), and mesenchymal stem cells (MSCs) are also potential candidates for cartilage regeneration for the OA treatment. Both ESCs and iPSCs have intrinsically pluripotent features and can differentiate into other cell types, including chondrocytes. Several studies have shown that chondrocytes can be differentiated using

ESCs and iPSCs [49,50]. However, there is a risk that both stem cells can form teratoma and immunogenicity [51].

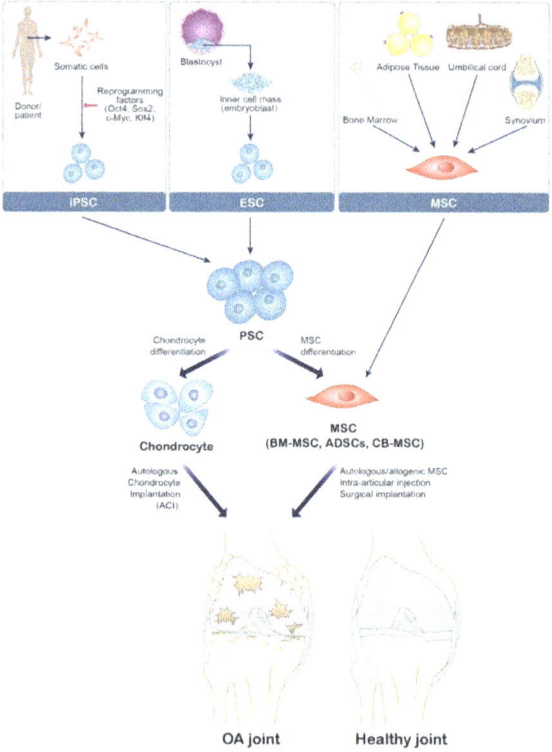

Figure 3. Schematic diagram of various cell-based therapy for osteoarthritis. Embryonic stem cells (ESC), induced pluripotent stem cells (iPSC), and mesenchymal stem cells (MSC) are potential candidates for cartilage regeneration for the OA treatment. MSC can be isolated from bone marrow, adipose tissue, umbilical cord, synovium. Pluripotent stem cells (PSC), including ESC and iPSC, are considered sources for the derivation of chondrocytes and MSC.

Mesenchymal stem cells (MSCs) are not as pluripotent as ESCs and iPSCs; however, may be considered the most ideal among the various types of stem cells for OA treatment [52]. MSCs have advantages that can be obtained in various ways such as bone marrow [53], adipose tissue [54], and umbilical cord [55]. As MSCs express and secrete various growth factors and cytokines and have anti-inflammatory activity, numerous studies have been conducted for the treatment of OA [56–58]. In addition to these sources of stem cells, infrapatellar fat pad (IFP) have been recently considered as a source of stem cells for cartilage regeneration in OA due to their increased chondrogenic capacity [59,60]. However, IFP-derived stem cells seem to be primed by the pathological environment and exert a protective role in the inflammatory environment [60]. Thus, further research is needed to clarify this point.

Human ESCs can be used as a raw material for cell therapy for the treatment of OA. It has been demonstrated that MSCs differentiated from ESCs have similar efficacy to those extracted from somatic tissues, such as bone marrow, and can treat various autoimmune and inflammatory diseases [61–63]. ESC-derived MSCs are more advantageous for use as a cell therapy than natural MSCs extracted from bone marrow tissues (Table 3). ESCs can be produced on a large scale from human ESC raw materials that can be supplied endlessly.

Table 3. Summary of advantages and disadvantages of ESCs, iPSCs, and MSCs.

	Advantages	Disadvantages
Embryonic stem cells (ESCs)	Unlimited self-renewal Unlimited proliferation Pluripotent Potentially unlimited supply	Ethical concerns Tumorigenic potential Difficulty in vitro work Difficulty in controlling differentiation
Induced pluripotent stem cells (iPSCs)	Autologous origin Extensive sources Unlimited self-renewal Unlimited proliferation Pluripotent No ethical issues	Security Tumorigenic potential Inefficiency Instability Unclear mechanism Difficulty in controlling differentiation
Mesenchymal stem cells (MSCs)	High chondrogenic potential Good expansion ability Easily accessible, reliable for isolation	A limited number of cells More affected by donor age Donor site pain

Additionally, human ESCs may be a potential treatment option as a source of consistently homogeneous cells with high chondrogenic ability [49]. Sources of expandable cartilage precursors may have broad potential to advance articular cartilage therapy and disease modeling and act as therapeutics that can promote cartilage regeneration or prevent degeneration [64,65]. Although several studies have attempted to treat OA using specialized cells derived from human ESCs [66–68], clinical studies on effective human ESC-derived cell therapeutics in patients with OA are still needed.

9.2. Induced Pluripotent Stem Cells for Osteoarthritis

The general differences in ESCs and iPSCs and the use of human PSCs for disease modeling have been extensively discussed in the literature [69–71]. The development of iPSCs opens up new horizons for the development of new research tools for OA that do not yet have a clear treatment [72]. The iPSCs, reprogrammed from somatic cells [73,74], provided a new opportunity to create a virtually unlimited number of patient-specific stem cells for OA for drug discovery. Thus, chondrogenic differentiation of iPSCs from patients with symptoms of OA may enable many studies of cartilage tissue [42,50,75]. Clinical studies using iPSCs in cell therapy for OA are still in the basic stage with an understanding of the cartilage regeneration mechanisms. iPSCs have proliferative and differentiation capabilities similar to those of other stem cells; however, do not have immune rejection reactions and ethical problems [76]. Additionally, studies on a new method of producing iPSCs without the use of viral vectors have been actively conducted in recent years to reduce the risk of tumorigenicity [77,78]. Nevertheless, there is still limited data on the effects of iPSCs on cartilage formation and OA, and further studies are needed (Table 4).

9.3. Mesenchymal Stem Cells for Osteoarthritis

MSCs are pluripotent progenitor cells derived from a population of adult stem cells that can be isolated from numerous tissues, including bone marrow, peripheral blood, adipose tissue, synovium, placenta, and umbilical cord [79,80]. Human MSCs are defined as cells that adhere to plastics; are positive for CD105, CD73 cell surface markers; negative for CD45, CD34 cell surface markers; and differentiate into osteoblasts, chondrocytes, and adipocytes [81]. Additionally, MSCs have unique immunomodulatory properties and can reduce inflammation and support other cells, enhancing angiogenesis, cell survival, and differentiation [82,83].

The primary isolated stromal cell represents the best option for OA treatment [84]. Bone marrow-derived stromal cells are the most common clinical source of MSC [85,86]. Although the main source has been on the use of bone marrow-derived stromal cells, some researchers have chosen to use adipose tissue-derived stromal cells as an alternative cell line [87,88]. These are harvested from bone marrow concentrate containing hematopoietic stem cells, endothelial progenitor cells, and related cytokines and growth factors [89]. Currently, translational medical research targeting MSCs for OA in the clinical trial database is promising. Clinical trials using MSCs in knee OA are actively underway. However, careful evaluation of clinical outcome data is necessary. The results appear to focus primarily on safety and efficacy [90]. Several studies have reported clinical trials for IA injection of MSCs in patients with OAs (Table 4) [91–101].

Table 4. Detailed clinical studies of ESCs, iPSCs, and MSCs in OA.

Cell Source	No. of Participants	Mean Follow-Up (Months)	Delivery Methods	Clinical Outcome	Reference/NCT
ESCs	N/A	N/A	N/A	N/A	N/A
iPSCs	N/A	N/A	N/A	N/A	N/A
BMSCs	45	75	Two-stage surgical approaches	No risk of serious complications	[91] N/A
BMSCs	4	12	IA injection	Improved pain, walking, and stairs climbing	[92] 00550524
BMSCs	56	24	IA injection	Better clinical outcomes and MRI in MSCs group	[93] N/A
BMSCs	12	12	IA injection	Improvement of cartilage quality on MRI	[94,95] 03956719
BMSCs	3	60	IA injection	Better than the baseline level	[96] 00550524
AMSCs (ASF)	18	6	IA injection	Better clinical outcomes	[97] 01300598
AMSCs (ASF)	100	26	IA injection	Improved pain VAS scores	[98] N/A
AMSCs (GSF)	40	29	IA injection and surgical implantation	Improved clinical outcomes	[99] N/A
AMSCs (ASF)	18	6	IA injection	Improved clinical outcomes	[100] 01585857
AMSCs (ASF)	24	6	IA injection	Improved clinical outcomes	[101] 02658344

ESCs; embryonic stem cells, iPSCs; induced pluripotent stem cells, BMSCs; bone-marrow-derived mesenchymal stem cells, AMSCs; adipose-derived mesenchymal stem cells, ASF; abdominal subcutaneous fat, GSF; gluteal subcutaneous fat, IA injection; intra-articular injection, MRI; magnetic resonance imaging, VAS; visual analog scale, NCT; The national clinical trial number, N/A; Not Assigned.

It has been suggested that the secretion of trophic factors, where exosomes play an important role, contributes to the MSC-based therapeutic mechanism of OA [102,103]. The paracrine secretion of MSC-derived exosomes may play a role in the repair of joint tissue as well as MSC-based treatments for other disorders. Recent studies have shown that MSC-derived exosomes may inhibit OA development and have summarized findings on exosomes derived from various MSCs and their effectiveness in OA therapy [104,105].

10. Noncellular Therapy

Gene Therapy

Gene therapy with genes encoding cartilage growth factor and anti-inflammatory cytokines is of interest in treating OA. Gene transfer was conducted in two ways: (1) In vivo injection/intravenous administration into the joint; (2) Ex vivo exposition/cell harvesting from the patient to the vector, and returning the modified cells to the joint [106].

In 2017, in vitro TGF-β1 gene therapy using retrovirus was approved for allogeneic chondrocytes [107]. In 2018, Kim et al. reported the clinical efficacy of TissueGene-C (TG-C), a cell and gene therapy for human knee OA consisting of non-transfected and transduced chondrocytes transduced using retrovirus to overexpress TGF-β1. They concluded that TG-C was associated with a statistically significant improvement in function and pain in patients with knee OA [108].

11. Conclusions

OA is the most common chronic joint disease, associated with obesity, aging, and socioeconomic impact. The mechanism works complex, local and systemic factors modulate clinical and structural representation, sometimes resulting in a common end-course of joint destruction. Treatment goals are to relieve symptoms and improve quality of life. Conventional management, surgery, and experimental therapy are summarized above.

Author Contributions: All authors were involved in drafting the manuscript or revising it critically for important intellectual content, and all authors approved the final version to be published. All authors have read and agreed to the published version of the manuscript.

Funding: This work was supported by a grant from the Korea Healthcare Technology R&D project, Ministry for Health, Welfare & Family Affairs, Republic of Korea (HI20C0495). This work was supported by a National Research Foundation of Korea (NRF) grant funded by the Korean government (MSIT) (No. 2020R1A2C3004123).

Institutional Review Board Statement: Not applicable.

Informed Consent Statement: Not applicable.

Data Availability Statement: No new data were created or analyzed in this study. Data sharing is not applicable to this article.

Conflicts of Interest: The authors declare no conflict of interest.

References

1. Zhang, Y.; Jordan, J.M. Epidemiology of osteoarthritis. *Clin. Geriatr. Med.* **2010**, *26*, 355–369. [CrossRef]
2. Kellgren, J.H.; Jeffrey, M.R.; Ball, J. *The Epidemiology of Chronic Rheumatism: A Symposium*; Blackwell Scientific Publications: Oxford, UK, 1963.
3. Hannan, M.T.; Felson, D.T.; Pincus, T. Analysis of the discordance between radiographic changes and knee pain in osteoarthritis of the knee. *J. Rheumatol.* **2000**, *27*, 1513–1517.
4. Mora, J.C.; Przkora, R.; Cruz-Almeida, Y. Knee osteoarthritis: Pathophysiology and current treatment modalities. *J. Pain Res.* **2018**, *11*, 2189–2196. [CrossRef] [PubMed]
5. Rubin, B.R. Management of osteoarthritic knee pain. *J. Am. Osteopath. Assoc.* **2005**, *105*, S23–S28.
6. Martel-Pelletier, J.; Barr, A.J.; Cicuttini, F.M.; Conaghan, P.G.; Cooper, C.; Goldring, M.B.; Goldring, S.R.; Jones, G.; Teichtahl, A.J.; Pelletier, J.P. Osteoarthritis. *Nat. Rev. Dis. Primers* **2016**, *2*, 16072. [CrossRef]
7. Netter, F.H.; Hansen, J.T. *Atlas of Human Anatomy*, 3rd ed.; Icon Learning Systems: Teterboro, NJ, USA, 2003.
8. Houard, X.; Goldring, M.B.; Berenbaum, F. Homeostatic mechanisms in articular cartilage and role of inflammation in osteoarthritis. *Curr. Rheumatol. Rep.* **2013**, *15*, 375. [CrossRef] [PubMed]
9. Troeberg, L.; Nagase, H. Proteases involved in cartilage matrix degradation in osteoarthritis. *Biochim. Biophys. Acta* **2012**, *1824*, 133–145. [CrossRef] [PubMed]
10. Glasson, S.S.; Askew, R.; Sheppard, B.; Carito, B.; Blanchet, T.; Ma, H.L.; Flannery, C.R.; Peluso, D.; Kanki, K.; Yang, Z.; et al. Deletion of active ADAMTS5 prevents cartilage degradation in a murine model of osteoarthritis. *Nature* **2005**, *434*, 644–648. [CrossRef] [PubMed]
11. Little, C.B.; Barai, A.; Burkhardt, D.; Smith, S.M.; Fosang, A.J.; Werb, Z.; Shah, M.; Thompson, E.W. Matrix metalloproteinase 13-deficient mice are resistant to osteoarthritic cartilage erosion but not chondrocyte hypertrophy or osteophyte development. *Arthritis Rheum.* **2009**, *60*, 3723–3733. [CrossRef] [PubMed]
12. Gosset, M.; Berenbaum, F.; Levy, A.; Pigenet, A.; Thirion, S.; Cavadias, S.; Jacques, C. Mechanical stress and prostaglandin E2 synthesis in cartilage. *Biorheology* **2008**, *45*, 301–320. [CrossRef] [PubMed]
13. Suri, S.; Gill, S.E.; Massena de Camin, S.; Wilson, D.; McWilliams, D.F.; Walsh, D.A. Neurovascular invasion at the osteochondral junction and in osteophytes in osteoarthritis. *Ann. Rheum. Dis.* **2007**, *66*, 1423–1428. [CrossRef]
14. Macchi, V.; Stocco, E.; Stecco, C.; Belluzzi, E.; Favero, M.; Porzionato, A.; De Caro, R. The infrapatellar fat pad and the synovial membrane: An anatomo-functional unit. *J. Anat.* **2018**, *233*, 146–154. [CrossRef] [PubMed]

15. Belluzzi, E.; Macchi, V.; Fontanella, C.G.; Carniel, E.L.; Olivotto, E.; Filardo, G.; Sarasin, G.; Porzionato, A.; Granzotto, M.; Pozzuoli, A.; et al. Infrapatellar Fat Pad Gene Expression and Protein Production in Patients with and without Osteoarthritis. *Int. J. Mol. Sci* **2020**, *21*, 6016. [CrossRef] [PubMed]

16. Kim, H.A.; Cho, M.L.; Choi, H.Y.; Yoon, C.S.; Jhun, J.Y.; Oh, H.J.; Kim, H.Y. The catabolic pathway mediated by Toll-like receptors in human osteoarthritic chondrocytes. *Arthritis Rheumatol.* **2006**, *54*, 2152–2163. [CrossRef] [PubMed]

17. Wu, C.W.; Morrell, M.R.; Heinze, E.; Concoff, A.L.; Wollaston, S.J.; Arnold, E.L.; Singh, R.; Charles, C.; Skovrun, M.L.; FitzGerald, J.D.; et al. Validation of American College of Rheumatology classification criteria for knee osteoarthritis using arthroscopically defined cartilage damage scores. *Semin. Arthritis Rheum.* **2005**, *35*, 197–201. [CrossRef] [PubMed]

18. Schiphof, D.; van Middelkoop, M.; de Klerk, B.M.; Oei, E.H.; Hofman, A.; Koes, B.W.; Weinans, H.; Bierma-Zeinstra, S.M. Crepitus is a first indication of patellofemoral osteoarthritis (and not of tibiofemoral osteoarthritis). *Osteoarthr. Cartil.* **2014**, *22*, 631–638. [CrossRef] [PubMed]

19. Ike, R.; O'Rourke, K.S. Compartment-directed physical examination of the knee can predict articular cartilage abnormalities disclosed by needle arthroscopy. *Arthritis Rheumatol.* **1995**, *38*, 917–925. [CrossRef] [PubMed]

20. Kohn, M.D.; Sassoon, A.A.; Fernando, N.D. Classifications in Brief: Kellgren-Lawrence Classification of Osteoarthritis. *Clin. Orthop. Relat. Res.* **2016**, *474*, 1886–1893. [CrossRef]

21. Hernandez-Vaquero, D.; Fernandez-Carreira, J.M. Relationship between radiological grading and clinical status in knee osteoarthritis. A multicentric study. *BMC Musculoskelet Disord.* **2012**, *13*, 194. [CrossRef] [PubMed]

22. Oosthuizen, C.R.; Takahashi, T.; Rogan, M.; Snyckers, C.H.; Vermaak, D.P.; Jones, G.G.; Porteous, A.; Maposa, I.; Pandit, H. The Knee Osteoarthritis Grading System for Arthroplasty. *J. Arthroplast.* **2019**, *34*, 450–455. [CrossRef]

23. Heidari, B. Knee osteoarthritis prevalence, risk factors, pathogenesis and features: Part I. *Casp. J. Intern. Med.* **2011**, *2*, 205–212.

24. McAlindon, T.E.; Bannuru, R.R.; Sullivan, M.C.; Arden, N.K.; Berenbaum, F.; Bierma-Zeinstra, S.M.; Hawker, G.A.; Henrotin, Y.; Hunter, D.J.; Kawaguchi, H.; et al. OARSI guidelines for the non-surgical management of knee osteoarthritis. *Osteoarthr. Cartil.* **2014**, *22*, 363–388. [CrossRef] [PubMed]

25. Kolasinski, S.L.; Neogi, T.; Hochberg, M.C.; Oatis, C.; Guyatt, G.; Block, J.; Callahan, L.; Copenhaver, C.; Dodge, C.; Felson, D.; et al. 2019 American College of Rheumatology/Arthritis Foundation Guideline for the Management of Osteoarthritis of the Hand, Hip, and Knee. *Arthritis Rheumatol.* **2020**, *72*, 220–233. [CrossRef]

26. Jevsevar, D.S.; Brown, G.A.; Jones, D.L.; Matzkin, E.G.; Manner, P.A.; Mooar, P.; Schousboe, J.T.; Stovitz, S.; Sanders, J.O.; Bozic, K.J.; et al. The American Academy of Orthopaedic Surgeons evidence-based guideline on: Treatment of osteoarthritis of the knee, 2nd edition. *J. Bone Jt. Surg. Am.* **2013**, *95*, 1885–1886. [CrossRef]

27. Bannuru, R.R.; Schmid, C.H.; Kent, D.M.; Vaysbrot, E.E.; Wong, J.B.; McAlindon, T.E. Comparative effectiveness of pharmacologic interventions for knee osteoarthritis: A systematic review and network meta-analysis. *Ann. Intern. Med.* **2015**, *162*, 46–54. [CrossRef]

28. Ayhan, E.; Kesmezacar, H.; Akgun, I. Intraarticular injections (corticosteroid, hyaluronic acid, platelet rich plasma) for the knee osteoarthritis. *World J. Orthop.* **2014**, *5*, 351–361. [CrossRef]

29. Georgiev, T. Multimodal approach to intraarticular drug delivery in knee osteoarthritis. *Rheumatol. Int.* **2020**, *40*, 1763–1769. [CrossRef]

30. Rozental, T.D.; Sculco, T.P. Intra-articular corticosteroids: An updated overview. *Am. J. Orthop.* **2000**, *29*, 18–23.

31. Osteoarthritis: National Clinical Guideline for Care and Management in Adults, London. 2008. Available online: https://www.ncbi.nlm.nih.gov/books/NBK48984/ (accessed on 3 March 2021).

32. McArthur, B.A.; Dy, C.J.; Fabricant, P.D.; Valle, A.G. Long term safety, efficacy, and patient acceptability of hyaluronic acid injection in patients with painful osteoarthritis of the knee. *Patient Prefer. Adherence* **2012**, *6*, 905–910.

33. Strauss, E.J.; Hart, J.A.; Miller, M.D.; Altman, R.D.; Rosen, J.E. Hyaluronic acid viscosupplementation and osteoarthritis: Current uses and future directions. *Am. J. Sports Med.* **2009**, *37*, 1636–1644. [CrossRef]

34. Pourcho, A.M.; Smith, J.; Wisniewski, S.J.; Sellon, J.L. Intraarticular platelet-rich plasma injection in the treatment of knee osteoarthritis: Review and recommendations. *Am. J. Phys. Med. Rehabil.* **2014**, *93*, S108–S121. [CrossRef] [PubMed]

35. Leopold, S.S. Minimally invasive total knee arthroplasty for osteoarthritis. *N. Engl. J. Med.* **2009**, *360*, 1749–1758. [CrossRef] [PubMed]

36. Skou, S.T.; Roos, E.M.; Laursen, M.B.; Rathleff, M.S.; Arendt-Nielsen, L.; Simonsen, O.; Rasmussen, S. A Randomized, Controlled Trial of Total Knee Replacement. *N. Engl. J. Med.* **2015**, *373*, 1597–1606. [CrossRef] [PubMed]

37. Sun, C.; Zhang, X.; Lee, W.G.; Tu, Y.; Li, H.; Cai, X.; Yang, H. Infrapatellar fat pad resection or preservation during total knee arthroplasty: A meta-analysis of randomized controlled trials. *J. Orthop. Surg. Res.* **2020**, *15*, 297. [CrossRef]

38. Rodriguez-Merchan, E.C.; Gomez-Cardero, P. Unicompartmental knee arthroplasty: Current indications, technical issues and results. *Efort Open Rev.* **2018**, *3*, 363–373. [CrossRef] [PubMed]

39. Liu, X.; Chen, Z.; Gao, Y.; Zhang, J.; Jin, Z. High Tibial Osteotomy: Review of Techniques and Biomechanics. *J. Healthc. Eng.* **2019**, *2019*, 8363128. [CrossRef]

40. Chua, M.J.; Hart, A.J.; Mittal, R.; Harris, I.A.; Xuan, W.; Naylor, J.M. Early mobilisation after total hip or knee arthroplasty: A multicentre prospective observational study. *PLoS ONE* **2017**, *12*, e0179820. [CrossRef]

41. Richter, D.L.; Schenck, R.C., Jr.; Wascher, D.C.; Treme, G. Knee Articular Cartilage Repair and Restoration Techniques: A Review of the Literature. *Sports Health* **2016**, *8*, 153–160. [CrossRef]

42. Nam, Y.; Rim, Y.A.; Lee, J.; Ju, J.H. Current Therapeutic Strategies for Stem Cell-Based Cartilage Regeneration. *Stem Cells Int.* **2018**, *2018*, 8490489. [CrossRef]

43. Nazempour, A.; Van Wie, B.J. Chondrocytes, Mesenchymal Stem Cells, and Their Combination in Articular Cartilage Regenerative Medicine. *Ann. Biomed. Eng.* **2016**, *44*, 1325–1354. [CrossRef] [PubMed]

44. Mobasheri, A.; Kalamegam, G.; Musumeci, G.; Batt, M.E. Chondrocyte and mesenchymal stem cell-based therapies for cartilage repair in osteoarthritis and related orthopaedic conditions. *Maturitas* **2014**, *78*, 188–198. [CrossRef]

45. Viste, A.; Piperno, M.; Desmarchelier, R.; Grosclaude, S.; Moyen, B.; Fessy, M.H. Autologous chondrocyte implantation for traumatic full-thickness cartilage defects of the knee in 14 patients: 6-year functional outcomes. *Orthop. Traumatol. Surg. Res.* **2012**, *98*, 737–743. [CrossRef] [PubMed]

46. Caron, M.M.; Emans, P.J.; Coolsen, M.M.; Voss, L.; Surtel, D.A.; Cremers, A.; van Rhijn, L.W.; Welting, T.J. Redifferentiation of dedifferentiated human articular chondrocytes: Comparison of 2D and 3D cultures. *Osteoarthr. Cartil.* **2012**, *20*, 1170–1178. [CrossRef]

47. Zakrzewski, W.; Dobrzynski, M.; Szymonowicz, M.; Rybak, Z. Stem cells: Past, present, and future. *Stem Cell Res. Ther.* **2019**, *10*, 68. [CrossRef]

48. Siegel, G.; Schafer, R.; Dazzi, F. The immunosuppressive properties of mesenchymal stem cells. *Transplantation* **2009**, *87*, S45–S49. [CrossRef] [PubMed]

49. Craft, A.M.; Ahmed, N.; Rockel, J.S.; Baht, G.S.; Alman, B.A.; Kandel, R.A.; Grigoriadis, A.E.; Keller, G.M. Specification of chondrocytes and cartilage tissues from embryonic stem cells. *Development* **2013**, *140*, 2597–2610. [CrossRef]

50. Lach, M.; Trzeciak, T.; Richter, M.; Pawlicz, J.; Suchorska, W.M. Directed differentiation of induced pluripotent stem cells into chondrogenic lineages for articular cartilage treatment. *J. Tissue Eng.* **2014**, *5*, 2041731414552701. [CrossRef] [PubMed]

51. de Almeida, P.E.; Ransohoff, J.D.; Nahid, A.; Wu, J.C. Immunogenicity of pluripotent stem cells and their derivatives. *Circ. Res.* **2013**, *112*, 549–561. [CrossRef]

52. Kim, S.H.; Ha, C.W.; Park, Y.B.; Nam, E.; Lee, J.E.; Lee, H.J. Intra-articular injection of mesenchymal stem cells for clinical outcomes and cartilage repair in osteoarthritis of the knee: A meta-analysis of randomized controlled trials. *Arch. Orthop. Trauma Surg.* **2019**, *139*, 971–980. [CrossRef] [PubMed]

53. Gnecchi, M.; Melo, L.G. Bone marrow-derived mesenchymal stem cells: Isolation, expansion, characterization, viral transduction, and production of conditioned medium. *Methods Mol. Biol* **2009**, *482*, 281–294.

54. Gruber, H.E.; Deepe, R.; Hoelscher, G.L.; Ingram, J.A.; Norton, H.J.; Scannell, B.; Loeffler, B.J.; Zinchenko, N.; Hanley, E.N.; Tapp, H. Human adipose-derived mesenchymal stem cells: Direction to a phenotype sharing similarities with the disc, gene expression profiling, and coculture with human annulus cells. *Tissue Eng. Part. A* **2010**, *16*, 2843–2860. [CrossRef] [PubMed]

55. Ishige, I.; Nagamura-Inoue, T.; Honda, M.J.; Harnprasopwat, R.; Kido, M.; Sugimoto, M.; Nakauchi, H.; Tojo, A. Comparison of mesenchymal stem cells derived from arterial, venous, and Wharton's jelly explants of human umbilical cord. *Int. J. Hematol.* **2009**, *90*, 261–269. [CrossRef] [PubMed]

56. Shariatzadeh, M.; Song, J.; Wilson, S.L. The efficacy of different sources of mesenchymal stem cells for the treatment of knee osteoarthritis. *Cell Tissue Res.* **2019**, *378*, 399–410. [CrossRef]

57. Iijima, H.; Isho, T.; Kuroki, H.; Takahashi, M.; Aoyama, T. Effectiveness of mesenchymal stem cells for treating patients with knee osteoarthritis: A meta-analysis toward the establishment of effective regenerative rehabilitation. *NPJ Regen. Med.* **2018**, *3*, 15. [CrossRef] [PubMed]

58. Wang, A.T.; Feng, Y.; Jia, H.H.; Zhao, M.; Yu, H. Application of mesenchymal stem cell therapy for the treatment of osteoarthritis of the knee: A concise review. *World J. Stem Cells* **2019**, *11*, 222–235. [CrossRef]

59. do Amaral, R.; Almeida, H.V.; Kelly, D.J.; O'Brien, F.J.; Kearney, C.J. Infrapatellar Fat Pad Stem Cells: From Developmental Biology to Cell Therapy. *Stem Cells Int.* **2017**, *2017*, 6843727. [CrossRef] [PubMed]

60. Stocco, E.; Barbon, S.; Piccione, M.; Belluzzi, E.; Petrelli, L.; Pozzuoli, A.; Ramonda, R.; Rossato, M.; Favero, M.; Ruggieri, P.; et al. Infrapatellar Fat Pad Stem Cells Responsiveness to Microenvironment in Osteoarthritis: From Morphology to Function. *Front. Cell Dev. Biol.* **2019**, *7*, 323. [CrossRef]

61. Kim, S.; Kim, T.M. Generation of mesenchymal stem-like cells for producing extracellular vesicles. *World J. Stem Cells* **2019**, *11*, 270–280. [CrossRef] [PubMed]

62. Lian, Q.; Lye, E.; Suan Yeo, K.; Khia Way Tan, E.; Salto-Tellez, M.; Liu, T.M.; Palanisamy, N.; El Oakley, R.M.; Lee, E.H.; Lim, B.; et al. Derivation of clinically compliant MSCs from CD105+, CD24- differentiated human ESCs. *Stem Cells* **2007**, *25*, 425–436. [CrossRef]

63. Hwang, N.S.; Varghese, S.; Lee, H.J.; Zhang, Z.; Ye, Z.; Bae, J.; Cheng, L.; Elisseeff, J. In vivo commitment and functional tissue regeneration using human embryonic stem cell-derived mesenchymal cells. *Proc. Natl. Acad. Sci. USA* **2008**, *105*, 20641–20646. [CrossRef]

64. Hwang, N.S.; Varghese, S.; Zhang, Z.; Elisseeff, J. Chondrogenic differentiation of human embryonic stem cell-derived cells in arginine-glycine-aspartate-modified hydrogels. *Tissue Eng.* **2006**, *12*, 2695–2706. [CrossRef] [PubMed]

65. Toh, W.S.; Lee, E.H.; Guo, X.M.; Chan, J.K.; Yeow, C.H.; Choo, A.B.; Cao, T. Cartilage repair using hyaluronan hydrogel-encapsulated human embryonic stem cell-derived chondrogenic cells. *Biomaterials* **2010**, *31*, 6968–6980. [CrossRef]

66. Jiang, B.; Fu, X.; Yan, L.; Li, S.; Zhao, D.; Wang, X.; Duan, Y.; Yan, Y.; Li, E.; Wu, K.; et al. Transplantation of human ESC-derived mesenchymal stem cell spheroids ameliorates spontaneous osteoarthritis in rhesus macaques. *Theranostics* **2019**, *9*, 6587–6600. [CrossRef] [PubMed]

67. Gibson, J.D.; O'Sullivan, M.B.; Alaee, F.; Paglia, D.N.; Yoshida, R.; Guzzo, R.M.; Drissi, H. Regeneration of Articular Cartilage by Human ESC-Derived Mesenchymal Progenitors Treated Sequentially with BMP-2 and Wnt5a. *Stem Cells Transl. Med.* **2017**, *6*, 40–50. [CrossRef]

68. Wang, Y.; Yu, D.; Liu, Z.; Zhou, F.; Dai, J.; Wu, B.; Zhou, J.; Heng, B.C.; Zou, X.H.; Ouyang, H.; et al. Exosomes from embryonic mesenchymal stem cells alleviate osteoarthritis through balancing synthesis and degradation of cartilage extracellular matrix. *Stem Cell Res. Ther.* **2017**, *8*, 189. [CrossRef]

69. Narsinh, K.H.; Plews, J.; Wu, J.C. Comparison of human induced pluripotent and embryonic stem cells: Fraternal or identical twins? *Mol. Ther.* **2011**, *19*, 635–638. [CrossRef] [PubMed]

70. Chin, M.H.; Mason, M.J.; Xie, W.; Volinia, S.; Singer, M.; Peterson, C.; Ambartsumyan, G.; Aimiuwu, O.; Richter, L.; Zhang, J.; et al. Induced pluripotent stem cells and embryonic stem cells are distinguished by gene expression signatures. *Cell Stem Cell* **2009**, *5*, 111–123. [CrossRef]

71. Puri, M.C.; Nagy, A. Concise review: Embryonic stem cells versus induced pluripotent stem cells: The game is on. *Stem Cells* **2012**, *30*, 10–14. [CrossRef]

72. Lietman, S.A. Induced pluripotent stem cells in cartilage repair. *World J. Orthop.* **2016**, *7*, 149–155. [CrossRef]

73. Takahashi, K.; Tanabe, K.; Ohnuki, M.; Narita, M.; Ichisaka, T.; Tomoda, K.; Yamanaka, S. Induction of pluripotent stem cells from adult human fibroblasts by defined factors. *Cell* **2007**, *131*, 861–872. [CrossRef] [PubMed]

74. Yu, J.; Vodyanik, M.A.; Smuga-Otto, K.; Antosiewicz-Bourget, J.; Frane, J.L.; Tian, S.; Nie, J.; Jonsdottir, G.A.; Ruotti, V.; Stewart, R.; et al. Induced pluripotent stem cell lines derived from human somatic cells. *Science* **2007**, *318*, 1917–1920. [CrossRef] [PubMed]

75. Wei, Y.; Zeng, W.; Wan, R.; Wang, J.; Zhou, Q.; Qiu, S.; Singh, S.R. Chondrogenic differentiation of induced pluripotent stem cells from osteoarthritic chondrocytes in alginate matrix. *Eur. Cell Mater.* **2012**, *23*, 1–12. [CrossRef] [PubMed]

76. Moradi, S.; Mahdizadeh, H.; Saric, T.; Kim, J.; Harati, J.; Shahsavarani, H.; Greber, B.; Moore, J.B., IV. Research and therapy with induced pluripotent stem cells (iPSCs): Social, legal, and ethical considerations. *Stem Cell Res. Ther.* **2019**, *10*, 341. [CrossRef]

77. Deng, X.Y.; Wang, H.; Wang, T.; Fang, X.T.; Zou, L.L.; Li, Z.Y.; Liu, C.B. Non-viral methods for generating integration-free, induced pluripotent stem cells. *Curr. Stem Cell Res. Ther.* **2015**, *10*, 153–158. [CrossRef]

78. Stadtfeld, M.; Nagaya, M.; Utikal, J.; Weir, G.; Hochedlinger, K. Induced pluripotent stem cells generated without viral integration. *Science* **2008**, *322*, 945–949. [CrossRef] [PubMed]

79. Pittenger, M.F.; Mackay, A.M.; Beck, S.C.; Jaiswal, R.K.; Douglas, R.; Mosca, J.D.; Moorman, M.A.; Simonetti, D.W.; Craig, S.; Marshak, D.R. Multilineage potential of adult human mesenchymal stem cells. *Science* **1999**, *284*, 143–147. [CrossRef]

80. Friedenstein, A.J.; Chailakhyan, R.K.; Gerasimov, U.V. Bone marrow osteogenic stem cells: In vitro cultivation and transplantation in diffusion chambers. *Cell Tissue Kinet.* **1987**, *20*, 263–272. [CrossRef]

81. Lv, F.J.; Tuan, R.S.; Cheung, K.M.; Leung, V.Y. Concise review: The surface markers and identity of human mesenchymal stem cells. *Stem Cells* **2014**, *32*, 1408–1419. [CrossRef] [PubMed]

82. Pittenger, M.F.; Discher, D.E.; Peault, B.M.; Phinney, D.G.; Hare, J.M.; Caplan, A.I. Mesenchymal stem cell perspective: Cell biology to clinical progress. *NPJ Regen. Med.* **2019**, *4*, 22. [CrossRef]

83. Wang, M.; Yuan, Q.; Xie, L. Mesenchymal Stem Cell-Based Immunomodulation: Properties and Clinical Application. *Stem Cells Int.* **2018**, *2018*, 3057624. [CrossRef] [PubMed]

84. Kim, C.; Keating, A. Cell Therapy for Knee Osteoarthritis: Mesenchymal Stromal Cells. *Gerontology* **2019**, *65*, 294–298. [CrossRef]

85. Berebichez-Fridman, R.; Montero-Olvera, P.R. Sources and Clinical Applications of Mesenchymal Stem Cells: State-of-the-art review. *Sultan Qaboos Univ. Med. J.* **2018**, *18*, e264–e277. [CrossRef] [PubMed]

86. Ramakrishnan, A.; Torok-Storb, B.; Pillai, M.M. Primary marrow-derived stromal cells: Isolation and manipulation. *Methods Mol. Biol.* **2013**, *1035*, 75–101.

87. Koh, Y.G.; Choi, Y.J. Infrapatellar fat pad-derived mesenchymal stem cell therapy for knee osteoarthritis. *Knee* **2012**, *19*, 902–907. [CrossRef]

88. Koh, Y.G.; Jo, S.B.; Kwon, O.R.; Suh, D.S.; Lee, S.W.; Park, S.H.; Choi, Y.J. Mesenchymal stem cell injections improve symptoms of knee osteoarthritis. *Arthroscopy* **2013**, *29*, 748–755. [CrossRef]

89. Cianca, J.C.; Jayaram, P. Musculoskeletal Injuries and Regenerative Medicine in the Elderly Patient. *Phys. Med. Rehabil. Clin. N. Am.* **2017**, *28*, 777–794. [CrossRef] [PubMed]

90. Shah, K.; Zhao, A.G.; Sumer, H. New Approaches to Treat Osteoarthritis with Mesenchymal Stem Cells. *Stem Cells Int.* **2018**, *2018*, 5373294. [CrossRef]

91. Wakitani, S.; Okabe, T.; Horibe, S.; Mitsuoka, T.; Saito, M.; Koyama, T.; Nawata, M.; Tensho, K.; Kato, H.; Uematsu, K.; et al. Safety of autologous bone marrow-derived mesenchymal stem cell transplantation for cartilage repair in 41 patients with 45 joints followed for up to 11 years and 5 months. *J. Tissue Eng. Regen. Med.* **2011**, *5*, 146–150. [CrossRef]

92. Davatchi, F.; Abdollahi, B.S.; Mohyeddin, M.; Shahram, F.; Nikbin, B. Mesenchymal stem cell therapy for knee osteoarthritis. Preliminary report of four patients. *Int. J. Rheum. Dis.* **2011**, *14*, 211–215. [CrossRef]

93. Wong, K.L.; Lee, K.B.; Tai, B.C.; Law, P.; Lee, E.H.; Hui, J.H. Injectable cultured bone marrow-derived mesenchymal stem cells in varus knees with cartilage defects undergoing high tibial osteotomy: A prospective, randomized controlled clinical trial with 2 years follow-up. *Arthroscopy* **2013**, *29*, 2020–2028. [CrossRef]

94. Orozco, L.; Munar, A.; Soler, R.; Alberca, M.; Soler, F.; Huguet, M.; Sentis, J.; Sanchez, A.; Garcia-Sancho, J. Treatment of knee osteoarthritis with autologous mesenchymal stem cells: A pilot study. *Transplantation* **2013**, *95*, 1535–1541. [CrossRef] [PubMed]

95. Orozco, L.; Munar, A.; Soler, R.; Alberca, M.; Soler, F.; Huguet, M.; Sentis, J.; Sanchez, A.; Garcia-Sancho, J. Treatment of knee osteoarthritis with autologous mesenchymal stem cells: Two-year follow-up results. *Transplantation* **2014**, *97*, e66–e68. [CrossRef]
96. Davatchi, F.; Sadeghi Abdollahi, B.; Mohyeddin, M.; Nikbin, B. Mesenchymal stem cell therapy for knee osteoarthritis: 5 years follow-up of three patients. *Int. J. Rheum. Dis.* **2016**, *19*, 219–225. [CrossRef]
97. Jo, C.H.; Lee, Y.G.; Shin, W.H.; Kim, H.; Chai, J.W.; Jeong, E.C.; Kim, J.E.; Shim, H.; Shin, J.S.; Shin, I.S.; et al. Intra-articular injection of mesenchymal stem cells for the treatment of osteoarthritis of the knee: A proof-of-concept clinical trial. *Stem Cells* **2014**, *32*, 1254–1266. [CrossRef] [PubMed]
98. Pak, J.; Chang, J.J.; Lee, J.H.; Lee, S.H. Safety reporting on implantation of autologous adipose tissue-derived stem cells with platelet-rich plasma into human articular joints. *BMC Musculoskelet Disord.* **2013**, *14*, 337. [CrossRef] [PubMed]
99. Kim, Y.S.; Kwon, O.R.; Choi, Y.J.; Suh, D.S.; Heo, D.B.; Koh, Y.G. Comparative Matched-Pair Analysis of the Injection Versus Implantation of Mesenchymal Stem Cells for Knee Osteoarthritis. *Am. J. Sports Med.* **2015**, *43*, 2738–2746. [CrossRef]
100. Pers, Y.M.; Rackwitz, L.; Ferreira, R.; Pullig, O.; Delfour, C.; Barry, F.; Sensebe, L.; Casteilla, L.; Fleury, S.; Bourin, P.; et al. Adipose Mesenchymal Stromal Cell-Based Therapy for Severe Osteoarthritis of the Knee: A Phase I Dose-Escalation Trial. *Stem Cells Transl. Med.* **2016**, *5*, 847–856. [CrossRef]
101. Lee, W.S.; Kim, H.J.; Kim, K.I.; Kim, G.B.; Jin, W. Intra-Articular Injection of Autologous Adipose Tissue-Derived Mesenchymal Stem Cells for the Treatment of Knee Osteoarthritis: A Phase IIb, Randomized, Placebo-Controlled Clinical Trial. *Stem Cells Transl. Med.* **2019**, *8*, 504–511. [CrossRef]
102. Mianehsaz, E.; Mirzaei, H.R.; Mahjoubin-Tehran, M.; Rezaee, A.; Sahebnasagh, R.; Pourhanifeh, M.H.; Mirzaei, H.; Hamblin, M.R. Mesenchymal stem cell-derived exosomes: A new therapeutic approach to osteoarthritis? *Stem Cell Res. Ther.* **2019**, *10*, 340. [CrossRef]
103. Ni, Z.; Zhou, S.; Li, S.; Kuang, L.; Chen, H.; Luo, X.; Ouyang, J.; He, M.; Du, X.; Chen, L. Exosomes: Roles and therapeutic potential in osteoarthritis. *Bone Res.* **2020**, *8*, 25. [CrossRef] [PubMed]
104. Cosenza, S.; Ruiz, M.; Toupet, K.; Jorgensen, C.; Noel, D. Mesenchymal stem cells derived exosomes and microparticles protect cartilage and bone from degradation in osteoarthritis. *Sci. Rep.* **2017**, *7*, 16214. [CrossRef] [PubMed]
105. Ha, D.H.; Kim, H.K.; Lee, J.; Kwon, H.H.; Park, G.H.; Yang, S.H.; Jung, J.Y.; Choi, H.; Lee, J.H.; Sung, S.; et al. Mesenchymal Stem/Stromal Cell-Derived Exosomes for Immunomodulatory Therapeutics and Skin Regeneration. *Cells* **2020**, *9*, 1157. [CrossRef] [PubMed]
106. Roseti, L.; Desando, G.; Cavallo, C.; Petretta, M.; Grigolo, B. Articular Cartilage Regeneration in Osteoarthritis. *Cells* **2019**, *8*, 1305. [CrossRef]
107. Evans, C.H.; Ghivizzani, S.C.; Robbins, P.D. Gene Delivery to Joints by Intra-Articular Injection. *Hum. Gene Ther.* **2018**, *29*, 2–14. [CrossRef] [PubMed]
108. Kim, M.K.; Ha, C.W.; In, Y.; Cho, S.D.; Choi, E.S.; Ha, J.K.; Lee, J.H.; Yoo, J.D.; Bin, S.I.; Choi, C.H.; et al. A Multicenter, Double-Blind, Phase III Clinical Trial to Evaluate the Efficacy and Safety of a Cell and Gene Therapy in Knee Osteoarthritis Patients. *Hum. Gene Ther. Clin. Dev.* **2018**, *29*, 48–59. [CrossRef]

International Journal of
Molecular Sciences

Review

From Pathogenesis to Therapy in Knee Osteoarthritis: Bench-to-Bedside

Elena Rezuş [1,†], Alexandra Burlui [1,*,†], Anca Cardoneanu [1,†], Luana Andreea Macovei [1,†], Bogdan Ionel Tamba [2,*] and Ciprian Rezuş [3]

[1] Department of Rheumatology and Physiotherapy, "Grigore T. Popa" University of Medicine and Pharmacy, 700115 Iaşi, Romania; elena.rezus@umfiasi.ro (E.R.); anca.cardoneanu@umfiasi.ro (A.C.); luana.macovei@umfiasi.ro (L.A.M.)

[2] Advanced Center for Research and Development in Experimental Medicine (CEMEX), "Grigore T. Popa" University of Medicine and Pharmacy, 700454 Iaşi, Romania

[3] Department of Internal Medicine, "Grigore T. Popa" University of Medicine and Pharmacy, 700115 Iaşi, Romania; ciprian.rezus@umfiasi.ro

* Correspondence: maria-alexandra.burlui@umfiasi.ro (A.B.); bogdan.tamba@umfiasi.ro (B.I.T.)

† These authors contributed equally to this work.

Abstract: Osteoarthritis (OA) is currently the most widespread musculoskeletal condition and primarily affects weight-bearing joints such as the knees and hips. Importantly, knee OA remains a multifactorial whole-joint disease, the appearance and progression of which involves the alteration of articular cartilage as well as the synovium, subchondral bone, ligaments, and muscles through intricate pathomechanisms. Whereas it was initially depicted as a predominantly aging-related and mechanically driven condition given its clear association with old age, high body mass index (BMI), and joint malalignment, more recent research identified and described a plethora of further factors contributing to knee OA pathogenesis. However, the pathogenic intricacies between the molecular pathways involved in OA prompted the study of certain drugs for more than one therapeutic target (amelioration of cartilage and bone changes, and synovial inflammation). Most clinical studies regarding knee OA focus mainly on improvement in pain and joint function and thus do not provide sufficient evidence on the possible disease-modifying properties of the tested drugs. Currently, there is an unmet need for further research regarding OA pathogenesis as well as the introduction and exhaustive testing of potential disease-modifying pharmacotherapies in order to structure an effective treatment plan for these patients.

Keywords: osteoarthritis; knee joint; disease modifying drugs; cartilage; bone remodeling; inflammation; platelet-rich plasma; mesenchymal stem cells; ozone; hyaluronic acid

check for
updates

Citation: Rezuş, E.; Burlui, A.; Cardoneanu, A.; Macovei, L.A.; Tamba, B.I.; Rezuş, C. From Pathogenesis to Therapy in Knee Osteoarthritis: Bench-to-Bedside. *Int. J. Mol. Sci.* **2021**, *22*, 2697. https://doi.org/10.3390/ijms22052697

Academic Editor: Masato Sato

Received: 31 January 2021
Accepted: 4 March 2021
Published: 7 March 2021

Publisher's Note: MDPI stays neutral with regard to jurisdictional claims in published maps and institutional affiliations.

1. Introduction

Osteoarthritis (OA) is a chronic musculoskeletal condition that primarily affects weight-bearing joints (such as the knees, hips, and spine) yet may involve the hands as well as other non-weight-bearing articular sites [1–5]. Genetic predisposition has been deemed relevant, however, more so in the hands and hips rather than in knee OA [1–6]. Moreover, certain racial and gender-related differences were also reported [6,7]. Nevertheless, OA remains a multifactorial whole-joint disease, the appearance and progression of which involves the alteration of articular cartilage as well as the synovium, subchondral bone, ligaments, and muscles through intricate pathogenic mechanisms [1–3].

Whereas it was initially depicted as a predominantly aging-related and mechanically driven condition given its clear association with old age, high body mass index (BMI), and joint malalignment, more recent research identified and described a plethora of further factors contributing to knee OA pathogenesis [6–10].

Expert opinion in OA proposes case stratification, describing four phenotypes of the disease largely based on pathogenesis: mechanical, metabolic, osteoporotic, and inflammatory. Nonetheless, patient stratification could lead to more precise identification of the potential therapeutic targets yet demands a comprehensive evaluation pretreatment [5]. Novel findings on the mechanisms underlying the development of knee OA prompted the search for potential disease-modifying OA drugs (DMOADs) able to counteract the molecular pathways involved in cartilage degradation, inflammation, and bone remodeling (Figure 1). However, most therapeutic agents with potential disease-modifying properties have not yet proven their efficacy in slowing the progression of knee OA in clinical trials [6–8].

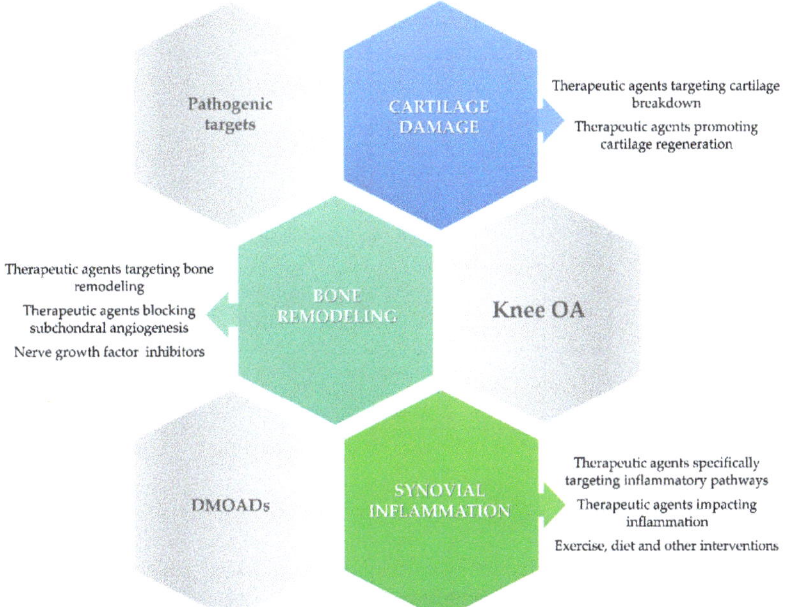

Figure 1. Therapeutic interventions targeting cartilage breakdown, bone remodeling, and inflammation in knee osteoarthritis (OA).

The present review aims to discuss the potentially disease-modifying therapeutic options targeting cartilage destruction, subchondral bone remodeling, and synovial inflammation in knee OA according to recent findings.

2. Therapeutic Approach to Cartilage Damage in Knee Osteoarthritis

The approval of DMOADs, according to regulatory guidelines from the United States Food and Drug Administration (FDA) and the European Medicines Agency (EMA) would need to meet the following conditions: slower loss in knee or hip joint space width (JSW) on x-ray and an appropriate symptomatic improvement [11]. Presently, there is no approved DMOAD despite the large amount of research conducted on the subject [12]. The currently available literature describes two subsets of potential DMOADs regarding cartilage damage in knee OA with respect to their specific actions: therapeutic agents targeting cartilage degeneration and drugs that support cartilage regeneration (Figure 2).

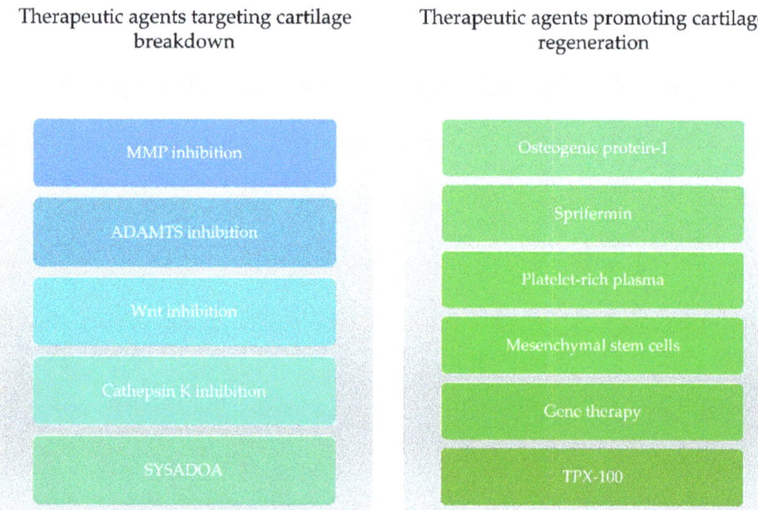

Figure 2. Therapeutic options targeting cartilage damage in knee OA.

2.1. Matrix Metalloproteinase Inhibition

Matrix metalloproteinases (MMPs) and aggrecanases are known as the main proteinases responsible for matrix degradation in OA [13,14]. MMPs, a family of zinc-dependent enzymes, are recognized for their involvement in the degeneration of the extracellular matrix (ECM) [15]. They can be classified into the following groups: collagenases (MMP-1 and MMP-13), gelatinases (MMP-2 and MMP-9), stromelysins (MMP-3), metalloelastase (MMP-12), matrilysin (MMP-7), and membrane-type MMPs (MT-MMPs) [16].

In 2007, Krzeski et al. published a study of PG-116,800 (PG-530,742), an MMP-inhibitor, but no significant changes were observed in joint space width (JSW) of the knee or Western Ontario and McMaster Universities Osteoarthritis Index (WOMAC) scores at 1 year [17], yet important musculoskeletal side effects (named the "musculoskeletal syndrome" (MSS)) such as pain, loss of range of motion in large joints, joint swelling, stiffness, soft tissue pain, and Dupuytren's contracture were observed in the treatment group. The exact cause of these reactions is not yet fully understood, but it is thought that chelation of zinc is involved in the process, seeing as the medication contains zinc-binding groups [18].

Given its notable pathogenic role in both OA as well as cartilage homeostasis, MMP-13 has been studied as a potential therapeutic target. ALS 1-0635 is a molecule that is free from zinc-chelating functional groups. Baragi et al. found a 67% deceleration in cartilage decline compared to the placebo in experimental animals in vivo [19].

Doxycycline is a broad-spectrum tetracycline antibiotic that also inhibits MMP activity, especially collagenase and gelatinase [20]. Nevertheless, a Cochrane review from 2012 stipulated that "the small benefit in terms of joint space narrowing is of questionable clinical relevance and outweighed by safety problems" [21].

2.2. ADAMTS Inhibition

In addition to MMPs, aggrecanases are also implicated in cartilage metabolism, acting as blockers of the aggrecan, the main proteoglycan of articular cartilage [22]. Aggrecan is a proteoglycan that incorporates chondroitin sulphate and keratan sulphate and is connected to a protein core. It links to another molecule, hyaluronan, thus creating stable large-molecular-weight molecules binding with a separate globular link [23]. In OA, the loss of aggrecan is an early event in the degradation of articular cartilage and results in the decrement of functional and structural ECM integrity followed by an irreversible loss of collagen [24].

Based on the fact that ADAMTS-4 (aggrecanase 1) and ADAMTS-5 (aggrecanase 2) are able to cleave proteoglycans (such as aggrecan—the main component of articular cartilage), both of these molecules have been studied as potentially disease-modifying targets in OA. A humanized anti-ADAMTS-5 antibody (GSK2394002) was assessed in a study conducted by Larkin et al. The authors demonstrated the efficacy of this drug in preventing constitutional destruction as well as in decreasing mechanical pain. Nevertheless, systemic administration resulted in significant side effects (increased mean arterial pressure and cardiac ischemia) and therefore was not approved as a DMOAD [25]. CRB007, a chimeric murine/human IgG4 anti-ADAMTS-5 monoclonal antibody, displayed disease-modifying properties in animal models of OA in the study published by Chiusaroli et al. However, further extensive research is needed [26].

Huang et al. described several small molecular aggrecanase inhibitors that demonstrated chondroprotective activity in patients with OA. One of these compounds is AGG-523, a per os ADAMTS-4 and -5 inhibitor [27]. Animal models of OA demonstrated a reduced level of aggrecan fragments in joints [28]. AGG-523 was part of two Phase I studies, but these trials were discontinued due to unknown reasons [27].

A derivative of 5-(1H-pyrazol-4-yl) methylene)-2-thioxothiazolidin-4-one has been suggested to exhibit important activity against ADAMTS-5. Together with a hyaluronic acid hydrogel (HAX), this molecule was injected into the knees of rats. The results of the study indicate that ADAMTS-5 could be a promising target for disease-modifying OA treatment [29]. A 2017 study reported the effects of per os GLPG1972/S201086, a potent inhibitor of ADAMTS-5. GLPG1972/S201086 confirmed the important protective effect on cartilage and subchondral bone in posttraumatic OA by remarkably diminishing cartilage proteoglycan loss, cartilage impairment, and subchondral bone sclerosis. Furthermore, the safety parameters evaluated indicated that the drug was well tolerated [30]. A Phase II trial assessing the efficacy and safety of 3 doses of per os GLPG1972/ S201086 once daily in patients with knee OA was completed in July 2020, yet the final results have yet to be reported [31].

2.3. Wnt Inhibition

Wang et al. described Wnt as a glycoprotein with extracellular position, for which signaling engages 19 Wnt genes and receptors that are capable of managing canonical β-catenin-dependent and noncanonical β-catenin-independent signaling pathways [32]. The abovementioned pathways are responsible for various biological aspects involving the cartilage, thus confirming their pathological role in OA [33,34].

Small molecules such as XAV-939 and SM04690 were discovered and studied as potential inhibitors of Wnt signaling in patients with knee OA. Certain Phase I, II, and IIb studies highlighted the positive effects of SM04690 in terms of pain reduction, functional impairment, and JSW in addition to a seemingly good safety profile. These data suggest that SM04690 could be considered for use as a disease-modifying agent in patients with moderate to severe symptomatic knee OA [35–37].

Takada et al. and Grossmann et al., in their respective studies of StAx-35R (a staple β-catenin-binding domain of Axin) and SAH-Bcl9 (a staple peptide derived from the Bcl9 homology domain-2), described the involvement and role in β-catenin transcriptional activity [38].

Another molecule involved in Wnt/β-catenin signaling is LRP5 (lipoprotein receptor-associated protein 5), which stimulates catabolic factors responsible for OA cartilage destruction and inhibits the anabolic factor type II collagen. A study on LRP5-knockdown mice confirmed the benefits on cartilage [39,40].

Lorecivivint inhibits the Wnt signaling pathway while also suppressing CLK2 (CDC-like kinase 2) and DYRK1A-mediated (dual-specificity tyrosine phosphorylation-regulated kinase 1A) phosphorylation of SIRT1 and FOXO1, both involved in Wnt/β-catenin activity. The drug has been shown to be relatively safe and well-tolerated, with important outcomes regarding cartilage destruction and inflammation, by decreasing catabolic enzymes and blocking the inflammatory process [41]. Studies on patients with knee OA have demonstrated the favorable effects of lorecivivint on reducing symptoms (notably pain), and improving physical function and patient global assessment [42].

2.4. Cathepsin K Inhibition

Cathepsin K is a lysosomal cysteine protease found in activated osteoclasts and chondrocytes (as well as other cell types) and has been shown to be involved in cartilage degradation (by destroying types I and II collagen and aggrecan found in the cartilage) and bone resorption [43,44].

Balicatib (AAE581), a cathepsin K inhibitor, was investigated in the treatment of patients with knee OA. The important side effects (skin rashes and dermal fibrosis) surpassed the beneficial outcome on both cartilage and bone, therefore leading to suspension of the study [45].

MIV-711 is described as being a selective and reversible agent that inhibits cathepsin K activity. MIV-711 holds an important role in decreasing serum CTX-I (a marker of bone turnover) and CTX-II levels (a marker of cartilage turnover), thus being implicated in both bone resorption and cartilage impairment [46]. MIV-711 was investigated in a multicenter, randomized, placebo-controlled, double-blind, three-arm parallel, Phase IIa study in patients with knee OA. The results showed that the reduction in medial femoral cartilage thickness was considerably reduced at 26 weeks. However, the medial tibia cartilage loss was not significantly changed. There was no substantial difference in pain reduction and quality of life scores or biomarker CTX-I and CTX-II values. The most important adverse events described were musculoskeletal symptoms, rashes, and infections. During the 6-month follow-up period, Conaghan et al. reported not finding significant benefits regarding OA-related symptoms. However, there is a need for further confirmation of MIV-711 as a DMOAD in OA through long-term trials [47].

2.5. Osteogenic Protein-1

BMP-7 (bone morphogenetic protein-7 or osteogenic protein-7), a member of the (TGF)-β superfamily, is a growth factor and is considered a possible therapeutic target in the process of restoration of damaged cartilage. Research analyzed BMP-7 (eptotermin-alpha) in the treatment of knee OA. A Phase I trial reported the amelioration of clinical symptoms and no dose-limiting toxicity. A Phase II trial and a Phase I trial also investigated BMP-7, but results have not yet been published [48].

2.6. Sprifermin

Sprifermin (AS902330) is a recombinant human fibroblast growth factor-18 (rhFGF18). According to published data, sprifermin has positive effects on cartilage by stimulating cell multiplication and ECM constituent production [49,50]. On intraarticular administration, it binds FGFR3 receptors from the cartilage [51,52].

The study of Dahlberg et al. (first in-human trial, randomized, double-blind, placebo-controlled study) investigated patients who proposed joint arthroplasty due to severe OA of the knee. However, no significant differences in terms of symptom improvement were described between the treatment group and the placebo cohort [53].

Preliminary results at 3 years of an important study of sprifermin were presented by Hochberg et al. in 2019. The FORWARD study, a 5-year Phase II, dose-ranging, randomized trial investigated the effect of intraarticular injections of sprifermin administered every 6 or 12 months versus the placebo. The consequence of drug injection was a lower mean cartilage thickness loss compared to natural evolution (total femorotibial joint as well as in the medial, lateral, central medial, and central-lateral regions). Importantly, the FORWARD study followed specifically structural progression, not other clinical issues associated with knee OA development and progression [54].

2.7. Platelet-Rich Plasma

Platelet-rich plasma (PRP) incorporates granules containing growth factors (transforming growth factor-β (TGF-b), platelet-derived growth factor (PDGF), insulin-like growth factor (IGF_, vascular endothelial growth factor (VEGF), and fibroblast growth factor (FGF)), cytokines, chemokines, and other mediators implicated in mesenchymal stem cell (MSC) proliferation and production of ECM and collagen, thus playing an important part in cartilage restoration [55,56].

Several randomized clinical trials have been completed in patients with knee OA, with intraarticular PRP being compared with hyaluronic acid. In terms of efficacy, the results of these studies revealed a favorable outcome in reducing symptoms and increasing mobility in these patients (with a mean duration of 12 months). Furthermore, no serious adverse effects were reported, suggesting a good safety profile for PRP [57–59]. However, there were a lot of discrepancies between these studies specifically in the protocols, methods, and patient characteristics; thus, it is challenging to compare all the data obtained [57].

2.8. Mesenchymal Stem Cells

Another promising treatment for cartilage repair seems to be MSCs derived from bone marrow, adipose tissue, or the umbilical cord. Studies have demonstrated that, compared with MSC injection, cell implantation has better results in patients with knee OA. Furthermore, there is a similar safety profile between autologous and allogenic MSCs [60,61].

A systematic review published by Jevotovsky et al. in 2018 evaluating 61 studies with MSCs as OA treatment concluded that MSC therapy has a favorable effect on OA patients but that there is limited high-quality evidence as well as a lack of long-term follow-up [62].

In 2015, Vega et al. presented a 1-year clinical trial that included patients with knee OA treated with intraarticular injections of allogeneic bone marrow MSCs. The results of this study showed considerable improvement in pain and functional indices, with a significant decrease in poor-quality cartilage areas and sustained augmentation of cartilage quality in the affected regions (quantified by magnetic resonance imaging (MRI) T2 mapping) [63].

An interesting aspect was represented by the possibility of treating MSCs before administration in order to boost the modulatory effect in OA. An example is combining MSCs with an inhibitor of signal transducer and activator of transcription 3. Under these conditions, the pro-inflammatory state was shown to be improved and the results of the treatment were more encouraging [64].

2.9. Gene Therapy

Gene therapy could be a promising alternative to OA treatment, mainly because of its long-term results on OA cartilage. Studies underline that intraarticular gene transfer treatment is a feasible option for patients suffering from OA compared to systemic administration with respect to safety, bioavailability, and direct targeting of the pathological site [65,66].

2.10. TPX-100

TPX-100, a peptide derived from matrix extracellular phosphoglycoprotein (MEPE), has expressed positive function when inducing articular cartilage regeneration in animal models. It is considered that this molecule facilitates cartilage formation only in areas with defects, without ectopic bone tissue formation [30].

A Phase II clinical study evaluating TPX-100 in patients with patellofemoral OA indicated significant amelioration and clinically efficient improvement in KOOS (Knee injury and Osteoarthritis Outcome Score) and WOMAC scores. Furthermore, promising effects were obtained concerning tibiofemoral cartilage thickness and volume at 6 and 12 months. Importantly, this performance was preserved for 30 months. The use of symptomatic treatment (nonsteroidal anti-inflammatory drugs (NSAIDs) and analgesics) diminished significantly across the study [67].

2.11. Symptomatic Slow-Acting Drugs for Osteoarthritis

A group of drugs known as SYSADOA (symptomatic slow-acting drugs for OA) have been used in patients with OA in order to improve clinical symptoms (pain and morning stiffness) as well as to slow progression of the disease. Among SYSADOA, chondroitin sulphate, glucosamine, and diacerein are the most widely studied. Using quantitative MRI (qMRI), clinical trials have demonstrated the contribution of chondroitin sulphate in decreasing cartilage volume loss [7,8]. Chondroitin sulphate influences proteoglycan metabolism and plays an important role in the balance of anabolic and catabolic processes at the ECM level. Chondroitin sulphate also decreases a series of proinflammatory factors with destructive properties in subchondral bone osteoblasts [9,10].

3. Therapeutic Approach to Bone Remodeling in Knee Osteoarthritis

Subchondral bone remodeling plays a very important role in OA, mediating and preceding cartilage damage [68]. The structural changes in the subchondral bone are different depending on the stages of OA (Figure 3). Thus, in the early stage, there is an increase in subchondral bone turnover characterized by thinning of the subchondral bone plate and by increasing porosity associated with impairment of the trabeculae: thickness decreasing and separation increasing [69]. The late stage of OA is characterized by thickening of the plate and trabecular layers, by decreasing bone marrow spacing, and by sclerosis of the subchondral bone. Even if bone thickening occurs, there is insufficient bone mineralization due to a decreased calcium and collagen ratio followed by increased bone turnover [70–75].

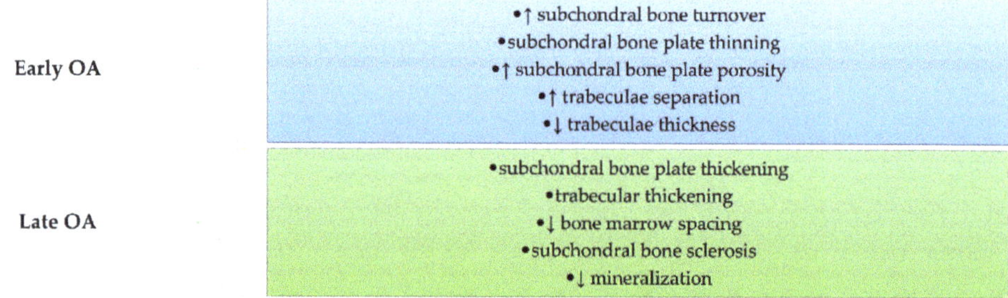

Figure 3. The main structural changes in the subchondral bone occurs in OA depending on the stage of the disease (early or late). In the early stage, there is an increase in subchondral bone turnover characterized by thinning of the subchondral bone plate and by increasing porosity associated with impairment of the trabeculae: thickness decreasing and separation increasing. The late stage of OA is characterized by thickening of the plate and trabecular layers, by decreasing bone marrow spacing, by sclerosis of the subchondral bone, and by decreased mineralization.

Furthermore, the changes in subchondral bone are related to several signaling pathways such as the Wnt/β-Catenin, TGF-β/Smad, RANK/RANKL/OPG (Receptor activator of nuclear factor kappa-B, RANK ligand, and osteoprotegerin), and MAPK (mitogen-activated protein kinase) signaling pathways [76–84]. Moreover, identification of the molecular pathways involved in OA-related bone remodeling has led to certain therapeutic agents targeting the mechanisms considered as possible DMOADs [85–90].

Regarding subchondral bone changes, therapies aim to block bone resorption caused by osteoclasts as well as other mechanisms such as novel angiogenesis or particular neuronal factors responsible for the onset of pain (Figure 4).

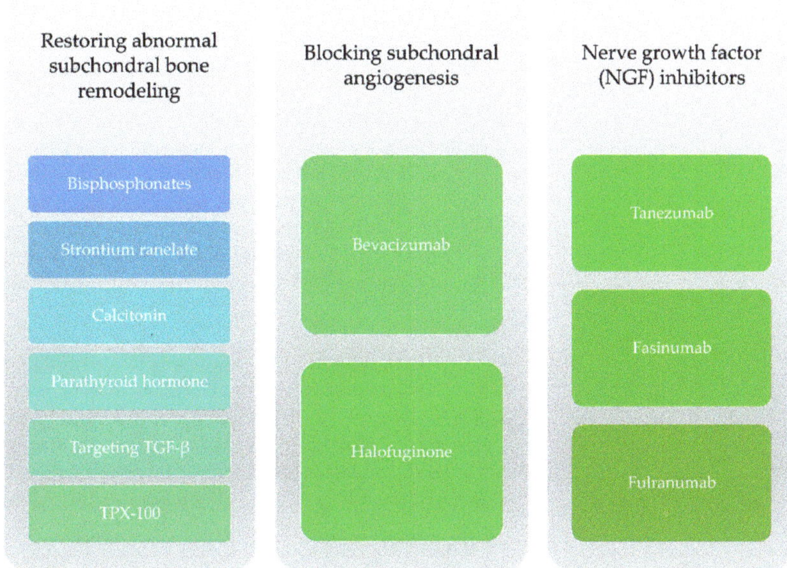

Figure 4. Therapeutic options targeting subchondral bone remodeling for knee OA.

3.1. Bisphosphonates

Bisphosphonates are antiresorptive drugs that can slow bone remodeling through the inhibition of osteoclasts. There are many published data that have shown pain relief, reduced bone destruction, and an improvement in joint structure following these treatments, but the clinical results are not very clear [91,92].

Alendronate, by inhibiting subchondral bone loss, caused an improvement in the structure of articular cartilage [93]. Other data showed an improvement of WOMAC pain score, an increase in bone mineral density, and a decrease in markers of bone destruction after alendronate administration for 2 years in patients with OA [91].

Risedronate use has not been shown to be clinically and radiologically effective in a group of 2483 patients with knee OA after 2 years of follow-up [94]. On the other hand, Spector et al. showed an improvement in pain and a decrease in subchondral remodeling following risedronate treatment [95].

The favorable effect of zoledronic acid was highlighted in a study that included 59 patients with knee OA. After 6 months of treatment, a significant reduction in pain was observed as well as a reduction in bone marrow lesions detected by magnetic resonance imaging [96]. Less favorable results (in relieving pain and loss of cartilage volume) after 24 months of treatment with zoledronic acid in knee OA cases were published by Aitken [91,92].

Clodronate, a non-amino bisphosphonate, appears to have a favorable effect in patients with knee OA, acting by increasing the secretion of SOX9, the transcription factor responsible for progenitor stem cell chondrogenic commitment. A 12-week, randomized, placebo-controlled study that included 80 cases of knee OA showed an improvement in pain and decreases in WOMAC score and in the need for analgesic medication after once weekly intraarticular injection of 2 mg clodronate [97].

3.2. Strontium Ranelate

Strontium ranelate can restore the abnormal subchondral bone remodeling by decreasing the activity of osteoclasts and by favoring the mineralization of new bone. After 3 years of treatment, patients with knee OA had an improvement in joint space narrowing compared to placebo [98]. At a dose of 1800 mg/day, there was a decrease in chondrocyte apoptosis and an improvement in cartilage matrix [99]. At a higher dose—2000 mg/day—strontium ranelate showed better effects in terms of loss of cartilage volume and regarding the improvement of bone marrow lesions [100].

3.3. Calcitonin

Calcitonin, a hormone secreted by parafollicular thyroid cells, binds to specific receptors on osteoclasts and inhibits their activity, thus reducing damage to the subchondral bone in OA [101,102]. In patients with knee OA, the use of oral calcitonin has been shown to have an effect in relieving pain and in increasing cartilage volume compared to placebo [103]. Due to the high risk of neoplasms, the use of calcitonin for short periods of time is recommended [104].

3.4. Cathepsin K Inhibition

Cathepsin K inhibitors are antiresorptive drugs that inhibit osteolytic protease of osteoclasts. There are several animal studies that have shown the beneficial role of these molecules in stopping the progression of OA [105,106]. In studies, there are 2 tested molecules: an oral inhibitor called MIV-711, for which structural changes in the joints were not considered, and balicatib (AAE581) used in the knee OA, in which narrowing of the articular space and the cartilage volume followed [107,108].

3.5. Parathyroid Hormone

Teriparatide, a recombinant human parathyroid hormone (PTH), is a bone anabolic therapy that regulates endochondral ossification and inhibits chondrocyte hypertrophy [109]. Systemic administration of teriparatide can inhibit cartilage degradation and abnormal chondrocyte differentiation, can stimulate regeneration of the matrix, and can improve the structure of the subchondral bone [110,111]. There are several ongoing studies that aim to highlight the beneficial effects of this molecule on both subchondral bone and joint cartilage, being assimilated as chondroregenerative therapy.

3.6. Trasforming Growth Factor β Inhibition

TGF-β gives positive feedback with the Wnt signaling pathway, favoring the differentiation of chondrocytes and osteoblasts [111]. In OA, TGF-β is secreted in excess by osteoblasts and causes the development of osteophytes [112]. Using a murine model, studies have highlighted the role of neutralizing TGF-β antibodies in stopping OA progression [113]. Other data have shown that implantation of an anti-TGF-β antibody (1D11) in alginic acid microbeads in the subchondral bone or deletion of the TGF-β type II receptor (TβRII) inhibits Smad2/3 phosphorylation in osteoblastic precursors, protecting the osteochondral unit [114].

3.7. TPX-100

MEPE is a bone protein secreted by osteocytes, with a negative role in regulating bone mineralization and involved in remodeling the subchondral bone [115]. A 23-aminoacid derived from MEPE (TPX-100) has been shown to be effective in cases of moderate femuropatellar OA, leading to significant improvement in WOMAC scores. However, there was no evidence of structural improvement after 12 months of treatment [116].

3.8. Subchondral Angiogenesis Inhibition

Subchondral angiogenesis acts as a bridge between the articular cartilage and the subchondral bone. Blocking neoangiogenesis, the data showed a decrease in subchondral bone loss and a reduction in cartilage degradation [117]. Numerous factors such as VEGF (vascular endothelial growth factor), TGF-β1, PDGF-BB (platelet-derived growth factor BB monomer), and SLIT3 participate in the process of angiogenesis in the subchondral bone [117]. Halofuginone stops the action of TGF-β1 by inhibiting Smad2/3, thus blocking subchondral angiogenesis [118]. Bevacizumab, an antibody against VEGF, has been shown to be effective in reducing the formation of subchondral blood vessels, thereby inhibiting chondrocyte hypertrophy [119].

3.9. Nerve Growth Factor (NGF) Inhibition

Abnormal remodeling of the subchondral bone is associated with the presence of particular neural factors that cause innervation of sensory nerve structures in cases of OA [120]. NGFs are secreted by preosteoclasts, being triggers of subchondral bone innervation in OA [121]. NGF inhibitors such as tanezumab, fasinumab, or fulranumab have been shown to be effective in ameliorating joint pain and function in patients with OA [122,123]. For tanezumab, the data showed a significant improvement in pain [122,123], while fasinumab is still under investigation [124].

4. Therapeutic Approach to Synovial Inflammation in Knee Osteoarthritis

The appearance of low-grade inflammation remains one of the most prominent features in OA pathogenesis. Proinflammatory cytokines such as IL-1, IL-6, and TNFα may be upregulated and may promote the release of ROS (reactive oxygen species), MMPs, and ADAMTS [125]. Though numerous emerging drugs were tested in degenerative joint conditions, a few studies focused specifically on their impact on knee OA-related inflammation (systemic or synovial fluid levels of inflammatory markers, the presence of joint effusion, or prolonged morning stiffness) [126–130]. Nevertheless, several drugs and dietary supplements as well as lifestyle interventions (diet and physical exercise) have been shown to exhibit disease-modifying properties in knee OA (Figure 5).

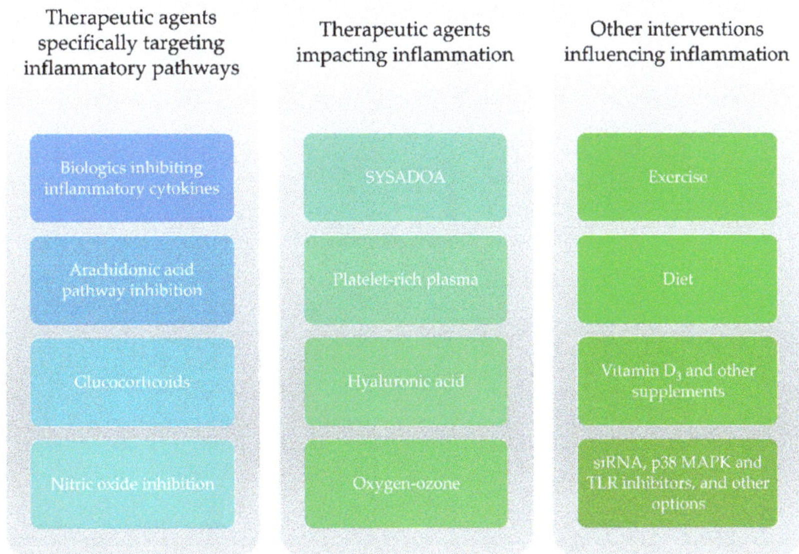

Figure 5. Drugs, dietary supplements, and other interventions contributing to a reduction in inflammation in knee OA.

4.1. Biologics Targeting Proinflammatory Cytokines

Whereas they are more commonly used in immune-inflammatory diseases such as rheumatoid arthritis, psoriasis, ankylosing spondylitis, juvenile idiopathic arthritis, and inflammatory bowel diseases, certain TNFα inhibitors have also been studied in knee OA [131]. An open-label study including 56 patients with knee OA compared the efficacy of a single intraarticular injection with 10 mg adalimumab (ADA) versus 25 mg hyaluronic acid (HA), obtaining better results in terms of pain and functional impairment in the TNF inhibitor subgroup compared to their counterparts (Table 1) [132]. However, the follow-up period was short (4 weeks) and all patients were under concomitant treatment with systemic NSAIDs (celecoxib 200 mg daily) [132].

Another research conducted on 39 patients suggested that intraarticular injections with etanercept (ETN) could be superior to hyaluronic acid for pain relief in moderate to severe knee OA. While this was true for WOMAC pain score improvement at 4 weeks, the differences in mean VAS (Visual Analogue Scale) levels between the two study groups failed to reach statistical significance at the end of the follow-up period [133].

A multicenter, randomized, double-blind, placebo-controlled study of 50 mg and 150 mg intraarticular anakinra (ANR) versus placebo found no significant benefits for ANR over 12 weeks [134]. Wang et al. conducted a placebo-controlled, randomized, double-blind, ascending dose research focused on anti-interleukin-1α/β dual variable domain immunoglobulin ABT-981 in patients with symptomatic knee OA, identifying lower levels of high-sensitivity C-reactive protein (hsCRP) as well as IL-1α and IL-1β in the treatment group [135].

Table 1. Proinflammatory cytokine blockers adalimumab, etanercept, anakinra, and ABT-981 in knee OA.

Reference	Compound	Intervention	Patients	Duration	Results
Wang [132]	Adalimumab (ETA)	10 mg intraarticular ADA + Celecoxib 200 mg/day versus 25 mg HA + Celecoxib 200 mg/day	ADA + Celecoxib (N = 28) HA + Celecoxib (N = 28)	4 weeks	The authors found a significant improvement in pain and functionality in the ADA group.
Ohtori et al. [133]	Etanercept (ETN)	10 mg intraarticular ETN versus 25 mg HA	ETN (N = 19) HA (N = 20)	4 weeks	An initial significant amelioration was obtained in the ETN group compared to HA (weeks 1 and 2), yet the results were not maintained at week 4.
Chevalier et al. [134]	Anakinra (ANR)	150 mg intraarticular ANR 50 mg intraarticular ANR versus Placebo	ANR 150 mg (N = 67) ANR 50 mg (N = 34) Placebo (N = 69)	12 weeks	A significant pain improvement was observed in the 150 mg ANR group compared to 50 mg ANR at day 4. Overall, intraarticular ANR did not demonstrate notable benefits, irrespective of the dose.
Wang et al. [135]	ABT-981	ABT-981 (various doses) versus Placebo	ABT-981 0.3 mg/kg fortnightly (N = 7) 1 mg/kg fortnightly (N = 7) 3 mg/kg fortnightly (N = 7) 3 mg/kg every 4 weeks (N = 7) Placebo (N = 8)	113 days (cohorts 1, 2, and 3) 127 days (cohort 4)	Mean hsCRP decreased through week 2 irrespective of ABT-981 dose/administration interval. While IL-1α and IL-1β were lower in the treatment group, serum vascular endothelial growth factor and MMP-9 did not demonstrate significant changes.

4.2. Arachidonic Acid Pathway Inhibition

The arachidonic acid pathway employs cyclooxygenase enzymes (COX) and prompts the release of proinflammatory prostaglandins. The latter are also involved in pain mediation, which is why NSAIDs are widely prescribed as symptomatic treatment in OA [136]. However, certain NSAIDs have additionally been considered for use as disease-modifying agents [137,138]. A study performed on knee OA patients examined the impact of celecoxib, ibuprofen, and diclofenac on synovial fluid proinflammatory cytokine (TNFα, IL-6, and IL-8) and VEGF expression [139]. Aside from significant pain relief, NSAID therapy (particularly higher doses) showed notable beneficial effects in reducing inflammatory markers in patients' synovial fluid, thus suggesting that these drugs could be regarded as having disease-modifying properties [139]. Nevertheless, the numerous adverse events associated with prolonged administration of systemic NSAIDs limit their use primarily in older individuals and persons with comorbid conditions [140].

The dual hindrance of COX2 and 5-lipooxygenase (LOX) diminished the release of prostaglandin E2, leukotriene B4, collagenase-1, cathepsin, MMP-13, and IL-1β [141]. Moreover, COX/LOX blockade diminished OA-related cartilage degradation and synovial hypertrophy in preclinical studies [138,141]. Licofelone is a COX/LOX inhibitor that displayed a protective effect on knee cartilage volume loss in a multicenter clinical trial [142].

4.3. Glucocorticoids

Glucocorticoids exhibit potent anti-inflammatory effects and have been shown to reduce pain and pain-related functional impairment in patients with OA. However, intraarticular or systemic corticosteroid administration has seldom been investigated with respect to its effect on clinical or paraclinical indicators of inflammation in knee OA (joint effusion and stiffness, synovial fluid inflammatory markers, and imaging tests) [143–145]. Furthermore, the risk–benefit ratio for using intraarticular glucocorticoids in knee OA remains a matter of discussion among researchers, clinicians, and medical organizations alike [146].

Nevertheless, a study evaluating a combination of triamcinolone acetonide and sodium hyaluronate was superior to intraarticular saline in improving joint stiffness, pain, and function in patients with knee OA with no or minimal to moderate joint space narrowing (Kellgren–Lawrence grades 1–3) [147]. A recent study investigating the effect of intraarticular injections with triamcinolone found that the latter reduced joint effusion in patients with knee OA [148].

4.4. Symptomatic Slow-Acting Drugs for Osteoarthritis

The DISSCO study examined the efficacy of diacerein (diacetylrhein, an anthraquinone derivative, and SYSADOA) compared to celecoxib on knee osteoarthritis. The multicenter randomized trial concluded that diacerein was non-inferior to celecoxib. The percentage of patients experiencing joint effusion diminished by half over time, however, without notable discrepancies between the two treatment groups [149]. In rat synovial cells, diacerein reduced the mRNA levels of inflammatory cytokines (TNFα, IL-1, and IL-6), COX2, cartilage-degrading enzymes (MMP-3 and MMP-13), and ADAMTS-5 in a dose-dependent manner while also raising those of certain anti-inflammatory factors (IL-4 and IL-10). Moreover, the analysis of biopsy samples taken from rats with monosodium iodoacetate-induced OA as well as imaging tests (computed tomography) showed that the administration of diacerein-loaded nanoparticles may protect against joint destruction [150].

Other SYSADOAs such as glucosamine and chondroitin sulphate have also been thought to display anti-inflammatory properties together with benefits regarding cartilage degradation in patients with knee OA, yet results remain discrepant across studies [7–10].

4.5. Nitric Oxide Inhibition

Nitric oxide (NO) together with inducible NO synthase (iNOS, an enzyme that participates in NO synthesis) are involved in OA-related inflammation, cartilage damage, and cellular injury. The release of proinflammatory cytokines such as TNFα, IL-1β, IL-17, and interferon γ (IFNγ) has been shown to upregulate the expression of iNOS, thus emphasizing the link between inflammation and oxidative stress [151,152].

Scientific evidence highlighting the involvement of iNOS in OA pathogenesis prompted testing of iNOS inhibitors in human subjects. A randomized double-blind placebo-controlled trial of the iNOS inhibitor cindunistat (50 or 200 mg daily) was conducted on patients with symptomatic knee OA. In the Kellgren–Lawrence grade 2 subgroup, cindunistat 50 mg daily was associated with a delay in joint space narrowing at 48 weeks compared to the placebo. Nevertheless, these results were not maintained at 96 weeks [153]. Furthermore, the Kellgren-Lawrence grade 3 cohort did not show a significant improvement with respect to slowing knee OA progression with iNOS inhibition, irrespective of cindunistat dose [153].

4.6. Platelet-Rich Plasma

PRP is an autologous formulation containing concentrated blood-derived platelets and growth factors that promotes synoviocyte differentiation and proliferation, with studies reporting significant benefits in OA of the knee [154,155]. Moreover, PRP is thought to be rich in bioactive molecules involved in tissue repair and regeneration and to wield anti-inflammatory properties by lowering synovial fluid proinflammatory marker levels [156–159]. Notably, some authors support the use of leukocyte-poor PRP (LP-PRP), arguing that the presence of white blood cells in intraarticular formulation may generate pro-inflammatory effects [160]. PRP has been deemed safe for use and has demonstrated promising results in different chronic rheumatic conditions and injury-related joint damage [161]. In knee OA, PRP displayed positive effects in improving pain, joint function, and quality of life according to recently published meta-analyses [157,162].

In young adults (18–30 years of age) diagnosed with knee OA, intraarticular PRP therapy led to a statistically significant decrease in IL-1β, IL-6, TNFα, IL-17A, RANKL, and IFNγ levels (Table 2) [163]. A study performed on eutrophic individuals (BMI between 22–25 kg/m^2) aged between 42 and 79 years comparing intraarticular PRP, HA, and the combination between PRP and HA found that IL-1β, TNFα, TIMP1, and MMP-3 values were diminished in the PRP group at 6 months posttreatment [164]. Nevertheless, the association of PRP and HA held the best results in reducing proinflammatory markers [164].

Table 2. Platelet-rich plasma (PRP) treatment in knee OA: anti-inflammatory properties.

Reference	Intervention	Patients	Duration	Results
Huang et al. [163]	2–14 mL intraarticular PRP weekly versus Placebo (10 mL saline solution) weekly	PRP (N = 310) Placebo (N = 56)	8 weeks	Significant improvements in plasma IL-1β, IL-6, TNFα, IL-17A, RANKL, and IFNγ were observed in the treatment group.
Xu et al. [164]	4 mL intraarticular PRP, 3 injections per knee, half-month interval 2 mL intraarticular HA, 3 injections per knee, half-month interval 4 mL PRP + 2 mL HA, 3 injections per knee, half-month interval	PRP (N = 40 knees) HA (N = 34 knees) PRP + HA (48 knees) N = 78 patients total (122 knees) N = 44 patients received bilateral injections	24 months	IL-1β, TNFα, TIMP1, and MMP-3 demonstrated significant decreases in the PRP group at 6 months posttreatment. Nevertheless, the PRP + HA group showed better results in this respect. Additionally, the PRP + HA cohort displayed IL-1β, TNFα, TIMP1, and MMP-3 inhibition at 12 months post-injection.

4.7. Hyaluronic Acid

A randomized, double-blind, placebo-controlled study of obese patients with knee OA examined the outcomes of a per os preparation containing HA (70%) and other glycosaminoglycans (GAGs) [165]. At 3 months, the treatment group exhibited a significant decrement in TNFα, IL-1α and IL-1β, IL-6, IL-17α, IFN, and GM-CSF (granulocyte-macrophage colony-stimulating factor) levels, while the placebo cohort demonstrated notably higher synovial fluid concentrations of proinflammatory cytokines as well as leptin [165].

A recent research comparing the efficacy of intraarticular HA to ozone (O_3) found that both reduced joint stiffness in patients with knee OA [166]. Other published studies aimed to investigate the possible differences between intraarticular HA and biologics, PRP, or O_3 (Table 3) [132,133,164–166]. Xu et al. reported positive results for a PRP + HA combination with respect to the reduction in proinflammatory cytokine values and MMP-3 [164].

Table 3. Hyaluronic acid (HA) treatment in knee OA: anti-inflammatory properties.

Reference	Intervention	Patients	Duration	Results
Nelson et al. [165]	80 mg oral preparation of HA (70%) + other GAGs versus Placebo	Oral preparation of HA + other GAGs (N = 21) Placebo (N = 19)	12 weeks	The HA-treated cohort demonstrated a notable decrease in TNFα, IL-1α, IL-1β, IL-6, IL-17α, IFN, and GM-CSF values. Furthermore, the placebo group exhibited significantly higher synovial fluid concentrations of inflammatory cytokines as well as leptin (a proinflammatory adipokine).
Wang [132]	25 mg HA + Celecoxib 200 mg/day versus 10 mg intraarticular ADA + Celecoxib 200 mg/day	HA + Celecoxib (N = 28) ADA + Celecoxib (N = 28)	4 weeks	Changes in joint stiffness did not exhibit statistically significant differences between HA + Celecoxib and ADA + Celecoxib.
Ohtori et al. [133]	25 mg HA versus 10 mg intraarticular ETN	HA (N = 20) ETN (N = 19)	4 weeks	The HA group displayed significantly weaker results in terms of joint stiffness improvement compared to ETN during the follow-up period.
Xu et al. [164]	2 mL intraarticular HA, 3 injections per knee, half-month interval 4 mL intraarticular PRP, 3 injections per knee, half-month interval 2 mL HA + 4 mL PRP, 3 injections per knee, half-month interval	HA (N = 34 knees) PRP (N = 40 knees) HA + PRP (48 knees)	24 months	The cohort that received the HA + PRP combination demonstrated IL-1β, TNFα, TIMP1, and MMP-3 inhibition at one-year posttreatment.
Raeissadat et al. [166]	3 weekly intraarticular injections of 20 mg/2 mL HA versus 3 weekly intraarticular injections of 30 µg/mL O_3 (10 mL)	HA (N = 74) O_3 (N = 67)	24 weeks	There was a significant reduction in joint stiffness in the group treated with HA. However, these results were not significantly different from the O_3-treated cohort.

4.8. Oxygen–Ozone

A study of intraarticular injections of oxygen–ozone (O_2–O_3) versus triamcinolone in knee OA determined that joint effusion was notably reduced in both treatment arms [148]. Whereas the decrement in WOMAC and VAS values were in favor of the O_2–O_3 group, the differences compared to corticosteroid-treated patients did not reach statistical significance with respect to the reduction in joint effusion on ultrasound [148]. Other published research reported finding a reduction in joint stiffness in knee OA patients following treatment with O_3 (Table 4) [148,166,167].

Table 4. O_3 and O_2–O_3 treatment in knee OA: anti-inflammatory properties.

Reference	Intervention	Patients	Duration	Results
Babaei-Ghazani et al. [148]	15 µg/mL intraarticular O_2–O_3 (10 mL) versus 40 mg intraarticular triamcinolone (1 mL)	O_2–O_3 (N = 31) Triamcinolone (N = 31)	12 weeks	The authors found an important reduction in joint effusion on ultrasound in both treatment arms at 3 months post-injection.
Lopes de Jesus et al. [167]	20 µg/mL intraarticular O_3 (10 mL) versus Placebo (10 mL air)	O_3 (N = 61) Placebo (N = 35)	8 weeks	There was a significant improvement in joint stiffness at 8 weeks in the treatment arm.
Raeissadat et al. [166]	3 weekly injections of 30 µg/mL intraarticular O_3 (10 mL) versus 3 weekly injections of 20 mg/2 mL HA	O_3 (N = 67) HA (N = 74)	24 weeks	Both groups demonstrated a notable amelioration in joint stiffness yet without significant discrepancies between the two treatment arms.

4.9. Exercise, Diet, and Supplements

Diet and lifestyle changes have also been shown to reduce inflammation in knee OA. The IDEA (Intensive Diet and Exercise for Arthritis) study enrolled over 400 overweight and obese individuals (BMI between 27–40.5 kg/m^2) with symptomatic knee OA. The research subjects were randomized to three intervention groups (exercise, diet, or diet and exercise) and followed the program for 18 months [168]. Both diet as well as the combination of diet and exercise significantly lowered serum inflammatory marker levels [168].

Another research performed on women with knee OA examining the role of resistance training and 1000 mg daily nanocurcumin (turmeric) showed that the latter reduced synovial fluid NO concentrations by over 26% and collagenase II by circa 4.7% [169]. However, the authors commented that the short duration of the study (16 weeks) could explain the lack of statistically significant differences between the 3 intervention groups (resistance exercise, curcumin supplementation, and exercise and supplementation) [169].

Vitamin D-deficient individuals were found to be at risk of developing knee OA [160]. The impact of vitamin D supplementation on pain relief, the improvement in functional parameters, and cartilage loss remain matters of debate [170,171]. However, an MRI study analyzing the effect of 50,000 IU monthly vitamin D_3 supplementation in patients with knee OA found that the placebo group exhibited a significant increase in effusion-synovitis volume over two years compared to the treatment cohort [172].

It has been stated that old age may be associated with intestinal dysbiosis and increased gut permeability, which could promote a "leakage" of gut microbes into tissues, thus leading to a chronic proinflammatory state [173,174]. Tsai et al. aimed to characterize the immunological signature of the microbiome in osteoarthritic knee synovium, finding marked discrepancy in microbial abundance in patients compared to healthy controls [175]. Whereas intestinal dysbiosis has been linked to synovial inflammation in knee OA, the impact of probiotic use on disease progression have not been studied extensively in clinical trials [176,177]. Certain experimental studies reported promising results with respect to the administration of *Lactobacillus acidophilus* in rats. In this regard, *Lactobacillus acidophilus* prompted the diminishment in proinflammatory marker expression together with an increase in anti-inflammatory cytokines in monosodium iodoacetate-induced OA. Moreover, *Lactobacillus acidophilus* exhibited anti-nociceptive effects in experimental animals [178].

It has been stated that dietary polyphenols and herbal medicine may also exhibit anti-inflammatory properties and reduce iNOS expression in OA [151,179]. Scoparone is a compound extracted from *Artemisia capillaris* that may exhibit potent anti-inflammatory effects. A recent study proved that scoparone could repress IL-1β-induced activation of the PI3K/Akt/NF-κB pathway as well as COX2 and iNOS [180]. In murine models of OA, acteoside (withdrawn from *Ligustrum purpurascens* kudingcha tea) reduced the expression of TNFα, IL-6, and IFNγ and hindered the IL-1β-related activation of the JAK/STAT (janus kinase/signal transducer and activator of transcription) pathway [181].

Avocado soybean unsaponifiables were studied in knee OA, showing positive results. Moreover, avocado soybean unsaponifiables diminished the expression of iNOS, MMP-13, and TNFα in rat monosodium iodoacetate-induced OA [182].

4.10. Other Therapeutic Options Exhibiting Antiinflammatory Properties

The inhibition of TNFα and pan-cytokine small interference RNA (siRNA) led to a decrease in proinflammatory cytokines (IL-1, IL-6, IL-8, and TNFα), with promising findings being recorded in collagenase-induced murine models of OA [183–185].

P38 mitogen-activated protein kinase (MAPK), toll-like receptors (TLR2 and TLR4), and leptin have been considered potential targets for the development of new DMOAD options based on their experimentally shown involvement in OA-related inflammation [138]. Though in vitro and animal studies reported certain encouraging results, the efficacy of siRNA, p38 MAPK, TLR, and inflammatory adipokine blockade remains in need of further testing in clinical trials of knee OA [138].

Carboxymethyl chitosan (a chitosan soluble derivative) displays a significant physicochemical likeness to cartilage proteoglycans. In rat chondrocytes, increasing doses of carboxymethyl chitosan demonstrated iNOS inhibition and upregulated anti-inflammatory cytokine expression (IL-10) [186].

5. Conclusions

OA is a chronic and pathogenically multifaceted disease that remains a major cause of disability worldwide. Cartilage degradation, inflammation, and bone remodeling are presently regarded as therapeutic targets for the development of DMOADs. However, the pathogenic intricacies between the molecular pathways involved in OA prompted the study of certain drugs for more than one therapeutic target (amelioration of cartilage and bone changes, and inflammation). Most clinical studies regarding knee OA focus mainly on improvement in pain or joint function and thus do not provide sufficient evidence on the possible disease-modifying properties of the tested drugs. Currently, there is an unmet need for further research regarding OA pathogenesis as well as the introduction and exhaustive testing of potential disease-modifying pharmacotherapies in order to structure an effective treatment plan for these patients.

Funding: The present research received no external funding.

Data Availability Statement: Not applicable.

Conflicts of Interest: The authors declare no conflict of interest.

References

1. Roemer, F.W.; Kwoh, C.K.; Hayashi, D.; Felson, D.T.; Guermazi, A. The role of radiography and MRI for eligibility assessment in DMOAD trials of knee OA. *Nat. Rev. Rheumatol.* **2018**, *14*, 372–380. [CrossRef]
2. Malfait, A.-M.; Tortorella, M.D. The "elusive DMOAD": Aggrecanase inhibition from laboratory to clinic. *Clin. Exp. Rheumatol.* **2019**, *37*, 130–134.

3. Castrogiovanni, P.; Di Rosa, M.; Ravalli, S.; Castorina, A.; Guglielmino, C.; Imbesi, R.; Vecchio, M.; Drago, F.; Szychlinska, M.A.; Musumeci, G. Moderate Physical Activity as a Prevention Method for Knee Osteoarthritis and the Role of Synoviocytes as Biological Key. *Int. J. Mol. Sci.* **2019**, *20*, 511. [CrossRef]
4. Yeap, S.S. Current DMOAD options for the treatment of osteoarthritis. *Clin. Exp. Rheumatol.* **2020**, *38*, 802.
5. Roman-Blas, J.A.; Bizzi, E.; Largo, R.; Migliore, A.; Herrero-Beaumont, G. An update on the up and coming therapies to treat osteoarthritis, a multifaceted disease. *Expert Opin. Pharmacother.* **2016**, *17*, 1745–1756. [CrossRef]
6. Rezuș, E.; Cardoneanu, A.; Burlui, A.; Luca, A.; Codreanu, C.; Tamba, B.I.; Stanciu, G.-D.; Dima, N.; Bădescu, C.; Rezuș, C. The Link Between Inflammaging and Degenerative Joint Diseases. *Int. J. Mol. Sci.* **2019**, *20*, 614. [CrossRef]
7. Martel-Pelletier, J.; Raynauld, J.-P.; Mineau, F.; Abram, F.; Paiement, P.; Delorme, P.; Pelletier, J.-P. Levels of serum biomarkers from a two-year multicentre trial are associated with treatment response on knee osteoarthritis cartilage loss as assessed by magnetic resonance imaging: An exploratory study. *Arthritis Res.* **2017**, *19*, 169. [CrossRef]
8. Vreju, F.A.; Ciurea, P.L.; Rosu, A.; Chisalau, B.A.; Parvanescu, C.D.; Firulescu, S.C.; Stiolica, A.T.; Barbulescu, A.L.; Dinescu, S.C.; Dumitrescu, C.I.; et al. The Effect of glucosamine, chondroitin and harpagophytum procumbens on femoral hyaline cartilage thickness in patients with knee osteoarthritis—An MRI versus ultrasonography study. *J. Mind Med. Sci.* **2019**, 162–168. [CrossRef]
9. Reginster, J.; Veronese, N. Highly purified chondroitin sulfate: A literature review on clinical efficacy and pharmaco-economic aspects in osteoarthritis treatment. *Aging Clin. Exp. Res.* **2020**, *33*, 37–47. [CrossRef] [PubMed]
10. Lane, N.; Shidara, K.; Wise, B. Osteoarthritis year in review 2016: Clinical. *Osteoarthr. Cartil.* **2017**, *25*, 209–215. [CrossRef] [PubMed]
11. United States Food and Drugs Administration. Available online: https://www.fda.gov/regulatory-information/search-fda-guidance-documents/osteoarthritis-structural-endpoints-development-drugs (accessed on 11 January 2021).
12. European Medicines Agency. Available online: https://www.ema.europa.eu/en/documents/scientific-guideline/guideline-clinical-investigation-medicinal-products-used-treatment-osteoarthritis_en.pdf (accessed on 3 January 2021).
13. Fosang, A.J.; Little, C.B. Drug Insight: Aggrecanases as therapeutic targets for osteoarthritis. *Nat. Clin. Pract. Rheumatol.* **2008**, *4*, 420–427. [CrossRef] [PubMed]
14. Murphy, G.; Nagase, H. Reappraising metalloproteinases in rheumatoid arthritis and osteoarthritis: Destruction or re-pair? *Nat. Clin. Pract. Rheumatol.* **2008**, *4*, 128–135. [CrossRef]
15. Chow, Y.Y.; Chin, K.-Y. The Role of Inflammation in the Pathogenesis of Osteoarthritis. *Mediat. Inflamm.* **2020**, *2020*, 1–19. [CrossRef] [PubMed]
16. Bauvois, B. New facets of matrix metalloproteinases MMP-2 and MMP-9 as cell surface transducers: Outside-in signaling and relationship to tumor progression. *Biochim. Biophys. Acta BBA Bioenerg.* **2012**, *1825*, 29–36. [CrossRef]
17. Krzeski, P.; Buckland-Wright, C.; Bálint, G.; Cline, G.A.; Stoner, K.; Lyon, R.; Beary, J.; Aronstein, W.S.; Spector, T.D. Development of musculoskeletal toxicity without clear benefit after administration of PG-116800, a matrix metalloproteinase inhibitor, to patients with knee osteoarthritis: A randomized, 12-month, double-blind, placebo-controlled study. *Arthritis Res. Ther.* **2007**, *9*, R109. [CrossRef] [PubMed]
18. Drake, C.G. Book Review of "Tumor Immunology and Cancer Vaccines". *Cancer Investig.* **2006**, *24*, 657. [CrossRef]
19. Baragi, V.M.; Becher, G.; Bendele, A.M.; Biesinger, R.; Bluhm, H.; Boer, J.; Deng, H.; Dodd, R.; Essers, M.; Feuerstein, T.; et al. A new class of potent matrix metalloproteinase 13 inhibitors for potential treatment of osteoarthritis: Evidence of histologic and clinical efficacy without musculoskeletal toxicity in rat models. *Arthritis Rheum.* **2009**, *60*, 2008–2018. [CrossRef]
20. Pinney, J.R.; Taylor, C.; Doan, R.; Burghardt, A.J.; Li, X.; Kim, H.T.; Ma, C.B.; Majumdar, S. Imaging longitudinal changes in articular cartilage and bone following doxycycline treatment in a rabbit anterior cruciate ligament transection model of osteoarthritis. *Magn. Reson. Imaging* **2012**, *30*, 271–282. [CrossRef]
21. da Costa, B.; Nüesch, E.; Reichenbach, S.; Jüni, P.; Rutjes, A. Doxycycline for osteoarthritis of the knee or hip. *Cochrane Database Syst. Rev.* **2012**. [CrossRef]
22. Dubail, J.; Apte, S.S. Insights on ADAMTS proteases and ADAMTS-like proteins from mammalian genetics. *Matrix Biol.* **2015**, *44–46*, 24–37. [CrossRef]
23. Kaushal, G.P.; Shah, S.V. The new kids on the block: ADAMTSs, potentially multifunctional metalloproteinases of the ADAM family. *J. Clin. Investig.* **2000**, *105*, 1335–1337. [CrossRef] [PubMed]
24. Little, C.B.; Meeker, C.T.; Golub, S.B.; Lawlor, K.E.; Farmer, P.J.; Smith, S.M.; Fosang, A.J. Blocking aggrecanase cleavage in the aggrecan interglobular domain abrogates cartilage erosion and promotes cartilage repair. *J. Clin. Investig.* **2008**, *118*, 3812. [CrossRef]
25. Larkin, J.; Lohr, T.A.; Elefante, L.; Shearin, J.; Matico, R.; Su, J.-L.; Xue, Y.; Liu, F.; Genell, C.; Miller, R.E.; et al. Translational development of an ADAMTS-5 antibody for osteoarthritis disease modification. *Osteoarthr. Cartil.* **2015**, *23*, 1254–1266. [CrossRef]
26. Chiusaroli, R.; Visentini, M.; Galimberti, C.; Casseler, C.; Mennuni, L.; Covaceuszach, S.; Lanza, M.; Ugolini, G.; Caselli, G.; Rovati, L.; et al. Targeting of ADAMTS5's ancillary domain with the recombinant mAb CRB0017 ameliorates disease progression in a spontaneous murine model of osteoarthritis. *Osteoarthr. Cartil.* **2013**, *21*, 1807–1810. [CrossRef]
27. Huang, Z.; Ding, C.; Li, T.; Yu, S. Current status and future prospects for disease modification in osteoarthritis. *Rheumatology* **2017**, *57* (Suppl. S4), iv108–iv123. [CrossRef]

28. Chockalingam, P.; Sun, W.; Rivera-Bermudez, M.; Zeng, W.; Dufield, D.; Larsson, S.; Lohmander, L.; Flannery, C.; Glasson, S.; Georgiadis, K.; et al. Elevated aggrecanase activity in a rat model of joint injury is attenuated by an aggrecanase specific inhibitor. *Osteoarthr. Cartil.* **2011**, *19*, 315–323. [CrossRef]

29. Chen, P.; Zhu, S.; Wang, Y.; Mu, Q.; Wu, Y.; Xia, Q.; Zhang, X.; Sun, H.; Tao, J.; Hu, H.; et al. The amelioration of cartilage degeneration by ADAMTS-5 inhibitor delivered in a hyaluronic acid hydrogel. *Biomaterials* **2014**, *35*, 2827–2836. [CrossRef] [PubMed]

30. Clement-Lacroix, P.; Little, C.; Meurisse, S.; Blanqué, R.; Mollat, P.; Brebion, F.; Gosmini, R.; De Ceuninck, F.; Botez, I.; Lepescheux, L.; et al. GLPG1972: A potent, selective, orally available adamts-5 inhibitor for the treatment of OA. *Osteoarthr. Cartil.* **2017**, *25*, S58–S59. [CrossRef]

31. Alcaraz, M.J.; Guillén, M.I.; Ferrándiz, M.L. Emerging therapeutic agents in osteoarthritis. *Biochem. Pharmacol.* **2019**, *165*, 4–16. [CrossRef]

32. Wang, Y.; Fan, X.; Xing, L.; Tian, F. Wnt signaling: A promising target for osteoarthritis therapy. *Cell Commun. Signal.* **2019**, *17*, 1–14. [CrossRef]

33. Liu, S.; Yang, H.; Shuan, L.; Zhang, M. Sirt1 regulates apoptosis and extracellular matrix degradation in resveratrol-treated osteoarthritis chondrocytes via the Wnt/β-catenin signaling pathways. *Exp. Ther. Med.* **2017**, *14*, 5057–5062. [CrossRef]

34. Zhou, Y.; Wang, T.; Hamilton, J.L.; Chen, D. Wnt/β-catenin Signaling in Osteoarthritis and in Other Forms of Arthritis. *Curr. Rheumatol. Rep.* **2017**, *19*, 1–8. [CrossRef] [PubMed]

35. Lietman, C.; Wu, B.; Lechner, S.; Shinar, A.; Sehgal, M.; Rossomacha, E.; Datta, P.; Sharma, A.; Gandhi, R.; Kapoor, M.; et al. Inhibition of Wnt/β-catenin signaling ameliorates osteoarthritis in a murine model of experimental osteoarthritis. *JCI Insight* **2018**, *3*. [CrossRef]

36. Deshmukh, V.; Hu, H.; Barroga, C.; Bossard, C.; Kc, S.; Dellamary, L.; Stewart, J.; Chiu, K.; Ibanez, M.; Pedraza, M.; et al. A small-molecule inhibitor of the Wnt pathway (SM04690) as a potential disease modifying agent for the treatment of osteoarthritis of the knee. *Osteoarthr. Cartil.* **2018**, *26*, 18–27. [CrossRef]

37. Seo, T.; Deshmukh, V.; Yazici, Y. AB0069 LORECIVIVINT (SM04690), an intra-articular, small-molecule CLK/DYRK1A inhibitor that modulates the WNT pathway, as a potential treatment for meniscal injuries. *Ann. Rheum. Dis.* **2020**, *79* (Suppl. S1), 1335. [CrossRef]

38. Takada, K.; Zhu, D.; Bird, G.; Sukhdeo, K.; Zhao, J.; Mani, M.; Lemieux, M.; Carrasco, D.; Ryan, J.; Horst, D.; et al. Targeted disruption of the BCL9/-catenin complex inhibits oncogenic WNT signaling. *Sci. Transl. Med.* **2012**, *4*, 148ra117. [CrossRef]

39. Shin, Y.; Huh, Y.H.; Kim, K.; Kim, S.; Park, K.H.; Koh, J.-T.; Chun, J.-S.; Ryu, J.-H. Low-density lipoprotein receptor–related protein 5 governs Wnt-mediated osteoarthritic cartilage destruction. *Arthritis Res. Ther.* **2014**, *16*, R37. [CrossRef] [PubMed]

40. Gao, S.-G.; Zeng, C.; Liu, J.-J.; Tian, J.; Cheng, C.; Zhang, F.-J.; Xiong, Y.-L.; Pan, D.; Xiao, Y.-B.; Lei, G.-H. Association between Wnt inhibitory factor-1 expression levels in articular cartilage and the disease severity of patients with osteoarthritis of the knee. *Exp. Ther. Med.* **2016**, *11*, 1405–1409. [CrossRef] [PubMed]

41. Deshmukh, V.; O'Green, A.; Bossard, C.; Seo, T.; Lamangan, L.; Ibanez, M.; Ghias, A.; Lai, C.; Do, L.; Cho, S.; et al. Modulation of the Wnt pathway through inhibition of CLK2 and DYRK1A by lorecivivint as a novel, potentially disease-modifying approach for knee osteoarthritis treatment. *Osteoarthr. Cartil.* **2019**, *27*, 1347–1360. [CrossRef]

42. Yazici, Y.; McAlindon, T.; Gibofsky, A.; Lane, N.; Swearingen, C.; Tambiah, J. Radiographic outcomes were concordant with outcome measures in rheumatology-osteoarthritis research society international (OMERACT-OARSI) strict response: Post-hoc analysis from a phase 2 study of a WNT pathway inhibitor, SM04690, for knee osteoarthritis treatment. *Osteoarthr. Cartil.* **2018**, *26*, S244–S245. [CrossRef]

43. Ghouri, A.; Conaghan, P.G. Update on novel pharmacological therapies for osteoarthritis. *Ther. Adv. Musculoskelet. Dis.* **2019**, *11*, 1759720–1986449. [CrossRef] [PubMed]

44. Ghouri, A.; Conaghan, P.G. Prospects for Therapies in Osteoarthritis. *Calcif. Tissue Int.* **2020**, 1–12. [CrossRef] [PubMed]

45. Manicourt, D.; Beaulieu, A.; Garnero, P.; Peterfy, C.; Bouisset, F.; Haemmerle, S.; Mindeholm, L. 229 Effect of treatment with the Cathepsin-K inhibitor, balicatib, on cartilage volume and biochemical markers of bone and cartilage degradation in patients with painful knee osteoarthritis. *Osteoarthr. Cartil.* **2007**, *15*, C130. [CrossRef]

46. Lindström, E.; Rizoska, B.; Henderson, I.; Terelius, Y.; Jerling, M.; Edenius, C.; Grabowska, U. Nonclinical and clinical pharmacological characterization of the potent and selective cathepsin K inhibitor MIV-711. *J. Transl. Med.* **2018**, *16*, 1–14. [CrossRef] [PubMed]

47. Conaghan, P.G.; Bowes, M.A.; Kingsbury, S.R.; Brett, A.; Guillard, G.; Tunblad, K.; Rizoska, B.; Larsson, T.; Holmgren, Å.; Manninen, A.; et al. Six months' treatment with MIV-711, a novel Cathepsin K inhibitor induces osteoarthritis structure modification: Results from a randomized double-blind placebo-controlled phase IIA trial. *Osteoarthr. Cartil.* **2018**, *26*, S25–S26. [CrossRef]

48. Hunter, D.; Pike, M.; Jonas, B.; Kissin, E.; Krop, J.; McAlindon, T. Phase 1 safety and tolerability study of BMP-7 in symptomatic knee osteoarthritis. *BMC Musculoskelet. Disord.* **2010**, *11*, 232. [CrossRef]

49. Reker, D.; Kjelgaard-Petersen, C.F.; Siebuhr, A.S.; Michaelis, M.; Gigout, A.; Karsdal, M.A.; Ladel, C.; Bay-Jensen, A.C. Sprifermin (rhFGF18) modulates extracellular matrix turnover in cartilage explants ex vivo. *J. Transl. Med.* **2017**, *15*, 1–12. [CrossRef]

50. Lohmander, L.S.; Hellot, S.; Dreher, D.; Krantz, E.F.W.; Kruger, D.S.; Guermazi, A.; Eckstein, F. Intraarticular Sprifermin (Recombinant Human Fibroblast Growth Factor 18) in Knee Osteoarthritis: A Randomized, Double-Blind, Placebo-Controlled Trial. *Arthritis Rheumatol.* **2014**, *66*, 1820–1831. [CrossRef]

51. Roemer, F.W.; Aydemir, A.; Lohmander, L.S.; Crema, M.D.; Marra, M.D.; Muurahainen, N.; Felson, D.T.; Eckstein, F.; Guermazi, A. Structural effects of sprifermin in knee osteoarthritis: A post-hoc analysis on cartilage and non-cartilaginous tissue alterations in a randomized controlled trial. *BMC Musculoskelet. Disord.* **2016**, *17*, 1–7. [CrossRef]

52. Gigout, A.; Guehring, H.; Froemel, D.; Meurer, A.; Ladel, C.; Reker, D.; Bay-Jensen, A.; Karsdal, M.; Lindemann, S. Sprifermin (rhFGF18) enables proliferation of chondrocytes producing a hyaline cartilage matrix. *Osteoarthr. Cartil.* **2017**, *25*, 1858–1867. [CrossRef]

53. Dahlberg, L.; Flechsenhar, K.; Dreher, D.; Goûteux, S.; Jurvelin, J. A randomized, double-blind, placebo-controllled, multicenter study of Rhfgf18 administered intraarticularly using single or multiple ascending doses in parients with primary knee osteo-arthritis (OA), scheduled for total knee replacement. *Osteoarthr. Cartil.* **2011**, *19*, S143. [CrossRef]

54. Hochberg, M.C.; Guermazi, A.; Guehring, H.; Aydemir, A.; Wax, S.; Fleuranceau-Morel, P.; Bihlet, A.R.; Byrjalsen, I.; Andersen, J.R.; Eckstein, F. Effect of intra-articular sprifermin vs placebo on femorotibial joint cartilage thickness in patients with Osteoarthritis. *JAMA* **2019**, *322*, 1360–1370. [CrossRef] [PubMed]

55. Pers, Y.-M.; Rackwitz, L.; Ferreira, R.; Pullig, O.; Delfour, C.; Barry, F.; Sensebe, L.; Casteilla, L.; Fleury, S.; Bourin, P.; et al. Adipose Mesenchymal Stromal Cell-Based Therapy for Severe Osteoarthritis of the Knee: A Phase I Dose-Escalation Trial. *STEM CELLS Transl. Med.* **2016**, *5*, 847–856. [CrossRef]

56. Muchedzi, T.A.; Roberts, S.B. A systematic review of the effects of platelet rich plasma on outcomes for patients with knee osteoarthritis and following total knee arthroplasty. *Surgeon* **2018**, *16*, 250–258. [CrossRef] [PubMed]

57. Bennell, K.L.; Hunter, D.J.; Paterson, K.L. Platelet-Rich Plasma for the Management of Hip and Knee Osteoarthritis. *Curr. Rheumatol. Rep.* **2017**, *19*, 24. [CrossRef]

58. Huebner, K.; Frank, R.M.; Getgood, A. Ortho-Biologics for Osteoarthritis. *Clin. Sports Med.* **2019**, *38*, 123–141. [CrossRef] [PubMed]

59. Davies, P.S.E.; Graham, S.M.; Macfarlane, R.J.; Leonidou, A.; Mantalaris, A.; Tsiridis, E. Disease-modifying osteoarthritis drugs:in vitroandin vivodata on the development of DMOADs under investigation. *Expert Opin. Investig. Drugs* **2013**, *22*, 423–441. [CrossRef] [PubMed]

60. Toh, W.S.; Lai, R.C.; Hui, J.H.P.; Lim, S.K. MSC exosome as a cell-free MSC therapy for cartilage regeneration: Implications for osteoarthritis treatment. *Semin. Cell Dev. Biol.* **2017**, *67*, 56–64. [CrossRef]

61. Kim, Y.; Kwon, O.; Choi, Y.; Suh, D.; Heo, D.; Koh, Y. Comparative matched-pair analysis of the injection versus implantation of mesenchymal stem cells for knee osteoarthritis. *Am. J. Sports Med.* **2015**, *43*, 2738–2746. [CrossRef]

62. Jevotovsky, D.; Alfonso, A.; Einhorn, T.; Chiu, E. Osteoarthritis and stem cell therapy in humans: A systematic review. *Osteoarthr. Cartil.* **2018**, *26*, 711–729. [CrossRef] [PubMed]

63. Vega, A.; Martín-Ferrero, M.A.; Del Canto, F.; Alberca, M.; García, V.; Munar, A.; Orozco, L.; Soler, R.; Fuertes, J.J.; Huguet, M.; et al. Treatment of knee osteoarthritis with allogeneic bone marrow mesenchymal stem cells. *Transplantation* **2015**, *99*, 1681–1690. [CrossRef] [PubMed]

64. Lee, S.; Lee, S.H.; Na, H.S.; Kwon, J.Y.; Kim, G.; Jung, K.; Cho, K.; Kim, S.A.; Go, E.J.; Park, M.; et al. The therapeutic effect of STAT3 signaling-suppressed MSC on pain and articular cartilage damage in a rat model of monosodium iodoacetate-induced osteoarthritis. *Front. Immunol.* **2018**, *9*, 2881. [CrossRef]

65. Madry, H.; Cucchiarini, M. Advances and challenges in gene-based approaches for osteoarthritis. *J. Gene Med.* **2013**, *15*, 343–355. [CrossRef] [PubMed]

66. Wu, Y.; Lu, X.; Shen, B.; Zeng, Y. The Therapeutic Potential and Role of miRNA, lncRNA, and circRNA in Osteoarthritis. *Curr. Gene Ther.* **2019**, *19*, 255–263. [CrossRef] [PubMed]

67. Zhang, W.; Robertson, W.B.; Zhao, J.; Chen, W.; Xu, J. Emerging Trend in the Pharmacotherapy of Osteoarthritis. *Front. Endocrinol.* **2019**, *10*, 431. [CrossRef] [PubMed]

68. Neogi, T. Clinical significance of bone changes in osteoarthritis. *Ther. Adv. Musculoskelet. Dis.* **2012**, *4*, 259–267. [CrossRef]

69. Hügle, T.; Geurts, J. What drives osteoarthritis? synovial versus subchondral bone pathology. *Rheumatology* **2017**, *56*, 1461–1471.

70. Burr, D.B. Anatomy and physiology of the mineralized tissues: Role in the pathogenesis of osteoarthrosis. *Osteoarthr. Cartil.* **2004**, *12*, 20–30. [CrossRef]

71. Day, J.S.; Ding, M.; Van Der Linden, J.C.; Hvid, I.; Sumner, D.R.; Weinans, H. A decreased subchondral trabecular bone tissue elastic modulus is associated with pre-arthritic cartilage damage. *J. Orthop. Res.* **2001**, *19*, 914–918. [CrossRef]

72. Ni, G.-X.; Zhan, L.-Q.; Gao, M.-Q.; Lei, L.; Zhou, Y.-Z.; Pan, Y.-X. Matrix metalloproteinase-3 inhibitor retards treadmill running-induced cartilage degradation in rats. *Arthritis Res. Ther.* **2011**, *13*, R192. [CrossRef]

73. Sanchez, C.; DeBerg, M.; Piccardi, N.; Msika, P.; Reginster, J.-Y.; Henrotin, Y. Osteoblasts from the sclerotic subchondral bone downregulate aggrecan but upregulate metalloproteinases expression by chondrocytes. This effect is mimicked by interleukin-6, -1β and oncostatin M pre-treated non-sclerotic osteoblasts. *Osteoarthr. Cartil.* **2005**, *13*, 979–987. [CrossRef] [PubMed]

74. Prasadam, I.; Crawford, R.; Xiao, Y. Aggravation of ADAMTS and Matrix Metalloproteinase Production and Role of ERK1/2 Pathway in the Interaction of Osteoarthritic Subchondral Bone Osteoblasts and Articular Cartilage Chondrocytes—Possible Pathogenic Role in Osteoarthritis. *J. Rheumatol.* **2012**, *39*, 621–634. [CrossRef] [PubMed]

75. Bianco, D.; Todorov, A.; Čengić, T.; Pagenstert, G.; Schären, S.; Netzer, C.; Hügle, T.; Geurts, J. Alterations of Subchondral Bone Progenitor Cells in Human Knee and Hip Osteoarthritis Lead to a Bone Sclerosis Phenotype. *Int. J. Mol. Sci.* **2018**, *19*, 475. [CrossRef]
76. Zhang, R.-K.; Li, G.-W.; Zeng, C.; Lin, C.-X.; Huang, L.-S.; Huang, G.-X.; Zhao, C.; Feng, S.-Y.; Fang, H. Mechanical stress contributes to osteoarthritis development through the activation of transforming growth factor beta 1 (TGF-β1). *Bone Jt. Res.* **2018**, *7*, 587–594. [CrossRef] [PubMed]
77. Jung, Y.-K.; Han, M.-S.; Park, H.-R.; Lee, E.-J.; Jang, J.-A.; Kim, G.-W.; Lee, S.-Y.; Moon, D.; Han, S. Calcium-phosphate complex increased during subchondral bone remodeling affects earlystage osteoarthritis. *Sci. Rep.* **2018**, *8*, 487. [CrossRef]
78. Loeser, R.F.; Shanker, G. Autocrine stimulation by insulin-like growth factor 1 and insulin-like growth factor 2 mediates chondrocyte survival in vitro. *Arthritis Rheum.* **2000**, *43*, 1552–1559. [CrossRef]
79. Kennedy, O.D.; Laudier, D.M.; Majeska, R.J.; Sun, H.B.; Schaffler, M.B. Osteocyte apoptosis is required for production of osteoclastogenic signals following bone fatigue in vivo. *Bone* **2014**, *64*, 132–137. [CrossRef] [PubMed]
80. Li, J.; Xue, J.; Jing, Y.; Wang, M.; Shu, R.; Xu, H.; Xue, C.; Feng, J.; Wang, P.; Bai, D. SOST Deficiency Aggravates Osteoarthritis in Mice by Promoting Sclerosis of Subchondral Bone. *BioMed Res. Int.* **2019**, *2019*, 1–8. [CrossRef]
81. Ganesh, T.; Laughrey, L.E.; Niroobakhsh, M.; Lara-Castillo, N. Multiscale finite element modeling of mechanical strains and fluid flow in osteocyte lacunocanalicular system. *Bone* **2020**, *137*, 115328. [CrossRef]
82. Dai, G.; Xiao, H.; Liao, J.; Zhou, N.; Zhao, C.; Xu, W.; Xu, W.; Liang, X.; Huang, W. Osteocyte TGFβ1–Smad2/3 is positively associated with bone turnover parameters in sub-chondral bone of advanced osteoarthritis. *Int. J. Mol. Med.* **2020**, *46*, 167–178.
83. Zhou, X.; Cao, H.; Yuan, Y.; Wu, W. Biochemical Signals Mediate the Crosstalk between Cartilage and Bone in Osteoarthritis. *BioMed Res. Int.* **2020**, *2020*, 5720360. [CrossRef]
84. Hu, S.; Mao, G.; Zhang, Z.; Wu, P.; Wen, X.; Liao, W.; Zhang, Z. MicroRNA-320c inhibits development of osteoarthritis through downregulation of canonical Wnt signaling pathway. *Life Sci.* **2019**, *228*, 242–250. [CrossRef] [PubMed]
85. De Santis, M.; Di Matteo, B.; Chisari, E.; Cincinelli, G.; Angele, P.; Lattermann, C.; Filardo, G.; Vitale, N.D.; Selmi, C.; Kon, E. The Role of Wnt Pathway in the Pathogenesis of OA and Its Potential Therapeutic Implications in the Field of Regenerative Medicine. *BioMed Res. Int.* **2018**, *2018*, 1–8. [CrossRef]
86. Zhao, W.; Wang, T.; Luo, Q.; Chen, Y.; Leung, V.Y.L.; Wen, C.; Shah, M.F.; Pan, H.; Chiu, K.; Cao, X.; et al. Cartilage degeneration and excessive subchondral bone formation in spontaneous osteoarthritis involves altered TGF-β signaling. *J. Orthop. Res.* **2015**, *34*, 763–770. [CrossRef]
87. Zhang, L.; Wang, P.-E.; Ying, J.; Jin, X.; Luo, C.; Xu, T.; Xu, S.; Dong, R.; Xiao, L.; Tong, P.; et al. Yougui Pills Attenuate Cartilage Degeneration via Activation of TGF-β/Smad Signaling in Chondrocyte of Osteoarthritic Mouse Model. *Front. Pharmacol.* **2017**, *8*, 611. [CrossRef]
88. Upton, A.R.; Holding, C.A.; Dharmapatni, A.A.S.S.K.; Haynes, D.R. The expression of RANKL and OPG in the various grades of osteoarthritic cartilage. *Rheumatol. Int.* **2012**, *32*, 535–540. [CrossRef]
89. Zhong, M.; Carney, D.H.; Jo, H.; Boyan, B.D.; Schwartz, Z. Inorganic Phosphate Induces Mammalian Growth Plate Chondrocyte Apoptosis in a Mitochondrial Pathway Involving Nitric Oxide and JNK MAP Kinase. *Calcif. Tissue Int.* **2011**, *88*, 96–108. [CrossRef]
90. Nishii, T.; Tamura, S.; Shiomi, T.; Yoshikawa, H.; Sugano, N. Alendronate treatment for hip osteoarthritis: Prospective randomized 2-year trial. *Clin. Rheumatol.* **2013**, *32*, 1759–1766. [CrossRef]
91. Aitken, D.; Laslett, L.L.; Cai, G.; Hill, C.; March, L.; Wluka, A.E.; Wang, Y.; Blizzard, L.; Cicuttini, F.; Jones, G. A protocol for a multicentre, randomised, double-blind, placebo-controlled trial to compare the effect of annual infusions of zoledronic acid to placebo on knee structural change and knee pain over 24 months in knee osteoarthritis patients—ZAP2. *BMC Musculoskelet. Disord.* **2018**, *19*, 217. [CrossRef] [PubMed]
92. Zhu, S.; Chen, K.; Lan, Y.; Zhang, N.; Jiang, R.; Hu, J. Alendronate protects against articular cartilage erosion by inhibiting subchondral bone loss in ovariectomized rats. *Bone* **2013**, *53*, 340–349. [CrossRef] [PubMed]
93. Bingham, C.O., III; Buckland-Wright, J.C.; Garnero, P.; Cohen, S.B.; Dougados, M.; Adami, S.; Clauw, D.J.; Spector, T.D.; Pelletier, J.-P.; Raynauld, J.-P.; et al. Risedronate decreases biochemical markers of cartilage degradation but does not decrease symptoms or slow radiographic progression in patients with medial compartment osteoarthritis of the knee: Results of the two-year multinational knee osteoarthritis structural arthritis study. *Arthritis Rheum.* **2006**, *54*, 3494–3507. [CrossRef] [PubMed]
94. Spector, T.D.; Conaghan, P.G.; Buckland-Wright, J.C.; Garnero, P.; Cline, G.A.; Beary, J.F.; Valent, D.J.; Meyer, J.M. Effect of risedronate on joint structure and symptoms of knee osteoarthritis: Results of the BRISK randomized, controlled trial [ISRCTN01928173]. *Arthritis Res. Ther.* **2005**, *7*, R625–R633. [CrossRef] [PubMed]
95. Laslett, L.L.; Doré, D.A.; Quinn, S.J.; Boon, P.; Ryan, E.; Winzenberg, T.M.; Jones, G. Zoledronic acid reduces knee pain and bone marrow lesions over 1 year: A randomised controlled trial. *Ann. Rheum. Dis.* **2012**, *71*, 1322–1328. [CrossRef] [PubMed]
96. Rossini, M.; Adami, S.; Fracassi, E.; Viapiana, O.; Orsolini, G.; Povino, M.R.; Idolazzi, L.; Gatti, D. Effects of intra-articular clodronate in the treatment of knee osteoarthritis: Results of a double-blind, randomized placebo-controlled trial. *Rheumatol. Int.* **2015**, *35*, 255–263. [CrossRef]
97. Reginster, J.-Y.; Badurski, J.; Bellamy, N.; Bensen, W.; Chapurlat, R.; Chevalier, X.; Christiansen, C.; Genant, H.; Navarro, F.; Nasonov, E.; et al. Efficacy and safety of strontium ranelate in the treatment of knee osteoarthritis: Results of a double-blind, randomised placebo-controlled trial. *Ann. Rheum. Dis.* **2013**, *72*, 179–186. [CrossRef] [PubMed]

98. Yu, D.-G.; Ding, H.-F.; Mao, Y.-Q.; Liu, M.; Yu, B.; Zhao, X.; Wang, X.-Q.; Li, Y.; Liu, G.-W.; Nie, S.-B.; et al. Strontium ranelate reduces cartilage degeneration and subchondral bone remodeling in rat osteoarthritis model. *Acta Pharmacol. Sin.* **2013**, *34*, 393–402. [CrossRef] [PubMed]

99. Pelletier, J.-P.; Roubille, C.; Raynauld, J.-P.; Abram, F.; Dorais, M.; Delorme, P.; Martel-Pelletier, J. Disease-modifying effect of strontium ranelate in a subset of patients from the Phase III knee osteoarthritis study SEKOIA using quantitative MRI: Reduction in bone marrow lesions protects against cartilage loss. *Ann. Rheum. Dis.* **2015**, *74*, 422–429. [CrossRef]

100. Granholm, S.; Lundberg, P.; Lerner, U.H. Expression of the calcitonin receptor, calcitonin receptor-like receptor, and receptor activity modifying proteins during osteoclast differentiation. *J. Cell. Biochem.* **2008**, *104*, 920–933. [CrossRef]

101. Nielsen, R.; Bay-Jensen, A.-C.; Byrjalsen, I.; Karsdal, M. Oral salmon calcitonin reduces cartilage and bone pathology in an osteoarthritis rat model with increased subchondral bone turnover. *Osteoarthr. Cartil.* **2011**, *19*, 466–473. [CrossRef]

102. Karsdal, M.; Byrjalsen, I.; Alexandersen, P.; Bihlet, A.; Andersen, J.; Riis, B.; Bay-Jensen, A.; Christiansen, C. Treatment of symptomatic knee osteoarthritis with oral salmon calcitonin: Results from two phase 3 trials. *Osteoarthr. Cartil.* **2015**, *23*, 532–543. [CrossRef]

103. EMA. European Medicines Agency Recommends Limiting Long-Term Use of Calcitonin Medicines 2012. Available online: https://www.ema.europa.eu/en/documents/press-release/european-medicines-agency-recommends-limiting-long-term-use-calcitonin-medicines_en.pdf (accessed on 3 January 2021).

104. McDougall, J.J.; Schuelert, N.; Bowyer, J. Cathepsin K inhibition reduces CTXII levels and joint pain in the guinea pig model of spontaneous osteoarthritis. *Osteoarthr. Cartil.* **2010**, *18*, 1355–1357. [CrossRef] [PubMed]

105. Hayami, T.; Zhuo, Y.; Wesolowski, G.A.; Pickarski, M.; Duong, L.T. Inhibition of cathepsin K reduces cartilage degeneration in the anterior cruciate ligament transection rabbit and murine models of osteoarthritis. *Bone* **2012**, *50*, 1250–1259. [CrossRef]

106. Lindström, E.; Rizoska, B.; Tunblad, K.; Edenius, C.; Bendele, A.M.; Maul, D.; Larson, M.; Shah, N.; Otto, V.Y.; Jerome, C.; et al. The selective cathepsin K inhibitor MIV-711 attenuates joint pathology in experimental animal models of osteoarthritis. *J. Transl. Med.* **2018**, *16*, 56. [CrossRef] [PubMed]

107. Karsdal, M.; Michaelis, M.; Ladel, C.; Siebuhr, A.; Bihlet, A.; Andersen, J.; Guehring, H.; Christiansen, C.; Bay-Jensen, A.; Kraus, V. Disease-modifying treatments for osteoarthritis (DMOADs) of the knee and hip: Lessons learned from failures and opportunities for the future. *Osteoarthr. Cartil.* **2016**, *24*, 2013–2021. [CrossRef] [PubMed]

108. Vortkamp, A.; Lee, K.; Lanske, B.; Segre, G.V.; Kronenberg, H.M.; Tabin, C.J. Regulation of Rate of Cartilage Differentiation by Indian Hedgehog and PTH-Related Protein. *Science* **1996**, *273*, 613–622. [CrossRef]

109. Sampson, E.R.; Hilton, M.J.; Tian, Y.; Chen, D.; Schwarz, E.M.; Mooney, R.A.; Bukata, S.V.; O'Keefe, R.J.; Awad, H.; Puzas, J.E.; et al. Teriparatide as a Chondroregenerative Therapy for Injury-Induced Osteoarthritis. *Sci. Transl. Med.* **2011**, *3*, 101ra93. [CrossRef] [PubMed]

110. Bellido, M.; Lugo, L.; Roman-Blas, J.A.; Castañeda, S.; Calvo, E.; Largo, R.; Herrero-Beaumont, G. Improving subchondral bone integrity reduces progression of cartilage damage in experimental osteoarthritis preceded by osteoporosis. *Osteoarthr. Cartil.* **2011**, *19*, 1228–1236. [CrossRef]

111. Wu, M.; Chen, G.; Li, Y.-P. TGF-β and BMP signaling in osteoblast, skeletal development, and bone formation, homeostasis and disease. *Bone Res.* **2016**, *4*, 16009. [CrossRef]

112. Shen, J.; Li, S.; Chen, D. TGF-β signaling and the development of osteoarthritis. *Bone Res.* **2014**, *2*, 14002. [CrossRef]

113. Xie, L.; Tintani, F.; Wang, X.; Li, F.; Zhen, G.; Qiu, T.; Wan, M.; Crane, J.; Chen, Q.; Cao, X. Systemic neutralization of TGF-β attenuates osteoarthritis. *Ann. N. Y. Acad. Sci.* **2016**, *1376*, 53–64. [CrossRef]

114. Zhen, G.; Wen, C.; Jia, X.; Li, Y.; Crane, J.L.; Mears, S.C.; Askin, F.B.; Frassica, F.J.; Chang, W.; Yao, J.; et al. Inhibition of TGF-β signaling in mesenchymal stem cells of subchondral bone attenuates osteoarthritis. *Nat. Med.* **2013**, *19*, 704–712. [CrossRef] [PubMed]

115. Rowe, P.; Kumagai, Y.; Gutierrez, G.; Garrett, I.; Blacher, R.; Rosen, D.; Cundy, J.; Navvab, S.; Chen, D.; Drezner, M.; et al. MEPE has the properties of an osteoblastic phosphatonin and minhibin. *Bone* **2004**, *34*, 303–319. [CrossRef] [PubMed]

116. McGuire, D.; Lane, N.; Segal, N.; Metyas, S.; Barthel, H.; Miller, M.; Rosen, D.; Kumagai, Y. TPX-100 leads to marked, sustained improvements in subjects with knee osteoarthritis: Pre-clinical rationale and results of a controlled clinical trial. *Osteoarthr. Cartil.* **2018**, *26*, S243. [CrossRef]

117. Nagai, T.; Sato, M.; Kutsuna, T.; Kokubo, M.; Ebihara, G.; Ohta, N.; Mochida, J. Intravenous administration of anti-vascular endothelial growth factor humanized monoclonal antibody bevacizumab improves articular cartilage repair. *Arthritis Res. Ther.* **2010**, *12*, R178. [CrossRef] [PubMed]

118. Cui, Z.; Crane, J.; Xie, H.; Jin, X.; Zhen, G.; Li, C.; Xie, L.; Wang, L.; Bian, Q.; Qiu, T.; et al. Halofuginone attenuates osteoarthritis by inhibition of TGF-β activity and H-type vessel formation in subchondral bone. *Ann. Rheum. Dis.* **2015**, *75*, 1714–1721. [CrossRef]

119. Lu, J.; Zhang, H.; Cai, D.; Zeng, C.; Lai, P.; Shao, Y.; Fang, H.; Li, D.; Ouyang, J.; Zhao, C.; et al. Positive-Feedback Regulation of Subchondral H-Type Vessel Formation by Chondrocyte Promotes Osteoarthritis Development in Mice. *J. Bone Miner. Res.* **2018**, *33*, 909–920. [CrossRef] [PubMed]

120. Walsh, D.A.; McWilliams, D.F.; Turley, M.J.; Dixon, M.R.; Fransès, R.E.; Mapp, P.I.; Wilson, D. Angiogenesis and nerve growth factor at the osteochondral junction in rheumatoid arthritis and osteoarthritis. *Rheumatology* **2010**, *49*, 1852–1861. [CrossRef]

121. Xie, H.; Cui, Z.; Wang, L.; Xia, Z.; Hu, Y.; Xian, L.; Li, C.; Xie, L.; Crane, J.; Wan, M.; et al. PDGF-BB secreted by preosteoclasts induces angiogenesis during coupling with osteogenesis. *Nat. Med.* **2014**, *20*, 1270–1278. [CrossRef]

122. Schnitzer, T.J.; Ekman, E.F.; Spierings, E.L.H.; Greenberg, H.S.; Smith, M.D.; Brown, M.T.; West, C.R.; Verburg, K.M. Efficacy and safety of tanezumab monotherapy or combined with non-steroidal anti-inflammatory drugs in the treatment of knee or hip osteoarthritis pain. *Ann. Rheum. Dis.* **2015**, *74*, 1202–1211. [CrossRef]

123. Hochberg, M.C.; Tive, L.A.; Abramson, S.B.; Vignon, E.; Verburg, K.M.; West, C.R.; Smith, M.D.; Hungerford, D.S. When Is Osteonecrosis Not Osteonecrosis?: Adjudication of Reported Serious Adverse Joint Events in the Tanezumab Clinical Development Program. *Arthritis Rheumatol.* **2016**, *68*, 382–391. [CrossRef]

124. Clinicaltrials.gov. Study to Determine the Safety and the Efficacy of Fasinumab Compared to Placebo and Nonsteroidal Anti-Inflammatory Drugs (NSAIDs) for Treatment of Adults with Pain from Osteoarthritis of the Knee or Hip (FACT OA2). Re-generon Pharmaceuticals. 2018. Available online: https://clinicaltrials.gov/ct2/show/NCT03304379 (accessed on 5 January 2021).

125. Yang, C.Y.; Chanalaris, A.; Troeberg, L. ADAMTS and ADAM metalloproteinases in osteoarthritis—Looking beyond the 'usual suspects'. *Osteoarthr. Cartil.* **2017**, *25*, 1000–1009. [CrossRef]

126. Atukorala, I.; Kwoh, C.K.; Guermazi, A.; Roemer, F.W.; Boudreau, R.M.; Hannon, M.J.; Hunter, D.J. Synovitis in knee osteoarthritis: A precursor of disease? *Ann. Rheum. Dis.* **2016**, *75*, 390–395. [CrossRef]

127. Schnitzer, T.J.; Easton, R.; Pang, S.; Levinson, D.J.; Pixton, G.; Viktrup, L.; Davignon, I.; Brown, M.T.; West, C.R.; Verburg, K.M. Effect of tanezumab on joint pain, physical function, and patient global assessment of osteoarthritis among patients with osteoarthritis of the hip or knee: A randomized clinical trial. *JAMA* **2019**, *322*, 37–48. [CrossRef]

128. Dakin, P.; DiMartino, S.J.; Gao, H.; Maloney, J.; Kivitz, A.J.; Schnitzer, T.J.; Stahl, N.; Yancopoulos, G.D.; Geba, G.P. The Efficacy, Tolerability, and Joint Safety of Fasinumab in Osteoarthritis Pain: A Phase IIb/III Double-Blind, Placebo-Controlled, Randomized Clinical Trial. *Arthritis Rheumatol.* **2019**, *71*, 1824–1834. [CrossRef]

129. Kim, J.-R.; Yoo, J.J.; Kim, H.A. Therapeutics in Osteoarthritis Based on an Understanding of Its Molecular Pathogenesis. *Int. J. Mol. Sci.* **2018**, *19*, 674. [CrossRef]

130. Yu, S.P.-C.; Hunter, D.J. Emerging drugs for the treatment of knee osteoarthritis. *Expert Opin. Emerg. Drugs* **2015**, *20*, 361–378. [CrossRef] [PubMed]

131. Siebuhr, A.S.; Bay-Jensen, A.C.; Jordan, J.M.; Kjelgaard-Petersen, C.F.; Christiansen, C.; Abramson, S.B.; Attur, M.; Berenbaum, F.; Kraus, V.; Karsdal, M.A. Inflammation (or synovitis)-driven osteoarthritis: An opportunity for personalizing prognosis and treatment? *Scand. J. Rheumatol.* **2015**, *45*, 87–98. [CrossRef] [PubMed]

132. Wang, J. Efficacy and safety of adalimumab by intra-articular injection for moderate to severe knee osteoarthritis: An open-label randomized controlled trial. *J. Int. Med Res.* **2017**, *46*, 326–334. [CrossRef]

133. Ohtori, S.; Orita, S.; Yamauchi, K.; Eguchi, Y.; Ochiai, N.; Kishida, S.; Kuniyoshi, K.; Aoki, Y.; Nakamura, J.; Ishikawa, T.; et al. Efficacy of Direct Injection of Etanercept into Knee Joints for Pain in Moderate and Severe Knee Osteoarthritis. *Yonsei Med. J.* **2015**, *56*, 1379–1383. [CrossRef]

134. Chevalier, X.; Goupille, P.; Beaulieu, A.D.; Burch, F.X.; Bensen, W.G.; Conrozier, T.; Loeuille, D.; Kivitz, A.J.; Silver, D.; Appleton, B.E. Intraarticular injection of anakinra in osteoarthritis of the knee: A multicenter, randomized, double-blind, placebo-controlled study. *Arthritis Rheum.* **2009**, *61*, 344–352. [CrossRef] [PubMed]

135. Wang, S.X.; Abramson, S.B.; Attur, M.; Karsdal, M.A.; Preston, R.A.; Lozada, C.J.; Kosloski, M.P.; Hong, F.; Jiang, P.; Saltarelli, M.J.; et al. Safety, tolerability, and pharmacodynamics of an anti-interleukin-1α/β dual variable domain immuno-globulin in patients with osteoarthritis of the knee: A randomized phase 1 study. *Osteoarthr. Cartil.* **2017**, *25*, 1952–1961. [CrossRef]

136. Migliore, A.; Paoletta, M.; Moretti, A.; Liguori, S.; Iolascon, G. The perspectives of intra-articular therapy in the management of osteoarthritis. *Expert Opin. Drug Deliv.* **2020**, *17*, 1213–1226. [CrossRef]

137. Gibofsky, A.; Hochberg, M.C.; Jaros, M.J.; Young, C.L. Efficacy and safety of low-dose submicron diclofenac for the treatment of osteoarthritis pain: A 12 week, phase 3 study. *Curr. Med Res. Opin.* **2014**, *30*, 1883–1893. [CrossRef] [PubMed]

138. Philp, A.M.; Davis, E.T.; Jones, S.W. Developing anti-inflammatory therapeutics for patients with osteoarthritis. *Rheumatology* **2016**, *56*, 869–881. [CrossRef]

139. Gallelli, L.; Galasso, O.; Falcone, D.; Southworth, S.; Greco, M.; Ventura, V.; Romualdi, P.; Corigliano, A.; Terracciano, R.; Savino, R.; et al. The effects of nonsteroidal anti-inflammatory drugs on clinical outcomes, synovial fluid cytokine concentration and signal transduction pathways in knee osteoarthritis. A randomized open label trial. *Osteoarthr. Cartil.* **2013**, *21*, 1400–1408. [CrossRef] [PubMed]

140. Lim, C.C.; Ang, A.T.W.; Kadir, H.B.A.; Lee, P.H.; Goh, B.Q.; Harikrishnan, S.; Kwek, J.L.; Gan, S.S.W.; Choo, J.C.J.; Tan, N.C. Short-Course Systemic and Topical Non-Steroidal Anti-Inflammatory Drugs: Impact on Adverse Renal Events in Older Adults with Co-Morbid Disease. *Drugs Aging* **2021**, *38*, 147–156. [CrossRef]

141. Jovanovic, D.V.; Fernandes, J.C.; Martel-Pelletier, J.; Jolicoeur, F.C.; Reboul, P.; Laufer, S.; Tries, S.; Pelletier, J.P. In vivo dual inhibition of cyclooxygenase and lipoxygenase by ML-3000 reduces the progression of experimental osteoarthritis: Suppression of collagenase 1 and interleukin-1β synthesis. *Arthritis Rheum.* **2001**, *44*, 2320–2330. [CrossRef]

142. Raynauld, J.P.; Martel-Pelletier, J.; Bias, P.; Laufer, S.; Haraoui, B.; Choquette, D.; Beaulieu, A.D.; Abram, F.; Dorais, M.; Vignon, E.; et al. Protective effects of licofelone, a 5-lipoxygenase and cy-clo-oxygenase inhibitor, versus naproxen on cartilage loss in knee osteoarthritis: A first multicentre clinical trial using quanti-tative MRI. *Ann. Rheum. Dis.* **2009**, *68*, 938–947. [CrossRef] [PubMed]

143. Vaishya, R.; Pandit, R.; Agarwal, A.K.; Vijay, V. Intra-articular hyaluronic acid is superior to steroids in knee osteoarthritis: A comparative, randomized study. *J. Clin. Orthop. Trauma* **2017**, *8*, 85–88. [CrossRef]

144. He, W.-W.; Kuang, M.-J.; Zhao, J.; Sun, L.; Lu, B.; Wang, Y.; Ma, J.-X.; Ma, X.-L. Efficacy and safety of intraarticular hyaluronic acid and corticosteroid for knee osteoarthritis: A meta-analysis. *Int. J. Surg.* **2017**, *39*, 95–103. [CrossRef]
145. Mol, M.F.; Runhaar, J.; Bos, P.K.; Dorleijn, D.M.J.; Vis, M.; Gussekloo, J.; Bindels, P.J.E.; Bierma-Zeinstra, S.M.A. Effectiveness of intramuscular gluteal glucocorticoid injection versus intra-articular glucocorticoid injection in knee osteoarthritis: Design of a multicenter randomized, 24 weeks comparative parallel-group trial. *BMC Musculoskelet. Disord.* **2020**, *21*, 1–9. [CrossRef]
146. Samuels, J.; Pillinger, M.H.; Jevsevar, D.; Felson, D.; Simon, L.S. Critical appraisal of intra-articular glucocorticoid injections for symptomatic osteoarthritis of the knee. *Osteoarthr. Cartil.* **2021**, *29*, 8–16. [CrossRef]
147. Hangody, L.; Szody, R.; Lukasik, P.; Zgadzaj, W.; Lénárt, E.; Dokoupilova, E.; Bichovsk, D.; Berta, A.; Vasarhelyi, G.; Ficzere, A.; et al. Intraarticular injection of a cross-linked sodium hyaluronate combined with triamcinolone hexacetonide (Cingal) to provide symptomatic relief of osteoarthritis of the knee: A randomized, double-blind, placebo-controlled multicenter clinical trial. *Cartilage* **2018**, *9*, 276–283. [CrossRef] [PubMed]
148. Babaei-Ghazani, A.; Najarzadeh, S.; Mansoori, K.; Forogh, B.; Madani, S.P.; Ebadi, S.; Fadavi, H.R.; Eftekharsadat, B. The effects of ultrasound-guided corticosteroid injection compared to oxygen-ozone (O2-O3) injection in patients with knee osteo-arthritis: A randomized controlled trial. *Clin. Rheumatol.* **2018**, *37*, 2517–2527. [CrossRef]
149. Pelletier, J.-P.; Raynauld, J.-P.; Dorais, M.; Bessette, L.; Dokoupilova, E.; Morin, F.; Pavelka, K.; Paiement, P.; Martel-Pelletier, J.; DISSCO Trial Investigator Group. An international, multicentre, double-blind, randomized study (DISSCO): Effect of diacerein vs celecoxib on symptoms in knee osteoarthritis. *Rheumatology* **2020**, *59*, 3858–3868. [CrossRef]
150. Jung, J.H.; Kim, S.E.; Kim, H.-J.; Park, K.; Song, G.G.; Choi, S.J. A comparative pilot study of oral diacerein and locally treated diacerein-loaded nanoparticles in a model of osteoarthritis. *Int. J. Pharm.* **2020**, *581*, 119249. [CrossRef] [PubMed]
151. Leonidou, A.; Lepetsos, P.; Mintzas, M.; Kenanidis, E.; Macheras, G.; Tzetis, M.; Potoupnis, M.; Tsiridis, E. Inducible nitric oxide synthase as a target for osteoarthritis treatment. *Expert Opin. Ther. Targets* **2018**, *22*, 299–318. [CrossRef]
152. Ahmad, N.; Ansari, M.Y.; Haqqi, T.M. Role of iNOS in osteoarthritis: Pathological and therapeutic aspects. *J. Cell. Physiol.* **2020**, *235*, 6366–6376. [CrossRef]
153. Le Graverand, M.-P.H.; Clemmer, R.S.; Redifer, P.; Brunell, R.M.; Hayes, C.W.; Brandt, K.D.; Abramson, S.B.; Manning, P.T.; Miller, C.G.; Vignon, E. A 2-year randomised, double-blind, placebo-controlled, multicentre study of oral selective iNOS inhibitor, cindunistat (SD-6010), in patients with symptomatic osteoarthritis of the knee. *Ann. Rheum. Dis.* **2013**, *72*, 187–195. [CrossRef]
154. Jacob, G.; Shetty, V.; Shetty, S. A study assessing intra-articular PRP vs PRP with HMW HA vs PRP with LMW HA in early knee osteoarthritis. *J. Arthrosc. Jt. Surg.* **2017**, *4*, 65–71. [CrossRef]
155. Cook, C.S.; Smith, P.A. Clinical Update: Why PRP Should Be Your First Choice for Injection Therapy in Treating Osteoarthritis of the Knee. *Curr. Rev. Musculoskelet. Med.* **2018**, *11*, 583–592. [CrossRef]
156. Saif, D.S.; Hegazy, N.N.; Zahran, E.S. Evaluating the efficacy of intra-articular injections of platelet rich plasma (PRP) in rheumatoid arthritis patients and its impact on inflammatory cytokines, disease activity and quality of life. *Curr. Rheumatol. Rev.* **2020**, *16*, 1. [CrossRef] [PubMed]
157. Filardo, G.; Previtali, D.; Napoli, F.; Candrian, C.; Zaffagnini, S.; Grassi, A. PRP injections for the treatment of knee osteoar-thritis: A meta-analysis of randomized controlled trials. *Cartilage* **2020**. [CrossRef]
158. O'Connell, B.; Wragg, N.M.; Wilson, S.L. The use of PRP injections in the management of knee osteoarthritis. *Cell Tissue Res.* **2019**, *376*, 143–152. [CrossRef] [PubMed]
159. Yin, W.; Xu, H.; Sheng, J.; Xu, Z.; Xie, X.; Zhang, C. Comparative evaluation of the effects of platelet rich plasma formulations on extracellular matrix formation and the NF κB signaling pathway in human articular chondrocytes. *Mol. Med. Rep.* **2017**, *15*, 2940–2948. [CrossRef] [PubMed]
160. Simental-Mendía, M.; Vílchez-Cavazos, J.F.; Peña-Martínez, V.M.; Said-Fernández, S.; Lara-Arias, J.; Martínez-Rodríguez, H.G. Leukocyte-poor platelet-rich plasma is more effective than the conventional therapy with acetaminophen for the treatment of early knee osteoarthritis. *Arch. Orthop. Trauma Surg.* **2016**, *136*, 1723–1732. [CrossRef]
161. Tong, S.; Liu, J.; Zhang, C. Platelet-rich plasma inhibits inflammatory factors and represses rheumatoid fibroblast-like syno-viocytes in rheumatoid arthritis. *Clin. Exp. Med.* **2017**, *17*, 441–449. [CrossRef]
162. Migliorini, F.; Driessen, A.; Quack, V.; Sippel, N.; Cooper, B.; El Mansy, Y.; Tingart, M.; Eschweiler, J. Comparison between intra-articular infiltrations of placebo, steroids, hyaluronic and PRP for knee osteoarthritis: A Bayesian network meta-analysis. *Arch. Orthop. Trauma Surg.* **2020**, 1–18. [CrossRef]
163. Huang, G.; Hua, S.; Yang, T.; Ma, J.; Yu, W.; Chen, X. Platelet-rich plasma shows beneficial effects for patients with knee osteoarthritis by suppressing inflammatory factors. *Exp. Ther. Med.* **2018**, *15*, 3096–3102. [CrossRef]
164. Xu, Z.; He, Z.; Shu, L.; Li, X.; Ma, M.; Ye, C. Intra-articular platelet-rich plasma combined with hyaluronic acid injection for knee osteoarthritis is superior to platelet-rich plasma or hyaluronic acid alone in inhibiting inflammation and improving pain and function. *Arthrosc. J. Arthrosc. Relat. Surg.* **2020**. [CrossRef]
165. Nelson, F.R.; Zvirbulis, R.A.; Zonca, B.; Li, K.W.; Turner, S.M.; Pasierb, M.; Wilton, P.; Martinez-Puig, D.; Wu, W. The effects of an oral preparation containing hyaluronic acid (Oralvisc®) on obese knee osteoarthritis patients determined by pain, function, bradykinin, leptin, inflammatory cytokines, and heavy water analyses. *Rheumatol. Int.* **2015**, *35*, 43–52. [CrossRef]
166. Raeissadat, S.A.; Rayegani, S.M.; Forogh, B.; Abadi, P.H.; Moridnia, M.; Dehgolan, S.R. Intra-articular ozone or hyaluronic acid injection: Which one is superior in patients with knee osteoarthritis? A 6-month randomized clinical trial. *J. Pain Res.* **2018**, *11*, 111. [CrossRef] [PubMed]

167. Lopes de Jesus, C.C.; Dos Santos, F.C.; de Jesus, L.M.; Monteiro, I.; Sant'Ana, M.S.; Trevisani, V.F. Comparison between in-tra-articular ozone and placebo in the treatment of knee osteoarthritis: A randomized, double-blinded, placebo-controlled study. *PLoS ONE* **2017**, *12*, e0179185. [CrossRef]
168. Loeser, R.; Beavers, D.; Bay-Jensen, A.; Karsdal, M.; Nicklas, B.; Guermazi, A.; Hunter, D.; Messier, S. Effects of dietary weight loss with and without exercise on interstitial matrix turnover and tissue inflammation biomarkers in adults with knee osteoarthritis: The Intensive Diet and Exercise for Arthritis trial (IDEA). *Osteoarthr. Cartil.* **2017**, *25*, 1822–1828. [CrossRef]
169. Cheragh-Birjandi, S.; Moghbeli, M.; Haghighi, F.; Safdari, M.R.; Baghernezhad, M.; Akhavan, A.; Ganji, R. Impact of resistance exercises and nano-curcumin on synovial levels of collagenase and nitric oxide in women with knee osteoarthritis. *Transl. Med. Commun.* **2020**, *5*, 1–6. [CrossRef]
170. Manoy, P.; Yuktanandana, P.; Tanavalee, A.; Anomasiri, W.; Ngarmukos, S.; Tanpowpong, T.; Honsawek, S. Vitamin D Supplementation Improves Quality of Life and Physical Performance in Osteoarthritis Patients. *Nutrients* **2017**, *9*, 799. [CrossRef]
171. Gao, X.R.; Chen, Y.S.; Deng, W. The effect of vitamin D supplementation on knee osteoarthritis: A meta-analysis of ran-domized controlled trials. *Int. J. Surg.* **2017**, *46*, 14–20. [CrossRef]
172. Wang, X.; Cicuttini, F.; Jin, X.; Wluka, A.; Han, W.; Zhu, Z.; Blizzard, L.; Antony, B.; Winzenberg, T.; Jones, G.; et al. Knee effusion-synovitis volume measurement and effects of vitamin D supplementation in patients with knee osteoarthritis. *Osteoarthr. Cartil.* **2017**, *25*, 1304–1312. [CrossRef] [PubMed]
173. Thevaranjan, N.; Puchta, A.; Schulz, C.; Naidoo, A.; Szamosi, J.; Verschoor, C.P.; Loukov, D.; Schenck, L.P.; Jury, J.; Foley, K.P.; et al. Age-Associated Microbial Dysbiosis Promotes Intestinal Permeability, Systemic Inflammation, and Macrophage Dysfunction. *Cell Host Microbe* **2017**, *21*, 455–466.e4. [CrossRef] [PubMed]
174. Franceschi, C.; Campisi, J. Chronic inflammation (inflammaging) and its potential contribution to age-associated diseases. *J. Gerontol. Ser. A Biol. Sci. Med. Sci.* **2014**, *69* (Suppl. S1), S4–S9. [CrossRef]
175. Tsai, J.C.; Casteneda, G.; Lee, A.; Dereschuk, K.; Li, W.T.; Chakladar, J.; Lombardi, A.F.; Ongkeko, W.M.; Chang, E.Y. Identification and Characterization of the Intra-Articular Microbiome in the Osteoarthritic Knee. *Int. J. Mol. Sci.* **2020**, *21*, 8618. [CrossRef]
176. Szychlinska, M.A.; Di Rosa, M.; Castorina, A.; Mobasheri, A.; Musumeci, G. A correlation between intestinal microbiota dysbiosis and osteoarthritis. *Heliyon* **2019**, *5*, e01134. [CrossRef]
177. Silvestre, M.P.; Rodrigues, A.M.; Canhão, H.; Marques, C.; Teixeira, D.; Calhau, C.; Branco, J. Cross-Talk between Diet-Associated Dysbiosis and Hand Osteoarthritis. *Nutrients* **2020**, *12*, 3469. [CrossRef]
178. Lee, S.H.; Kwon, J.Y.; Jhun, J.; Jung, K.; Park, S.-H.; Yang, C.W.; Cho, Y.; Kim, S.J.; Cho, M.-L. Lactobacillus acidophilus ameliorates pain and cartilage degradation in experimental osteoarthritis. *Immunol. Lett.* **2018**, *203*, 6–14. [CrossRef]
179. Ahmad, N.; Ansari, M.Y.; Bano, S.; Haqqi, T.M. Imperatorin suppresses IL-1β-induced iNOS expression via inhibiting ERK-MAPK/AP1 signaling in primary human OA chondrocytes. *Int. Immunopharmacol.* **2020**, *85*, 106612. [CrossRef] [PubMed]
180. Lu, C.; Li, Y.; Hu, S.; Cai, Y.; Yang, Z.; Peng, K. Scoparone prevents IL-1β-induced inflammatory response in human osteo-arthritis chondrocytes through the PI3K/Akt/NF-κB pathway. *Biomed. Pharmacother.* **2018**, *106*, 1169–1174. [CrossRef]
181. Qiao, Z.; Tang, J.; Wu, W.; Liu, M. Acteoside inhibits inflammatory response via JAK/STAT signaling pathway in osteoarthritic rats. *BMC Complement. Altern. Med.* **2019**, *19*, 1–8. [CrossRef]
182. Al-Afify, A.S.; El-Akabawy, G.; El-Sherif, N.M.; El-Safty, F.E.-N.A.; El-Habiby, M.M. Avocado soybean unsaponifiables ameliorates cartilage and subchondral bone degeneration in mono-iodoacetate-induced knee osteoarthritis in rats. *Tissue Cell* **2018**, *52*, 108–115. [CrossRef]
183. Shin, H.J.; Park, H.; Shin, N.; Kwon, H.H.; Yin, Y.; Hwang, J.A.; Kim, S.I.; Kim, S.R.; Kim, S.; Joo, Y.; et al. p47phox siR-NA-Loaded PLGA nanoparticles suppress ROS/oxidative stress-induced chondrocyte damage in osteoarthritis. *Polymers* **2020**, *12*, 443. [CrossRef] [PubMed]
184. Hoshi, H.; Akagi, R.; Yamaguchi, S.; Muramatsu, Y.; Akatsu, Y.; Yamamoto, Y.; Sasaki, T.; Takahashi, K.; Sasho, T. Effect of inhibiting MMP13 and ADAMTS5 by intra-articular injection of small interfering RNA in a surgically induced osteoarthritis model of mice. *Cell Tissue Res.* **2017**, *368*, 379–387. [CrossRef] [PubMed]
185. Chen, L.X.; Lin, L.; Wang, H.J.; Wei, X.L.; Fu, X.; Zhang, J.Y.; Yu, C.L. Suppression of early experimental osteoarthritis by in vivo delivery of the adenoviral vector-mediated NF-kappaBp65-specific siRNA. *Osteoarthr. Cartil.* **2008**, *16*, 174–184. [CrossRef] [PubMed]
186. Kong, Y.; Zhang, Y.; Zhao, X.; Wang, G.; Liu, Q. Carboxymethyl-chitosan attenuates inducible nitric oxide synthase and promotes interleukin-10 production in rat chondrocytes. *Exp. Ther. Med.* **2017**, *14*, 5641–5646. [CrossRef] [PubMed]

International Journal of
Molecular Sciences

Article

E8002 Reduces Adhesion Formation and Improves Joint Mobility in a Rat Model of Knee Arthrofibrosis

Seiya Takada [1], Kentaro Setoyama [2,†], Kosuke Norimatsu [3,†], Shotaro Otsuka [1], Kazuki Nakanishi [3], Akira Tani [3], Tomomi Nakakogawa [3], Ryoma Matsuzaki [3], Teruki Matsuoka [3], Harutoshi Sakakima [3], Salunya Tancharoen [4], Ikuro Maruyama [1], Eiichiro Tanaka [5], Kiyoshi Kikuchi [1,5,6,*] and Hisaaki Uchikado [6,7,*]

[1] Department of Systems Biology in Thromboregulation, Kagoshima University Graduate School of Medical and Dental Sciences, 8-35-1 Sakuragaoka, Kagoshima 890-8520, Japan; karaagetantou0110@gmail.com (S.T.); k3360022@kadai.jp (S.O.); maruyama-i@eva.hi-ho.ne.jp (I.M.)

[2] Division of Laboratory Animal Science, Natural Science Center for Research and Education, Kagoshima University, 8-35-1 Sakuragaoka, Kagoshima 890-8520, Japan; seto@m.kufm.kagoshima-u.ac.jp

[3] Course of Physical Therapy, School of Health Sciences, Faculty of Medicine, Kagoshima University, 8-35-1 Sakuragaoka, Kagoshima 890-8544, Japan; k6745961@kadai.jp (K.N.); k9378361@kadai.jp (K.N.); k3694885@kadai.jp (A.T.); 1033.msr@gmail.com (T.N.); ryoma.m1026@outlook.jp (R.M.); k3988526@kadai.jp (T.M.); sakaki@health.nop.kagoshima-u.ac.jp (H.S.)

[4] Department of Pharmacology, Faculty of Dentistry, Mahidol University, Bangkok 10400, Thailand; salunya.tan@mahidol.edu

[5] Division of Brain Science, Department of Physiology, Kurume University School of Medicine, 67 Asahi-machi, Kurume, Fukuoka 830-0011, Japan; eacht@med.kurume-u.ac.jp

[6] Department of Neurosurgery, Kurume University School of Medicine, 67 Asahi-machi, Kurume, Fukuoka 830-0011, Japan

[7] Uchikado Neuro-Spine Clinic, 1-2-3 Naka, Hakata-ku, Fukuoka 812-0893, Japan

* Correspondence: kikuchi_kiyoshi@kurume-u.ac.jp (K.K.); uchikado@me.com (H.U.); Tel.: +81-942-31-7542 (K.K.); +81-92-477-2355 (H.U.); Fax: +81-942-31-7695 (K.K.); +81-92-477-2325 (H.U.)

† These authors contributed equally to this work.

Citation: Takada, S.; Setoyama, K.; Norimatsu, K.; Otsuka, S.; Nakanishi, K.; Tani, A.; Nakakogawa, T.; Matsuzaki, R.; Matsuoka, T.; Sakakima, H.; et al. E8002 Reduces Adhesion Formation and Improves Joint Mobility in a Rat Model of Knee Arthrofibrosis. *Int. J. Mol. Sci.* **2022**, *23*, 1239. https://doi.org/10.3390/ijms23031239

Academic Editor: Masato Sato

Received: 7 January 2022
Accepted: 21 January 2022
Published: 22 January 2022

Publisher's Note: MDPI stays neutral with regard to jurisdictional claims in published maps and institutional affiliations.

Abstract: Knee arthrofibrosis is a common complication of knee surgery, caused by excessive scar tissue, which results in functional disability. However, no curative treatment has been established. E8002 is an anti-adhesion material that contains L-ascorbic acid, an antioxidant. We aimed to evaluate the efficacy of E8002 for the prevention of knee arthrofibrosis in a rat model, comprising injury to the surface of the femur and quadriceps muscle 1 cm proximal to the patella. Sixteen male, 8-week-old Sprague Dawley rats were studied: in the Adhesion group, haemorrhagic injury was induced to the quadriceps and bone, and in the E8002 group, an adhesion-preventing film was implanted between the quadriceps and femur after injury. Six weeks following injury, the restriction of knee flexion owing to fibrotic scarring had not worsened in the E8002 group but had worsened in the Adhesion group. The area of fibrotic scarring was smaller in the E8002 group than in the Adhesion group ($p < 0.05$). In addition, the numbers of fibroblasts ($p < 0.05$) and myofibroblasts ($p < 0.01$) in the fibrotic scar were lower in the E8002 group. Thus, E8002 reduces myofibroblast proliferation and fibrotic scar formation and improves the range of motion of the joint in a model of knee injury.

Keywords: anti-adhesive membrane; knee adhesion; E8002; arthrofibrosis; myofibroblast

1. Introduction

Arthrofibrosis is a disease characterized by pain, loss of range of motion (ROM), warmth, exudation, swelling, and the development of fibrous scarring [1]. The postoperative incidence of arthrofibrosis has been reported to be 1–13% after total knee arthroplasty, 0–4% after ligament injury, such as to the anterior cruciate ligament, and 7% after high-energy knee fracture [2–4]. The formation of excessive scar tissue in the knee joint leads to a loss of ROM, tissue contracture, greater pain, and impairments in activities of daily living, such as standing up, walking, and climbing stairs [5–8]. In consequence, knee joint fibrosis represents a serious challenge for orthopaedic and rehabilitation departments worldwide.

Although the aetiology remains to be fully elucidated, one possible cause of arthrofibrosis is that tissue damage stimulates immune cells, inducing oxidative stress and the production of pro-inflammatory cytokines, such as platelet-derived growth factor (PDGF), transforming growth factor-beta (TGF-beta), and interleukin (IL)-1 [9,10]. This causes fibroblasts to differentiate into myofibroblasts, which produce excessive amounts of extracellular matrix (ECM), including collagen and elastin, which limits joint capsule flexibility and joint movement [9,10]. Although ECM is required for healing and wound repair, dysregulation of its production and degradation leads to pathological fibrosis [10,11]. The approaches to the treatment of knee fibrosis include manipulation under anaesthesia, debridement, and physical therapy [12], but the ideal measure would be to prevent the development of arthrofibrosis. Hydrogel microspheres, temperature-sensitive anti-adhesive poloxamer (TAP) hydrogel, hyaluronic acid, carboxymethylcellulose, decorin, chitosan, lovastatin, rapamycin, and hydroxycaptotensin have been shown to have anti-adhesive effects [13–20], and chitosan and TAP hydrogel have been shown to be effective in clinical trials [14,17]. However, these agents have yet to be commercialized, and no surgical means of preventing adhesions following knee surgery have been established.

E8002 is a three-layered anti-adhesion material that was previously known as nDM-14R [21]. Its central layer consists of pullulan, which is harmless, easily absorbed, and used in foods and pharmaceuticals [22], and the surface layers comprise l-lactide, glycolide, and ε-caprolactone copolymer, which is produced by ring-opening polymerization (Taki Chemical, Kakogawa, Japan), catalysed by tin octoate [$Sn(O_2C_8H_{15})_2$] [22]. The material comprising the central layer dissolves quickly under moist conditions, while that used for the surface layers dissolves slowly [22]. In addition, both the central and surface layers contain L-ascorbic acid [22]. Previously, we reported that E8002 inhibits peripheral nerve adhesion in a model of sciatic nerve injury and epidural scar formation after vertebroplasty in rats [22,23]. E8002 has been shown to act as a slowly dissolving temporary physical barrier that prevents the formation of inter-tissue adhesions, and to inhibit fibrous scar formation via the tissue plasminogen activator-mediated fibrinolytic action of L-ascorbic acid [22,23]. However, there have been no studies of the use of E8002 for the prevention of fibrosis in the knee joint. Therefore, in the present study, we evaluated the effects of E8002 in a rat model of knee joint fibrosis.

2. Results

2.1. E8002 Treatment Reduces the Post-Injury Limitation of Joint Flexion

To determine whether E8002 maintains postoperative knee mobility, we examined the range of motion of the stifles of rats over time. The rats were randomly allocated to an E8002 group ($n = 8$) or Adhesion group ($n = 8$) and the uninjured hindlimbs of eight randomly selected rats from the two groups were regarded as No Adhesion controls ($n = 4$ from each group). The acute angle formed by the femoral greater trochanter, femoral lateral epicondyle, and fibular external capsule was measured. The flexion restriction angle of the stifle in the No Adhesion group did not change significantly over the 6 weeks of the study (pre-injury: $38.33 \pm 1.26°$, 6 weeks: $39.10 \pm 1.06°$). The Adhesion group showed an increase in knee flexion restriction angle with time following injury (pre-injury: $37.31 \pm 1.87°$, 6 weeks: $45.52 \pm 1.14°$), whereas the E8002 group did not show an increase, despite the injury (pre-injury: $37.04 \pm 1.72°$, 6 weeks: $38.16 \pm 0.87°$) (Figure 1A,B). The results of two-way repeated-measures analysis of variance were $p = 0.056$ for the effect of group, $p = 0.001$ for the effect of time, and $p < 0.001$ for the interaction. Specifically, the knee flexion restriction angle of the E8002 group was lower than that of the Adhesion group after 6 weeks, and was comparable to that of the No Adhesion group (Tukey's range test: Adhesion group vs. No Adhesion group: $p < 0.001$, Adhesion group vs. E8002 group: $p < 0.001$, No Adhesion group vs. E8002 group: $p = 0.643$) (Figure 1B). Thus, the insertion of E8002 at the site of linear scarring following knee arthroplasty can reduce subsequent restriction of knee joint motion.

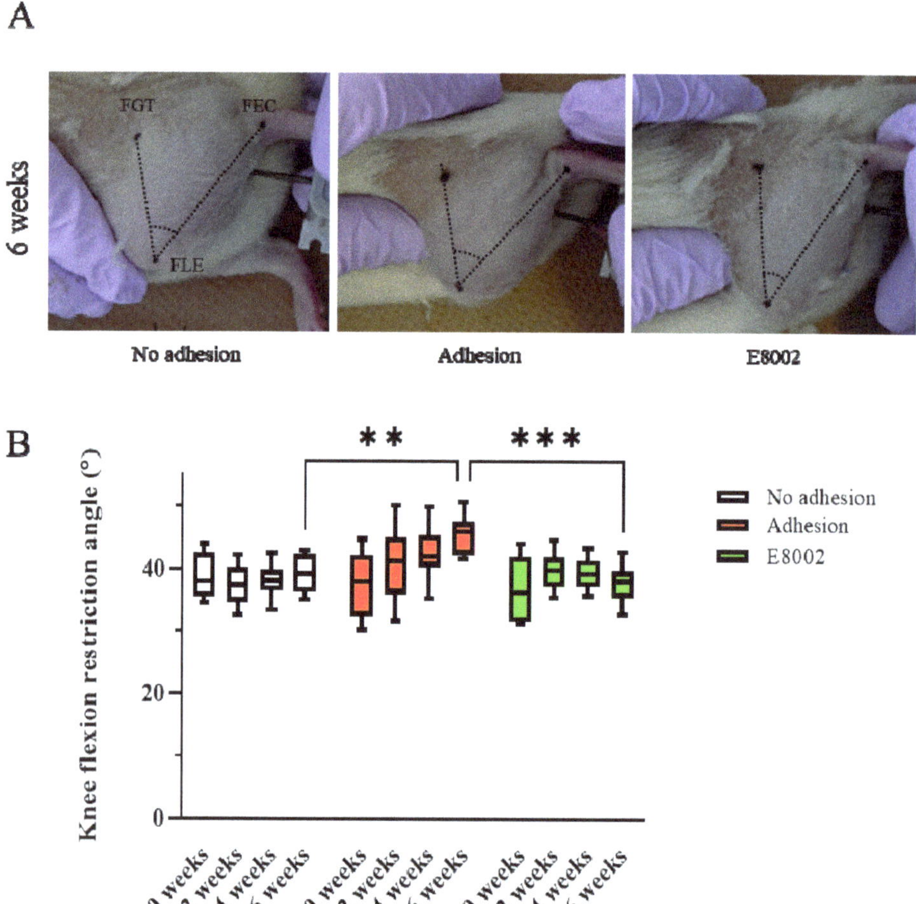

Figure 1. E8002 preserves joint mobility after stifle injury. (**A**) Restriction of joint flexion after stifle injury was compared between the No Adhesion, Adhesion, and E8002 groups over 6 weeks (n = 8 per group). The acute angle formed by the three points of the femoral greater trochanter, femoral lateral condyle, and tibial epicondyle was measured. (**B**) Knee flexion in each group at each time point. Data are mean $\pm$ SE. The results of two-way RM-ANOVA were p = 0.056 for the effect of group, $p = 0.001$ for the effect of time, and $p < 0.001$ for the interaction. Post hoc testing (Tukey's range test) indicated a significantly greater restriction of knee flexion at 6 weeks in the Adhesion group (Adhesion group vs. No Adhesion group: $p < 0.001$, Adhesion group vs. E8002 group: $p < 0.001$, No Adhesion group vs. E8002 group: p = 0.643). ** $p < 0.05$, *** $p < 0.01$.

2.2. E8002 Inhibits Myofibroblast Proliferation and Scar Tissue Formation

The cause of the E8002-associated reduction in knee flexion restriction angle was then investigated morphologically and histologically. Sagittal sections of the normal and injured limbs obtained 6 weeks after injury are shown in Figure 2A. The stifle joint epiphyses were normal in the No Adhesion group, whereas connective tissue proliferation was present in the Adhesion and E88002 groups. However, the area of scar tissue area in the

E8002 group was significantly smaller than that in the Adhesion group (Adhesion group: 2.73 ± 0.34 mm^2, E8002 group: 1.41 ± 0.23 mm^2, Student's t-test: $p < 0.01$) (Figure 2B).

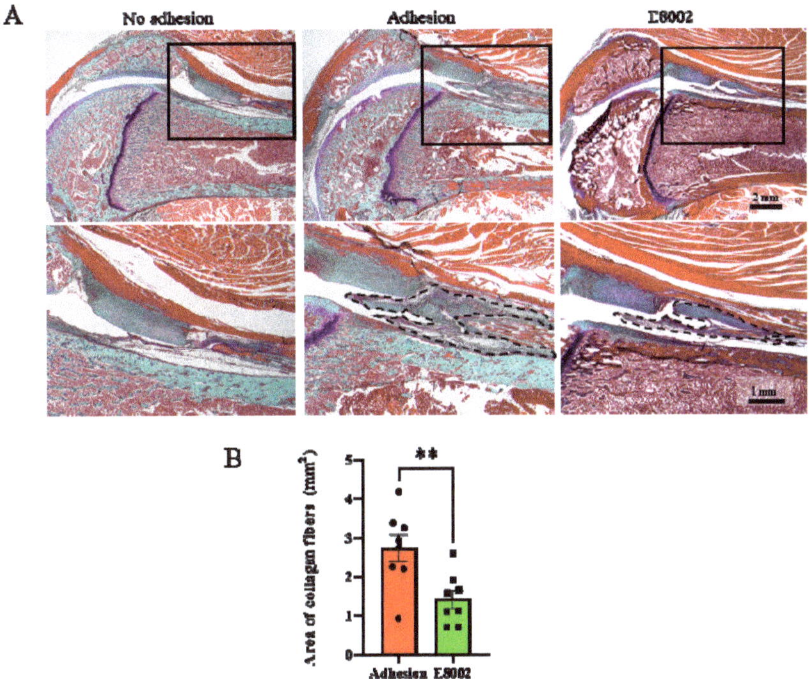

Figure 2. E8002 reduces the formation of fibrous scars after stifle injury. (**A**) Upper row: representative photographs of fibrous scars stained using aldehyde-fuchsin-Masson-Goldner stain. Lower row: higher magnification of the areas indicated by solid black boxes on the photomicrographs above. Black dotted lines outline the areas of fibrotic scarring. (**B**) E8002 significantly reduced the formation of fibrotic scars at the 6-week time point. Data are mean $\pm$ SE. ** $p < 0.01$ (Student's t-test).

Haematoxylin and eosin (HE)-stained images of the fibrous scars are shown in Figure 3A. Because no scarring developed in the No Adhesion group, we counted the numbers of haematoxylin-stained nuclei in the fibrous scars of the other two groups, which were considered to be the nuclei of fibroblasts [20]. There were significantly fewer nuclei in the E8002 group than in the Adhesion group (Adhesion group: 5495 ± 244 cells, E8002 group: 4540 ± 349 cells, Student's t-test: $p = 0.04$) (Figure 3A,B).

Immunostaining for α-smooth muscle actin (α-SMA) was next performed to show the presence of myofibroblasts in the scar tissue (Figure 3C). Myofibroblasts are a key mediator of excessive fibrotic scar formation [24,25]. The number of myofibroblasts was significantly lower in the E8002 group then in the Adhesion group (Adhesion group: 1344 ± 186 cells, E8002 group: 601 ± 182 cells, Mann–Whitney U test: $p = 0.02$) (Figure 3C,D). These data suggest that E8002 reduces excessive scar tissue formation by reducing fibroblast and myofibroblast proliferation.

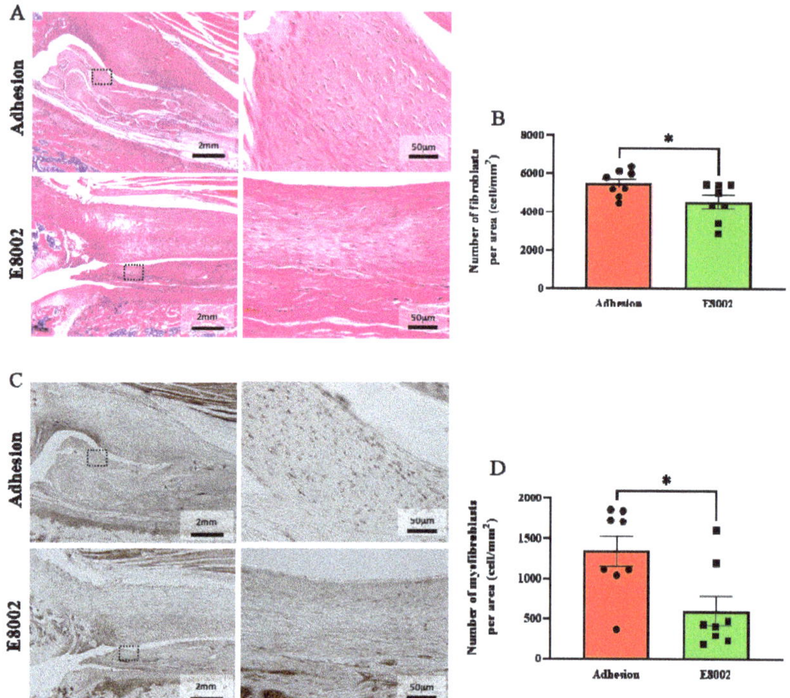

Figure 3. E8002 inhibits the proliferation of myofibroblasts. (**A**) Left: representative photomicrographs of fibroblasts in fibrotic scars (haematoxylin and eosin staining). Right: magnified images of the areas within the black dotted frames. (**B**) Comparison of the mean number of fibroblasts at three different locations. E8002 reduced the number of fibroblasts in the fibrotic scars 6 weeks after injury. (**C**) Left: representative photomicrographs of myofibroblasts in fibrotic scars (α-smooth muscle actin immunostaining. Right: magnified images of the areas within the black dotted frames. (**D**) Comparison of the mean number of myofibroblasts at three different locations. E8002 reduced the number of fibroblasts in the fibrotic scars 6 weeks after injury. Data are mean $\pm$ SE. * $p < 0.05$ (Figure 3B: Student's *t*-test; Figure 3D: Mann–Whitney U test).

3. Discussion

E8002 is a novel multilayer membrane that has been shown to have anti-adhesive effects in rat models of appendage adhesion, vertebral arch resection, and sciatic nerve adhesion [21–23]. In the present study, we have confirmed the efficacy of E8002 using a rat model of arthrofibrosis. Specifically, we have shown that E8002 reduces the limitation to the ROM of the stifle induced by surgical injury and that this may be mediated through decreases in the number of myofibroblasts and the amount of scar tissue.

Arthrofibrosis is a fibrotic joint disorder that begins as an inflammatory response to an insult, such as injury, surgery, or infection [9]. The treatments for arthrofibrosis include manipulation under anaesthesia and debridement, but these procedures may be associated with complications such as neurovascular disorders [2–4]. In addition, excessive physical therapy may trigger an inflammatory response and thereby worsen the arthrofibrosis [12,26,27]. Therefore, the implantation of E8002 may prevent the development of postoperative arthrofibrosis and facilitate 'moderate' physical therapy after treatment.

Myofibroblasts are important for wound healing but are usually absent in healthy tissues [28]. They are the primary effector cells of fibrosis, which involves the deposition of excess ECM and dense fibrous collagen [11,29,30]. Reactive oxygen species (ROS) and pro-inflammatory cytokines, such as TGF-β, IL-1β, and IL-6, play important roles in myofibroblast proliferation [28]. However, E8002 contains ascorbic acid, which is a potent antioxidant [31] and prevents ROS-induced oxidative damage to biological macromolecules, such as DNA, lipids, and proteins [32]. Ascorbic acid has been shown to cause the proliferation of myofibroblasts in collaboration with TGF-β in vitro [33], but the model used in this study did not mimic the prevalent microenvironment associated with tissue injury. However, in a double-blind, placebo-controlled, randomized trial, oral ascorbic acid was found to significantly reduce the incidence of knee arthrofibrosis 1 year following injury [34]. Moreover, the consumption of vitamin C has been reported to reduce complex regional pain syndrome after total knee arthroplasty [35]. Finally, we have previously shown that E8002 inhibits peripheral nerve adhesion by increasing the fibrinolytic effects of ascorbic acid [23]. The present findings are consistent with the fibrinolytic effect of E8002 [23] and suggest that E8002 may inhibit myofibroblast proliferation by reducing postoperative oxidative stress, rather than directly inhibiting myofibroblast proliferation, through the maintenance of an effective local concentration of ascorbic acid.

Positive effects of hydrogel microspheres, TAP hydrogel, hyaluronic acid and carboxymethylcellulose, decorin, chitosan, lovastatin, rapamycin, and hydroxycaptotensin have been identified in animals, but most of these have yet to be tested in human patients [13–20], and to the best of our knowledge, there have no studies of the use of these agents in clinical practice for the prevention of arthrofibrosis following knee joint surgery in humans. The reasons for the delay in the trial of these substances in clinical practice is unknown, but this serves to emphasize the necessity for the development of new methods for the control of arthrofibrosis. As has been frequently demonstrated previously, imperfect knee function is often a source of stress and may reduce the quality of life of patients [5–8]. As treatment technologies develop further, the level of joint function required by patients would be expected to increase, and therefore the control of arthrofibrosis using biological or synthetic materials and drugs will be of increasing importance. Therefore, long-term studies of knee joint mobility will be necessary in the future.

There were several limitations to the present study. First, we did not characterize the changes in the tissues over time or how E8002 affects the surrounding tissues. This is because the number of animals per group was restricted for ethical reasons. Second, the importance of oxidative stress in models of knee joint injury requires further study. However, it is not easy to directly assess oxidative stress caused by ROS or free radicals because ROS are generated immediately after injury and are likely to be undetectable after 6 weeks. Third, the anatomy and mechanics of the rat stifle differ to those of the human knee: rats are quadrupedal, and therefore, their forelimbs are also weight-bearing. Fourth, arthrofibrosis can have a number of different aetiologies, whereas the present model reflects only one of these. Finally, we have not evaluated the effects of the physical properties of E8002 on knee mechanics. There was a slight increase in knee flexion restriction in the E8002 group 2 weeks after the procedure; therefore, it is possible that it may have a small effect, but its softness and thinness would limit this, and because it gradually disappears from the joint, any effect is likely to be transient.

In conclusion, the results of the present study suggest that E8002 reduces myofibroblast proliferation and fibrotic scar formation and maintains the joint ROM in a rat model of knee injury. These findings suggest that a combination of physical and pharmacological means of reducing the postoperative formation of adhesions and scarring may be effective.

4. Materials and Methods

4.1. Animals and Experimental Groups

We studied 16 male, 8-week-old Sprague Dawley rats (250–300 g) that were purchased from Charles River Laboratories (Yokohama, Kanagawa, Japan). The rats were housed under a 12-h light/dark cycle and controlled temperature ($22.0 \pm 1.0\ °C$), with free access to food and water. They were randomly allocated to either an E8002 group ($n = 8$) or an Adhesion group (control group: $n = 8$). The healthy hindlimbs of eight randomly selected rats were classified as the No Adhesion group (*n* = 4, E8002; *n* = 4, Adhesion). The experimental protocol was approved by the Kagoshima University Animal Ethics Committee (approval number MD18025; date: 25 June 2018).

4.2. Knee Arthrofibrosis Model

A knee arthrofibrosis model was created, as described previously [36]. Briefly, rats were anesthetized with 1.5%–2.0% isoflurane (Pfizer Inc., Shibuya-ku, Tokyo, Japan) using an MK-A110 Small Animal Anesthetizer (Muromachi Kikai Co., Ltd., Chuo-ku, Tokyo, Japan). The left stifle of each was shaved using electric clippers, sterilized with 70% ethanol, and draped under aseptic conditions. After the skin was incised, the stifle was opened using a medial parapatellar approach and the medial and lateral sides of the femoral condyle were exposed. Three cuts were then made using a scalpel on the articular surface of the femur and quadriceps muscle 1 cm above the patella, causing bleeding. In the E8002 group, a rectangle of E8002 was inserted (width: 5 mm, height: 15 mm) between the femur and quadriceps (Figure 4), but in the Adhesion group, nothing was inserted. The muscle and skin of all the rats were then sutured using Monocryl 4-0 (Ethicon, Somerville, NJ, USA). The left leg was not fixed, so that it could be moved freely. Six weeks after the procedure, none of the rats showed evidence of wound infection.

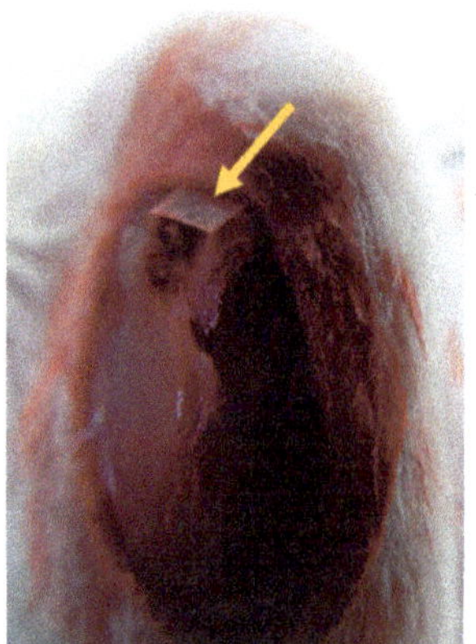
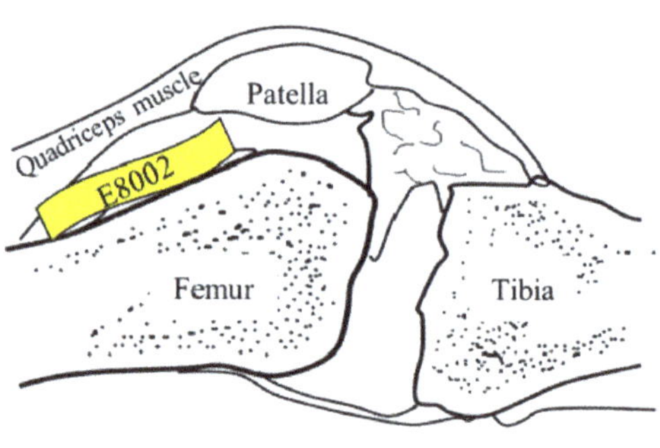

Figure 4. Site of insertion of the E8002 rectangles. The skin was incised to expose the surface of the stifle joint. The yellow arrows indicate the anti-adhesion E8002 membrane, which was inserted between the femur and the quadriceps muscle. The illustration on the right shows the site of insertion of the E8002 in the sagittal plane.

4.3. Joint Angle Measurement

The flexion angle of the stifle of each rat was measured every 2 weeks from prior to injury to 6 weeks afterwards under anaesthesia using 1.5%–2.0% isoflurane. The trunk and left leg were fixed, and a force of 0.5 N was applied to the distal tibia using a Wds-180A Compact Digital Indicator (Kyowa Electronic Instruments Co., Ltd., Chofu City, Tokyo, Japan). It was thought that the application of too much pressure would elongate the soft tissues around the joint and affect the ROM of the joint. Preliminary measurements were made at 1.0 N, 0.5 N, and 0.3 N, and 0.5 N. At 0.3 N, the reaction force was weak, and it was thought to be from the fatty tissue and muscles on the posterior surface of the thigh, not from the fibrous scar. 1N applied strong extension stress to the fibrous scar, and it was thought to be a range-of-motion expansion training. 0.5 N felt moderate resistance. Therefore, 0.5N was adopted. A similar force was used in a recent study [37]. The acute angle between the long axis of the femur and the tibia was measured, using the femoral greater trochanter, femoral lateral epicondyle, and external capsule as landmarks.

Two investigators measured the angles formed by these three points using Scion Image Software 4.0.3 (Scion Corp, Frederick, MD, USA) and the mean value was calculated.

4.4. Histological and Immunohistochemical Evaluation

After the final measurement of the joint angle, the stifle was dissected from the mid-point to the distal patella under terminal anaesthesia. The samples were immersed overnight in 4% paraformaldehyde in 0.1 M phosphate buffer (pH 7.4) at 4 °C, then decalcified using Kalchitox (hydrochloric acid 4.8%, disodium EDTA 0.1% aqueous solution; Fujifilm Wako Pure Chemical Industries, Osaka City, Osaka, Japan) for 72 h, and neutralized using 5% sodium sulphate solution for 24 h. They were then cut as symmetrically as possible along the long axis of the femur and paraffin-embedded. Blocks containing the medial femoral condyles were then cut into 4-μm sections ~100–150 μm medial to the mid-femur, using the patellar tendon transition as a landmark. Aldehyde-fuchsin-Masson-Goldner staining and HE staining were performed. The scar tissue on the aldehyde-fuchsin-Masson-Goldner-stained sections was magnified 4× and photographed using an Olympus DP21 microscope (Shinjuku-ku, Tokyo, Japan).

Sections were also immunostained using rabbit anti-α-SMA antibody to identify myofibroblasts (Cosmo Bio Co., Ltd. Koto-ku, Tokyo, Japan). Following deparaffinization and rehydration, the endogenous peroxidase activity was blocked by incubation in methanol containing 3.0% hydrogen peroxide for 10 min. The sections were then rinsed three times (5 min each) with PBS (pH 7.6) and blocked with 10% skimmed milk in PBS for 20 min. The sections were individually incubated at 4 °C overnight in the α-SMA antibody (1:200), washed in PBS three further times, and then incubated for 60 min with goat anti-rabbit IgG conjugated to a peroxidase-labelled dextran polymer (Dako EnVision+ System-HRP Labelled Polymer Anti-Rabbit; Agilent Technologies, Santa Clara, CA, USA), according to the manufacturer's instructions. Finally, the sections were rinsed with PBS and their immunoreactivity was visualized by diaminobenzidine staining.

4.5. Quantitative Analysis

The area of the scar tissue was analysed using Scion Image Software 4.0.3. The fibroblasts and myofibroblasts were photographed at three random locations using the DP21 microscope at 40× magnification. Each type of cell was counted, and the mean numbers were calculated.

4.6. Statistical Analysis

All data were subjected to the Shapiro–Wilk test, then two-way repeated-measures analysis of variance was used to determine the effect of E8002 on the change in knee flexion over time. Tukey's range test was used for post hoc testing. Student's *t*-test was used to compare the area of scarring and the number of fibroblasts between the E8002 and Adhesion groups, and the Mann–Whitney U test was used to compare the number of myofibroblasts. Two-sided *p*-values < 0.05 were considered to represent statistical significance. Data are expressed as mean ± standard error. The data were analysed using GraphPad prism 9.1.0 (221) (Graphpad Holdings, LLC, San Diego, CA, USA).

Author Contributions: Conceptualization, K.K. and H.U.; formal analysis, S.T. (Seiya Takada), H.S., S.T. (Salunya Tancharoen) and I.M.; investigation, S.T. (Seiya Takada), K.S., K.N. (Kosuke Norimatsu), S.O., K.N. (Kazuki Nakanishi), A.T., T.N., R.M., T.M. and K.K.; resources, E.T., K.K. and H.U.; writing—original draft preparation, S.T. (Seiya Takada); writing—review and editing, all authors. All authors have read and agreed to the published version of the manuscript.

Funding: This research was funded by the Japan Society for the Promotion of Science KAKENHI, grant numbers JP16K10746 and JP20K11640, awarded to Kiyoshi Kikuchi.

Institutional Review Board Statement: The animal study protocol was approved by the Ethics Committee of Kagoshima University (approval number: MD18025; date: 25 June 2018).

Informed Consent Statement: Not applicable.

Data Availability Statement: The data that support the findings of this study are available from the corresponding author (Kiyoshi Kikuchi) upon reasonable request.

Acknowledgments: E8002 was provided by Kawasumi Laboratories Inc. (Tokyo, Japan).

Conflicts of Interest: The authors declare no conflict of interest.

References

1. Millett, P.J.; Johnson, B.; Carlson, J.; Krishnan, S.; Steadman, J.R. Rehabilitation of the arthrofibrotic knee. *Am. J. Orthop.* **2003**, *32*, 531–538. [PubMed]
2. Egol, K.A.; Tejwani, N.C.; Capla, E.L.; Wolinsky, P.L.; Koval, K.J. Staged management of high-energy proximal tibia fractures (OTA types 41): The results of a prospective, standardized protocol. *J. Orthop. Trauma* **2005**, *19*, 448–455. [CrossRef]
3. Laskin, R.S.; Beksac, B. Stiffness after total knee arthroplasty. *J. Arthroplast.* **2004**, *19* (Suppl. 1), 41–46. [CrossRef] [PubMed]
4. Fisher, S.E.; Shelbourne, K.D. Arthroscopic treatment of symptomatic extension block complicating anterior cruciate ligament reconstruction. *Am. J. Sports Med.* **1993**, *21*, 558–564. [CrossRef] [PubMed]
5. Bong, M.R.; Di Cesare, P.E. Stiffness after total knee arthroplasty. *J. Am. Acad. Orthop. Surg.* **2004**, *12*, 164–171. [CrossRef]
6. Bhave, A.; Mont, M.; Tennis, S.; Nickey, M.; Starr, R.; Etienne, G. Functional problems and treatment solutions after total hip and knee joint arthroplasty. *J. Bone Joint Surg Am.* **2005**, *87* (Suppl. 2), 9–21.
7. Seyler, T.M.; Marker, D.R.; Bhave, A.; Plate, J.F.; Marulanda, G.A.; Bonutti, P.M.; Delanois, R.E.; Mont, M.A. Functional problems and arthrofibrosis following total knee arthroplasty. *J. Bone Joint Surg. Am.* **2007**, *89* (Suppl. 3), 59–69.
8. Scheffler, S.U.; Unterhauser, F.N.; Weiler, A. Graft remodeling and ligamentization after cruciate ligament reconstruction. *Knee Surg. Sports Traumatol. Arthrosc.* **2008**, *16*, 834–842. [CrossRef] [PubMed]
9. Usher, K.M.; Zhu, S.; Mavropalias, G.; Carrino, J.A.; Zhao, J.; Xu, J. Pathological mechanisms and therapeutic outlooks for arthrofibrosis. *Bone Res.* **2019**, *7*, 9. [CrossRef]
10. Watson, R.S.; Gouze, E.; Levings, P.P.; Bush, M.L.; Kay, J.D.; Jorgensen, M.S.; Dacanay, E.A.; Reith, J.W.; Wright, T.W.; Ghivizzani, S.C. Gene delivery of TGF-β1 induces arthrofibrosis and chondrometaplasia of synovium in vivo. *Lab. Investig.* **2010**, *90*, 1615–1627. [CrossRef]
11. Luo, Y.; Xie, X.; Luo, D.; Wang, Y.; Gao, Y. The role of halofuginone in fibrosis: More to be explored? *J. Leukoc. Biol.* **2017**, *102*, 1333–1345. [CrossRef] [PubMed]
12. Cheuy, V.A.; Foran, J.R.H.; Paxton, R.J.; Bade, M.J.; Zeni, J.A.; Stevens-Lapsley, J.E. Arthrofibrosis associated with total knee arthroplasty. *J. Arthroplast.* **2017**, *32*, 2604–2611. [CrossRef] [PubMed]
13. Han, Y.; Yang, J.; Zhao, W.; Wang, H.; Sun, Y.; Chen, Y.; Luo, J.; Deng, L.; Xu, X.; Cui, W.; et al. Biomimetic injectable hydrogel microspheres with enhanced lubrication and controllable drug release for the treatment of osteoarthritis. *Bioact. Mater.* **2021**, *6*, 3596–3607. [CrossRef] [PubMed]
14. Kim, J.K.; Park, J.Y.; Lee, D.W.; Ro, D.H.; Lee, M.C.; Han, H.S. Temperature-sensitive anti-adhesive poloxamer hydrogel decreases fascial adhesion in total knee arthroplasty: A prospective randomized controlled study. *J. Biomater. Appl.* **2019**, *34*, 386–395. [CrossRef] [PubMed]
15. Hayashi, M.; Sekiya, H.; Takatoku, K.; Kariya, Y.; Hoshino, Y. Experimental model of knee contracture in extension: Its prevention using a sheet made from hyaluronic acid and carboxymethylcellulose. *Knee Surg. Sports Traumatol. Arthrosc.* **2004**, *12*, 545–551. [CrossRef] [PubMed]
16. Fukui, N.; Fukuda, A.; Kojima, K.; Nakajima, K.; Oda, H.; Nakamura, K. Suppression of fibrous adhesion by proteoglycan decorin. *J. Orthop. Res.* **2001**, *19*, 456–462. [CrossRef]
17. Xu, R.S.; Hou, C.L.; Yin, C.H.; Wang, Y.S.; Chen, A.M. Clinical study on chitosan in prevention of knee adhesion after patellar operation. *Zhongguo Xiu Fu Chong Jian Wai Ke Za Zhi* **2002**, *16*, 240–241.
18. Li, X.; Sun, Y.; Chen, H.; Zhu, G.; Liang, Y.; Wang, Q.; Wang, J.; Yan, L. Hydroxycamptothecin induces apoptosis of fibroblasts and prevents intraarticular scar adhesion in rabbits by activating the IRE-1 signal pathway. *Eur. J. Pharmacol.* **2016**, *781*, 139–147. [CrossRef]
19. Zhao, S.; Sun, Y.; Li, X.; Wang, J.; Yan, L.; Chen, H.; Wang, D.; Dai, J.; He, J. Reduction of intraarticular adhesion of knee by local application of rapamycin in rabbits via inhibition of fibroblast proliferation and collagen synthesis. *J. Orthop. Surg. Res.* **2016**, *11*, 45. [CrossRef]
20. Wu, H.; Germanov, A.V.; Goryaeva, G.L.; Yachmenev, A.N.; Gordienko, D.I.; Kuzin, V.V.; Skoroglyadov, A.V. The Topical Application of Rosuvastatin in Preventing Knee Intra-Articular Adhesion in Rats. *Med. Sci. Monit.* **2016**, *22*, 1403–1409. [CrossRef]
21. Mukai, T.; Kamitani, S.; Shimizu, T.; Fujino, M.; Tsutamoto, Y.; Endo, Y.; Hanasawa, K.; Tani, T. Development of a novel, nearly insoluble antiadhesive membrane. *Eur. Surg. Res.* **2011**, *47*, 248–253. [CrossRef] [PubMed]
22. Kikuchi, K.; Setoyama, K.; Terashi, T.; Sumizono, M.; Tancharoen, S.; Otsuka, S.; Takada, S.; Nakanishi, K.; Ueda, K.; Sakakima, H.; et al. Application of a novel anti-adhesive membrane, E8002, in a rat laminectomy model. *Int. J. Mol. Sci.* **2018**, *19*, 1513. [CrossRef] [PubMed]

23. Kikuchi, K.; Setoyama, K.; Takada, S.; Otsuka, S.; Nakanishi, K.; Norimatsu, K.; Tani, A.; Sakakima, H.; Kawahara, K.I.; Hosokawa, K.; et al. E8002 inhibits peripheral nerve adhesion by enhancing fibrinolysis of l-ascorbic acid in a rat sciatic nerve model. *Int. J. Mol. Sci.* **2020**, *21*, 3972. [CrossRef] [PubMed]
24. Wynn, T.A.; Ramalingam, T.R. Mechanisms of fibrosis: Therapeutic translation for fibrotic disease. *Nat. Med.* **2012**, *18*, 1028–1040. [CrossRef] [PubMed]
25. Monument, M.J.; Hart, D.A.; Salo, P.T.; Befus, A.D.; Hildebrand, K.A. Neuroinflammatory mechanisms of connective tissue fibrosis: Targeting neurogenic and mast cell contributions. *Adv. Wound Care* **2015**, *4*, 137–151. [CrossRef]
26. Daluga, D.; Lombardi, A.V., Jr.; Mallory, T.H.; Vaughn, B.K. Knee manipulation following total knee arthroplasty. Analysis of prognostic variables. *J. Arthroplast.* **1991**, *6*, 119–128. [CrossRef]
27. Kurosaka, M.; Yoshiya, S.; Mizuno, K.; Yamamoto, T. Maximizing flexion after total knee arthroplasty: The need and the pitfalls. *J. Arthroplast.* **2002**, *17* (Suppl. 1), 59–62. [CrossRef]
28. Kendall, R.T.; Feghali-Bostwick, C.A. Fibroblasts in fibrosis: Novel roles and mediators. *Front. Pharmacol.* **2014**, *5*, 123. [CrossRef]
29. Pines, M. Halofuginone for fibrosis, regeneration and cancer in the gastrointestinal tract. *World J. Gastroenterol.* **2014**, *20*, 14778–14786. [CrossRef]
30. Hinz, B.; Phan, S.H.; Thannickal, V.J.; Galli, A.; Bochaton-Piallat, M.L.; Gabbiani, G. The myofibroblast: One function, multiple origins. *Am. J. Pathol.* **2007**, *170*, 1807–1816. [CrossRef]
31. Njus, D.; Kelley, P.M.; Tu, Y.J.; Schlegel, H.B. Ascorbic acid: The chemistry underlying its antioxidant properties. *Free Radic. Biol. Med.* **2020**, *159*, 37–43. [CrossRef] [PubMed]
32. Conway, F.J.; Talwar, D.; McMillan, D.C. The relationship between acute changes in the systemic inflammatory response and plasma ascorbic acid, alpha-tocopherol and lipid peroxidation after elective hip arthroplasty. *Clin. Nutr.* **2015**, *34*, 642–646. [CrossRef] [PubMed]
33. Piersma, B.; Wouters, O.Y.; de Rond, S.; Boersema, M.; Gjaltema, R.A.F.; Bank, R.A. Ascorbic acid promotes a TGFβ1-induced myofibroblast phenotype switch. *Physiol. Rep.* **2017**, *5*, e13324. [CrossRef]
34. Behrend, H.; Lengnick, H.; Zdravkovic, V.; Ladurner, A.; Rudin, D.; Erschbamer, M.; Joerger, M.; Kuster, M. Vitamin C demand is increased after total knee arthroplasty: A double-blind placebo-controlled-randomized study. *Knee Surg. Sports Traumatol. Arthrosc.* **2019**, *27*, 1182–1188. [CrossRef] [PubMed]
35. Jacques, H.; Jérôme, V.; Antoine, C.; Lucile, S.; Valérie, D.; Amandine, L.; Theofylaktos, K.; Olivier, B. Prospective randomized study of the vitamin C effect on pain and complex pain regional syndrome after total knee arthroplasty. *Int. Orthop.* **2021**, *45*, 1155–1162. [CrossRef]
36. Gao, Z.Y.; Wu, J.X.; Liu, W.B.; Sun, J.K. Reduction of adhesion formation after knee surgery in a rat model by botulinum toxin A. *Biosci. Rep.* **2017**, *37*, BSR20160460. [CrossRef] [PubMed]
37. Karahan, N.; Kaya, M.; Yılmaz, B.; Pepele Kurdal, D.; Midi, A.; Vurucu, O. The effect of collagenase clostridium histolyticum on adhesion reduction in a rat knee arthrofibrosis model. *Acta Orthop. Traumatol. Turc.* **2021**, *55*, 385–390. [CrossRef] [PubMed]

International Journal of
Molecular Sciences

Article

Local Administration of Low-Dose Nerve Growth Factor Antibody Reduced Pain in a Rat Osteoarthritis Model

Yuan Tian [1], Tomohiro Onodera [1,2,*], Mohamad Alaa Terkawi [1,2], Koji Iwasaki [1], Ryosuke Hishimura [3], Dawei Liang [1], Takuji Miyazaki [1] and Norimasa Iwasaki [1,2]

[1] Department of Orthopedic Surgery, Faculty of Medicine and Graduate School of Medicine, Hokkaido University, Kita-15, Nish-7, Kita-ku, Sapporo 060-8638, Japan; tyfree1001@gmail.com (Y.T.); materkawi@med.hokudai.ac.jp (M.A.T.); rockcape324@gmail.com (K.I.); 527079901ldw@gmail.com (D.L.); takuzimiyazaki@gmail.com (T.M.); niwasaki@med.hokudai.ac.jp (N.I.)

[2] Global Institution for Collaborative Research and Education (GI-CoRE), Hokkaido University, Sapporo 060-0808, Japan

[3] Department of Orthopedic Surgery, Hokkaido University Hospital, Kita14, Nishi5, Kita-Ku, Sapporo 060-8648, Japan; hishi_piero@yahoo.co.jp

* Correspondence: tomozou@med.hokudai.ac.jp; Tel.: +81-11-706-5935; Fax: +81-11-706-6054

Abstract: Systemic injection of a nerve growth factor (NGF) antibody has been proven to have a significant relevance in relieving osteoarthritis (OA) pain, while its adverse effects remain a safety concern for patients. A local low-dose injection is thought to minimize adverse effects. In this study, OA was induced in an 8-week-old male Sprague–Dawley (SD) rat joint by monoiodoacetate (MIA) injection for 2 weeks, and the effect of weekly injections of low-dose (1, 10, and 100 µg) NGF antibody or saline (control) was evaluated. Behavioral tests were performed, and at the end of week 6, all rats were sacrificed and their knee joints were collected for macroscopic and histological evaluations. Results showed that 100 µg NGF antibody injection relieved pain in OA rats, as evidenced from improved weight-bearing performance but not allodynia. In contrast, no significant differences were observed in macroscopic and histological scores between rats from different groups, demonstrating that intra-articular treatment does not worsen OA progression. These results suggest that local administration yielded a low effective NGF antibody dose that may serve as an alternative approach to systemic injection for the treatment of patients with OA.

Keywords: osteoarthritis; pain; nerve growth factor (NGF); intra-articular injection

Citation: Tian, Y.; Onodera, T.; Terkawi, M.A.; Iwasaki, K.; Hishimura, R.; Liang, D.; Miyazaki, T.; Iwasaki, N. Local Administration of Low-Dose Nerve Growth Factor Antibody Reduced Pain in a Rat Osteoarthritis Model. *Int. J. Mol. Sci.* **2021**, *22*, 2552. https://doi.org/10.3390/ijms22052552

Academic Editor: Henning Madry

Received: 27 January 2021
Accepted: 26 February 2021
Published: 4 March 2021

Publisher's Note: MDPI stays neutral with regard to jurisdictional claims in published maps and institutional affiliations.

1. Introduction

Osteoarthritis (OA) is the most common type of arthritis that affects more than 300 million people globally and contributes to an economic burden on both patients and society [1,2]. According to the recent Osteoarthritis Research Society International white paper, OA has been considered a serious disease because of the lack of any efficient treatment [3]. Clinically, the symptoms of OA include pain, joint stiffness, and disability that lead to a decline in patients' quality of life, the loss of social labor, and an economic burden on the whole society. Pain is particularly important in all clinical problems, as it is not only the cause of hospital visits for treatment but also the main reason underlying poor quality of life and social labor loss [4]. Current pharmacological treatment for OA pain using traditional analgesics, such as non-steroidal anti-inflammatory drugs and acetaminophen, is partly effective and accompanied with serious side-effects, such as disruption of the gastrointestinal mucosa ulceration, cardiovascular toxicity, and suppression of platelet aggregation [5,6]. Therefore, more effective treatments that relieve OA pain are warranted.

A humanized immunoglobulin G2 monoclonal nerve growth factor (NGF) antibody that has been used as an analgesic agent for OA has recently gained significant relevance in relieving OA-associated pain in a clinical trial [7]. Intravenous injections could effectively improve chronic pain and joint function in patients with OA at a dose of 5 or 10 mg every 8 weeks as compared with a placebo [8]. However, a clinical phase III study of tanezumab (NGF antibody) was held by the Food and Drug Administration in 2015 because of its adverse effect, as all patients presented with progressively worsening OA and subsequently required total joint replacement in one of 13 phase III studies [7]. Moreover, other adverse effects, such as paresthesia, arthralgia, pain in the extremities, and headaches were also observed after systemic administration of tanezumab, and these effects remain a safety concern for patients [9–11].

Considering the high effectiveness of NGF antibody treatment for relieving chronic pain, such as OA pain, researchers continue to focus on developing NGF antibody treatment. As OA only affects a limited number of joints, intra-articular injection therapy appears to be a more attractive alternative for patients than other treatments [12]. Local injection can largely decrease the risk of systemic exposure and the incidence of adverse effects. Moreover, local injection is thought to reduce the effective dosage, possibly preventing the aggravation of adverse effects and decreasing the economic burden on patients [13,14]. Although local administration of the NGF antibody, such as its intra-articular injection, might be a preferable way to maintain its effectiveness for the treatment of chronic pain and to reduce the incidence of adverse effects, the analgesic effects of local treatment with a low-dose NGF antibody on OA pain and related adverse effects on cartilage degeneration have not yet been investigated. The purpose of this study was to investigate the effect of low-dose intra-articular injections of the NGF antibody on OA joints using a rat model.

2. Results

2.1. The NGF Antibody Can Relieve the Pain and Improve the Weight-Bearing Performance but Not Allodynia

To observe the effect of analgesic local treatment on OA, a murine model of monoiodoacetate (MIA)-induced OA was employed [15]. To confirm the effect of local treatment with the NGF antibody, different doses (1, 10, and 100 µg) of the NGF antibody were intra-articularly injected into the right knees of rats once a week from the end of week 2. Behavioral tests were performed twice a week to observe the pain behavior in animals. The results of the behavioral tests showed that MIA injection impaired the weight-bearing performance (Table S1) and decreased the threshold, termed as allodynia (Table S2), as a sign of pain from the first week. Rats receiving saline injection did not show any pain behavioral changes. The intra-articular injection of 100 µg of the NGF antibody effectively relieved the pain in the OA model rat, as evidenced from improved weight-bearing performance (Figure 1a, saline, 1 µg, 10 µg, vs. 100 µg from week 3, Table S1), whereas 1 and 10 µg doses showed no pain-relieving effects. However, allodynia, which was also induced by MIA injection, did not improve after NGF antibody injection at any concentrations tested (Figure 1b, Table S2). These results show that only 100 µg of NGF antibody could relieve the OA pain induced by MIA injection, as confirmed from the improvement in weight-bearing asymmetry but not allodynia.

2.2. The NGF Antibody Injection Exerts No Negative Effect on the Cartilage

To observe the effect of the NGF antibody on the joint pathological progress, macroscopic and histological scores were evaluated. The results of macroscopic evaluations showed that MIA injection damaged the cartilage, imitating OA characterized with cartilage erosions (Figure 2a). No erosion was reported in sham knee joint cartilage treated with saline. The injection of the NGF antibody at all doses and saline had no evident adverse effects on the joints (Figure 2b). Consistent with these results, histological evaluations were performed based on hematoxylin and eosin (H&E; Figure 3a) and safranin O staining (Figure 3b). MIA inhibited the function of glyceraldehyde-3-phosphatase-induced cell death, resulting in disorganized cartilage structure, reduction in safranin-O staining, and destruction of tidemark integrity [16]. These pathological changes that appeared at the end of week 6 indicated the damage to the rat knee joint cartilage. H&E (Figure 3a) and safranin

O staining (Figure 3b) showed no significant difference in each MIA group, indicating that NGF antibody injection exhibited no negative effects on cartilage pathology (Figure 3c). However, during the progression of MIA-induced OA, NGF antibody injection did not obviously interrupt the pathological progression of OA.

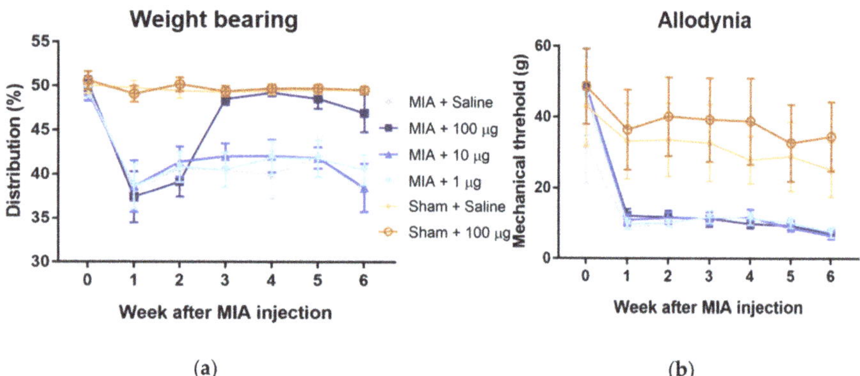

(**a**) (**b**)

Figure 1. Anti-nerve growth factor (NGF) antibody treatment relieved the pain in a rat osteoarthritis (OA) model, as evidenced by improved weight-bearing performance but not allodynia. (**a**) Monoiodoacetate (MIA) injection can induce weight-bearing asymmetry, whereas a saline injection did not show any significant changes in the rat's weight-bearing performance. Moreover, 100 µg of anti-NGF antibody treatment reduced pain in the rat OA model, as evident from the improvement in weight-bearing performance ($p < 0.0001$). Furthermore, 1 and 10 µg antibody treatment did not improve the weight-bearing performance after MIA injection. The results represent the 95% confidence intervals for six rats. (**b**) MIA injection significantly lowered rat hind paw withdrawal mechanical thresholds compared to saline injection. No significant improvement in allodynia was observed after the injection of the NGF antibody at any doses. The results represent the 95% confidence intervals for six rats.

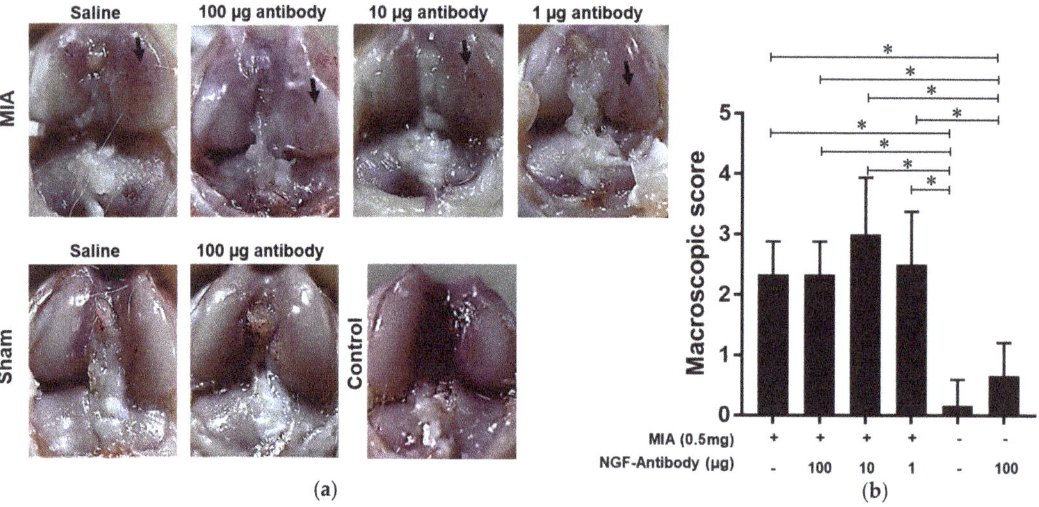

(**a**) (**b**)

Figure 2. Macroscopic evaluation of rat-affected knee joints indicates no differences among the MIA injection groups. (**a**) The macroscopic figures showed that MIA injection can induce cartilage degradation, whereas saline injection had no effect. Arrows indicate cartilage erosions. (**b**) Macroscopic score using the Likert scale showed that the nerve growth factor antibody injection had no evident effect. The results represent 95% confidence intervals for six rats. * indicates a significant difference, as determined by one-way analysis of variance, followed by the Tukey's multiple-comparison procedure ($p \leq 0.05$).

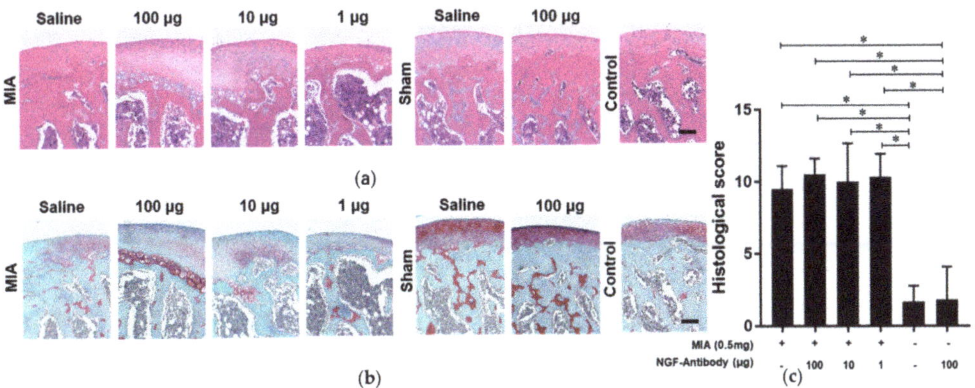

Figure 3. Histological evaluation of rat-affected knee joints consistent with macroscopic evaluation. (**a**) Hematoxylin and eosin (H&E) staining showed that the inflammation formed around the cartilage and chondrocytes was disorganized and no longer observed after MIA injection. The scale bar is 100 μm. (**b**) Results of safranin O staining showed that MIA injection induced cartilage degradation, characterized with cartilage irregularities and reduction in staining intensity. The scale bar is 100 μm. (**c**) The results of the Mankin score showed no significant difference among the MIA groups or between the sham groups, revealing that the treatment of the anti-NGF antibody did not exacerbate the pathological progression of OA joints. The results represent the 95% confidence intervals for six rats. * indicates a significant difference, as determined by one-way analysis of variance, followed by the Tukey's multiple-comparison procedure ($p \leq 0.05$).

Fluorescence staining for the NGF revealed the MIA injection-mediated increase in the concentration of NGF in the synovial tissue surrounding the affected joints (Figure 4). Injection of the NGF antibody neutralized the NGF in the tissue and, consequently, reduced the signal of NGF staining.

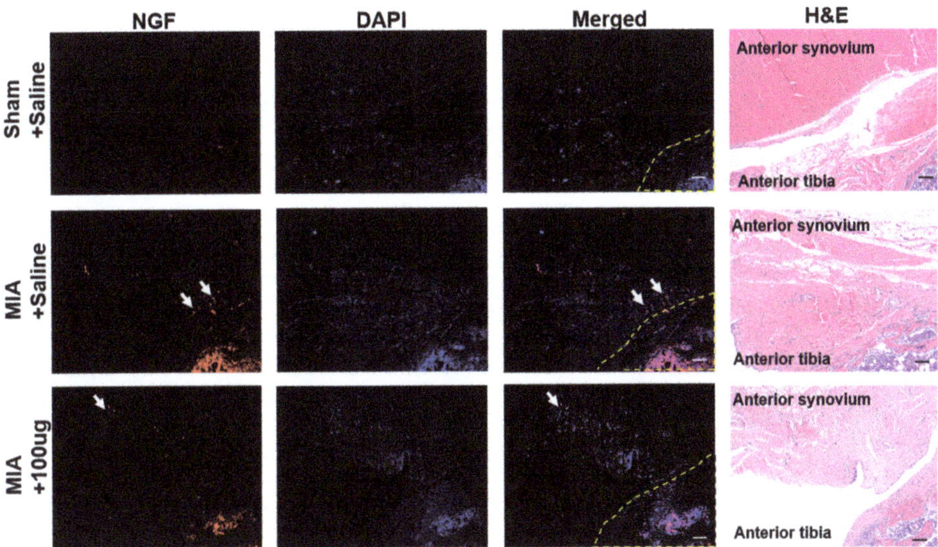

Figure 4. Fluorescent staining of knee joints indicates NGF concentration changes before and after treatment. MIA injection induced a NGF increase, which appears as a high NGF positive area. After the injection of 100 μg NGF, there was a decrease of NGF positive reaction. Arrows indicate the positive reaction for the NGF. The yellow dotted line indicates the area of the articular cartilage and subchondral bone tissue. DAPI; 4′,6-diamidino-2-phenylindole. The scale bar is 100 μm.

3. Discussion

NGF plays an important role in pain and can serve as a signal in inflammatory joint disease [17,18]. NGF antibody treatment has been proven to be effective to relieve chronic pain, such as OA pain. However, high-dose administration of the NGF antibody (5 mg/kg) is accompanied with adverse effects, including progressively worsening cartilage degeneration and potential nerve system side-effects, such as paresthesia, arthralgia, and headaches, and is considered problematic in OA [10,19]. Hochberg revealed the rapid progression in OA after systemic treatment with the NGF antibody and concluded that it is imperative to use low effective doses to control the risk of adverse effects [20]. Moreover, Bélanger et al. reviewed the evaluation of safety data of systemic treatment with the NGF antibody and found safety concerns both in clinical and nonclinical cases [21]. Therefore, a low effective dose should be considered to decrease the risk of side-effects and systemic exposure.

In this study, we found that a low dose of 100 µg NGF antibody could alleviate MIA-induced pain, suggesting that intra-articular administration of a low dose of NGF antibody may be an effective treatment for OA, specifically when a limited number of arthritic joints are effected, such as mono- or oligo-articular OA. Several studies have focused on the efficacy and safety of the NGF antibody for chronic pain using an animal OA model through systemic treatment. In general, systemic injection necessitates a 10 mg/kg dose, corresponding to 5 mg/injection [22,23]. Aso et al. found that the inhibition of tropomyosin receptor kinase (TrkA), a high-affinity receptor of the NGF, reduced pain in meniscal transection-operated rats after oral treatment with 30 mg/kg twice daily [24]. Additionally, systemic administration of the NGF antibody in rat models increased limbs edema [25,26]. In fact, intra-articular injection has been considered a more cost-effective treatment for OA than systemic injection [10,27]. Direct delivery of drugs necessitates a low yet effective dose, which can decrease the risk of side-effects and damage to other unimpaired tissues. Therefore, local treatment with the NGF antibody may be an appropriate strategy to reduce the dosage and thereby the incidence of side-effects and exposure of the whole body to the antibody. This is the first study to elaborate on the pain-relieving effect of local treatment with the NGF antibody and its effects on articular cartilage.

Systemic injection is known to improve both weight-bearing asymmetry and mechanical allodynia [22]. The intra-articular injection of the NGF antibody could only improve weight-bearing performance. Mechanical allodynia is a painful sensation stimulated by light touch. Although the mechanism of allodynia is incompletely understood, it is thought to involve alterations in mechano-transduction and sensory neurons of the central nervous system (CNS) [28,29]. NGF can bind to two receptors, a high-affinity TrkA receptor and a low-affinity p75 neurotrophic receptor (p75NTR). Upregulated levels of NGF and TrkA have been reported in the synovial fluid of some patients with arthritis [18,30]. The binding of NGF to TrkA results in the retrograde transport of the resulting complex to the cell body of sensory neurons located in the dorsal root ganglia (DRG) [31]. The increased expression of NGF under OA conditions contributes to changes in receptor sensitivity or DRG, which may result in central sensitization [32,33].

The limitation of this study is the insufficient analgesic effect on allodynia-related pathologies, such as complex regional pain syndrome after intra-articular injection of the NGF antibody, as suggested from our results. Whether or not the local injection of the NGF antibody can effectively reverse the CNS changes established during the NGF rising phase requires further exploration and experiments. Furthermore, although OA pain was relieved by NGF antibody injection, complications of OA, such as cartilage degeneration, still require treatment. Patients treated with the NGF antibody may have better ability for physical activities because of the absence of pain in the OA joint, thereby aggravating cartilage damage and eventually leading to early joint replacement. Moreover, the mechanism by which topical administration of the NGF antibody does not suppress allodynia without exacerbating OA is still unknown. Further studies are warranted to clarify the association between these two phenomena.

In conclusion, the intra-articular administration of a low dose of NGF antibody could reduce pain but not allodynia or, more importantly, cartilage degeneration in rat. Although doses and intervals need to be considered in humans due to a species mismatch, the local administration may serve as a safe alternative approach to systemic injection for pain relief treatment in OA patients.

4. Materials and Methods

4.1. Ethics Statement

All animal experiments were approved on 20 February 2018 by the Institute of Animal Care and Use Committee of the Hokkaido University Graduate School of Medicine (no. 17–0136).

4.2. MIA-Induced Rat OA Pain Model

Eight-week-old male Sprague–Dawley (SD) rats (CLEA Tokyo, Japan) were housed on a 12 h light/dark cycle and had free access to food and water. Rats were randomly divided into six groups, including four MIA injection groups and two sham groups (saline injection). Each group contained a sample size of six rats. Rats were anesthetized by intraperitoneal injections of 100 and 10 mg/kg ketamine and xylazine, respectively, before local injection. Later, 0.5 mg MIA (Nacalai Tesque, Kyoto, Japan) dissolved in 25 μL saline solution was injected once into the right knee joint capsules of rats from MIA groups through the infrapatellar ligament to induce OA-like pain [34]. The same volume of saline solution was injected into the right knee joints of sham rats. To confirm the effect of MIA and saline local injection, behavioral tests, such as weight-bearing and von Frey filament tests, were continuously performed before and after local injection for 6 weeks.

4.3. Behavioral Tests

To define the pain behavior performance of rat knee joints, direct pain behavior performance (for detecting weight-bearing asymmetry) and indirect pain behavior performance (for detecting allodynia) were tested in this study. Both behavioral evaluations were performed twice a week for each rat from the week before local injection. All data were collected and applied for statistical analysis after 6 weeks, and data were combined to show weekly changes in behavior tests.

Weight-bearing changes in the hind paw represented the weight distribution between the right (operated) and left (control) limbs as a direct index of joint pain in the osteoarthritic knee [35]. A static weight-bearing test device (Bioseb, Chaville, France) was used to determine the hind paw weight distribution. Rats were rested in an angled chamber so that each hind paw was placed on a separate force testing plate. The force exerted by each hind limb (measured in grams) was averaged over a 5-s period. Each data point was the mean of the three tests of 5-s reading. Results were presented as the distribution of weight-bearing between the left (control) limb and right (operated) limb, as calculated by the following equation: bearing weight of operated leg/bearing weight of both legs × 100%. Decreasing the distribution of weight-bearing can be considered a direct index of joint pain.

A von Frey filament (Shin factory, Fukuoka, Japan) was used to measure the mechanical threshold for indicating allodynia, which was induced by mechanical stimulation. Rats were placed in a chamber with a mesh bottom, which allowed access to the plantar surface of each hind paw. The animals were allowed to acclimatize in the chamber for 10 min before testing. The mechanical threshold of the ipsilateral hind paw was assessed using the modified up-down method [36]. A von Frey hair was perpendicularly applied to the plantar surface of the ipsilateral hind paw until the hair flexed and held in place for 3 s. The von Frey hair (range, 0.4–60.0 g) was applied in ascending order to observe the withdrawal reaction of rats. If a rapid withdrawal reaction was observed, the von Frey hair was subsequently applied in descending order until the withdrawal reaction was no longer observed. This method was repeated thrice at an interval of 10 min. Allodynia was confirmed from the decrease in the mechanical threshold as compared to the original mechanical threshold before injection of MIA or saline.

4.4. Treatment with the NGF Antibody

To observe the effect of the local administration of the NGF antibody on OA pain, saline and different doses (1, 10, and 100 µg) of the NGF antibody (Mochida Pharmaceutical Co., Ltd., Tokyo, Japan) were injected into the right knee joint capsules of rats from MIA groups through the infrapatellar ligament at week 2. The injection was administered once a week, four times until the end of week 6 (Figure 5). The left knee joints remained non-operated and were considered control groups.

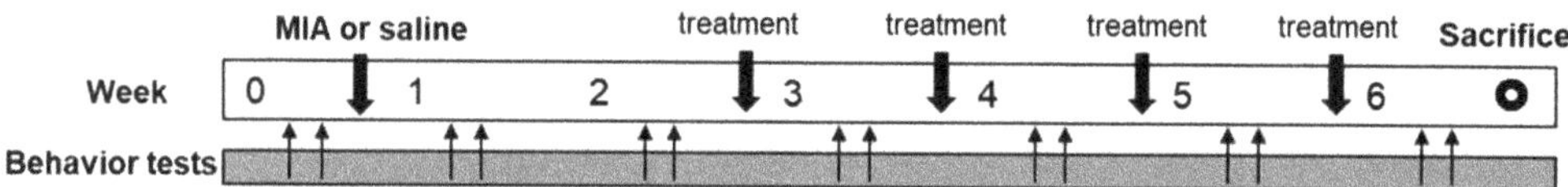

Figure 5. Timeline of local injection of the NGF antibody in an MIA-induced OA rat model.

4.5. Cartilage Degradation Evaluation

Rats were sacrificed at the end of week 6. The knee joints were collected and fixed in 10% formalin and subjected to cartilage degradation evaluations, including macroscopic and histological scoring, in a blinded manner.

Macroscopic scores were assessed after joint collection and evaluated using the Likert scale (Table 1) [37].

Table 1. Likert scale (Guingamp macroscopic lesions).

Grade
0 = normal appearance
1 = slight yellowish discoloration of the chondral surface
2 = little cartilage erosion in load-bearing areas
3 = large erosions extending down to the subchondral bone
4 = large erosions with large areas of subchondral bone exposure.

Regarding the performance of knee joint histological examination, knee joints were decalcified in decalcifying solution B (Wako, Osaka, Japan) for 3 days before the joints were embedded in paraffin. Sections of 5 µm were prepared and subjected to H&E and safranin O staining. The degeneration of the cartilage was microscopically examined using an all-in-one microscope (Keyence, Osaka, Japan) and scored using the modified Mankin scoring system (Table 2) [38].

4.6. Immunofluorescence Staining

To observe the inhibition of NGF in the knee joint capsule, 5-µm sections of knee joints were prepared and blocked with horse serum. The sections were subsequently incubated for overnight at 4 °C with a primary NGF antibody (1:500, Mochida Pharmaceutical Co. Ltd., Tokyo, Japan). The primary antibody was detected with a treatment with goat anti-mouse Alexa Fluor Plus 594 (Invitrogen, Waltham, MA, USA) for 60 min at 37 °C. The cell nuclei were stained with 4′,6-diamidino-2-phenylindole (DAPI, Invitrogen, Carlsbad, CA, USA). Finally, the sections were mounted, covered with cover slides, and examined using a fluorescence microscope (Keyence, Osaka, Japan).

Table 2. Modified Mankin scoring system.

Cartilage Structure	
Normal	0
Surface irregularities	1
Pannus and surface irregularities	2
Clefts to transitional zone	3
Clefts to radial zone	4
Clefts to calcified zone	5
Complete disorganization	6
Cartilage cells	
Normal	0
Pyknosis, lipid degeneration hypercellularity	1
Clusters	2
Hypocellularity	3
Safranin-O	
Normal	0
Slight reduction	1
Moderate reduction	2
Severe reduction	3
No staining	4
Tidemark integrity	
Intact	0
Destroyed	1

4.7. Statistical Analyses

For behavioral evaluation, a two-way analysis of variance (ANOVA), followed by Tukey's test, was used for the comparison of each group. For macroscopic and histological evaluations, one-way ANOVA, followed by Tukey's multiple-comparison procedure, was used to compare differences among groups (GraphPad Software, La Jolla, CA, USA). Results presented as mean $\pm$ standard error of the mean (SEM) were considered statistically significant at $p < 0.05$.

5. Conclusions

Based on the results of this study, the effective dose of the NGF antibody was significantly lower with intra-articular injection than that with systemic injection without acceleration in OA progression. Moreover, intra-articular injection might be an alternative approach to systemic injection for the treatment of patients with OA. Further experiments are necessary to improve the intra-articular injection treatment with the NGF antibody to treat cartilage degeneration and elucidate the mechanism of allodynia in OA, as well as the relationship between the NGF antibody and allodynia.

Supplementary Materials: The following are available online at https://www.mdpi.com/1422-0067/22/5/2552/s1.

Author Contributions: Study designs were performed by Y.T. and T.O. In vivo and in vitro experiments were performed by Y.T., R.H., T.M., and D.L. Data acquisition and statistical analysis were performed by Y.T. and T.O. Drafting of the manuscript was carried out by Y.T. and T.O. A critical review of the manuscript was conducted by K.I., M.A.T., and N.I. All authors have read and agreed to the published version of the manuscript.

Funding: This work was supported by KAKEN-Grant-in-Aid for Young Scientists under grant #18K16642.

Institutional Review Board Statement: All procedures for animal experiments were approved by the Institute of Animal Care and Use Committee of the Hokkaido University Graduate School of Medicine (no. 17-0136).

Informed Consent Statement: Not applicable.

Data Availability Statement: The data presented in this study are available in results section, figures and Supplementary Material.

Acknowledgments: The NGF antibody used in this study was provided by MOCHIDA PHARMACEUTICAL CO., LTD. Tokyo Japan.

Conflicts of Interest: The authors declare no conflict of interest.

References

1. Kloppenburg, M.; Berenbaum, F. Osteoarthritis year in review 2019: Epidemiology and therapy. *Osteoarthr. Cartil.* **2020**, *28*, 242–248. [CrossRef] [PubMed]
2. Chua, J.R.; Jamal, S.; Riad, M.; Castrejon, I.; Malfait, A.M.; Block, J.A.; Pincus, T. Disease burden in osteoarthritis is similar to that of rheumatoid arthritis at initial rheumatology visit and significantly greater six months later. *Arthritis Rheumatol.* **2019**, *71*, 1276–1284. [CrossRef] [PubMed]
3. March, L.; Cross, M.; Arden, N.; Hawker, G. *Osteoarthritis: A Serious Disease*; OARSI White Paper; OARSI: Mount Laurel, NJ, USA, 2016; p. 103.
4. Perrot, S. Osteoarthritis pain. *Best Pract. Res. Clin. Rheumatol.* **2015**, *29*, 90–97. [CrossRef]
5. Krebs, E.E.; Gravely, A.; Nugent, S.; Jensen, A.C.; DeRonne, B.; Goldsmith, E.S.; Kroenke, K.; Bair, M.J.; Noorbaloochi, S. Effect of opioid vs nonopioid medications on pain-related function in patients with chronic back pain or hip or knee osteoarthritis pain: The space randomized clinical trial. *J. Am. Med. Assoc.* **2018**, *319*, 872–882. [CrossRef] [PubMed]
6. Da Costa, B.R.; Reichenbach, S.; Keller, N.; Nartey, L.; Wandel, S.; Jüni, P.; Trelle, S. Effectiveness of non-steroidal anti-inflammatory drugs for the treatment of pain in knee and hip osteoarthritis: A network meta-analysis. *Lancet* **2017**, *390*, e21–e33. [CrossRef]
7. Lane, N.E.; Schnitzer, T.J.; Birbara, C.A.; Mokhtarani, M.; Shelton, D.L.; Smith, M.D.; Brown, M.T. Tanezumab for the treatment of pain from osteoarthritis of the knee. *N. Engl. J. Med.* **2010**, *363*, 1521–1531. [CrossRef] [PubMed]
8. Schnitzer, T.J.; Ekman, E.F.; Spierings, E.L.H.; Greenberg, H.S.; Smith, M.D.; Brown, M.T.; West, C.R.; Verburg, K.M. Efficacy and safety of tanezumab monotherapy or combined with non-steroidal anti-inflammatory drugs in the treatment of knee or hip osteoarthritis pain. *Ann. Rheum. Dis.* **2015**, *74*, 1202–1211. [CrossRef]
9. Kivitz, A.J.; Gimbel, J.S.; Bramson, C.; Nemeth, M.A.; Keller, D.S.; Brown, M.T.; West, C.R.; Verburg, K.M. M and safety of tanezumab versus naproxen in the treatment of chronic low back pain. *Pain* **2013**, *154*, 1009–1021. [CrossRef] [PubMed]
10. Katz, N.; Borenstein, D.G.; Birbara, C.; Bramson, C.; Nemeth, M.A.; Smith, M.D.; Brown, M.T. Efficacy and safety of tanezumab in the treatment of chronic low back pain. *Pain* **2011**, *152*, 2248–2258. [CrossRef] [PubMed]
11. Schnitzer, T.J.; Easton, R.; Pang, S.; Levinson, D.J.; Pixton, G.; Viktrup, L.; Davignon, I.; Brown, M.T.; West, C.R.; Verburg, K.M. Effect of Tanezumab on Joint Pain, Physical Function, and Patient Global Assessment of Osteoarthritis among Patients with Osteoarthritis of the Hip or Knee: A Randomized Clinical Trial. *JAMA* **2019**, *322*, 37–48. [CrossRef]
12. Jones, I.A.; Togashi, R.; Wilson, M.L.; Heckmann, N.; Vangsness, C.T., Jr. Intra-articular treatment options for knee osteoarthritis. *Nat. Rev. Rheumatol.* **2019**, *15*, 77–90. [CrossRef] [PubMed]
13. Evans, C.H.; Kraus, V.B.; Setton, L.A. Progress in intra-articular therapy. *Nat. Rev. Rheumatol.* **2014**, *10*, 11–22. [CrossRef] [PubMed]
14. Rosen, J.; Sancheti, P.; Fierlinger, A.; Niazi, F.; Johal, H.; Bedi, A. Cost-effectiveness of different forms of intra-articular injections for the treatment of osteoarthritis of the knee. *Adv. Ther.* **2016**, *33*, 998–1011. [CrossRef] [PubMed]
15. Udo, M.; Muneta, T.; Tsuji, K.; Ozeki, N.; Nakagawa, Y.; Ohara, T.; Saito, R.; Yanagisawa, K.; Koga, H.; Sekiya, I. Monoiodoacetic acid induces arthritis and synovitis in rats in a dose- and time-dependent manner: Proposed model-specific scoring systems. *Osteoarthr. Cartil.* **2016**, *24*, 1284–1291. [CrossRef] [PubMed]
16. Sabri, M.I.; Ochs, S.J. Inhibition of glyceraldehyde-3-phosphate dehydrogenase in mammalian nerve by iodoacetic acid. *J. Neurochem.* **1971**, *18*, 1509–1514. [CrossRef] [PubMed]
17. Kumar, V.; Mahal, B.A. NGF-The TrkA to successful pain treatment. *J. Pain Res.* **2012**, *5*, 279–287. [CrossRef] [PubMed]
18. Halliday, D.A.; Zettler, C.; Rush, R.A.; Scicchitano, R.; McNeil, J.D. Elevated nerve growth factor levels in the synovial fluid of patients with inflammatory joint disease. *Neurochem. Res.* **1998**, *23*, 919–922. [CrossRef]
19. Miller, R.E.; Block, J.A.; Malfait, A.M. What is new in pain modification in osteoarthritis? *Rheumatol. Oxf. Engl.* **2018**, *57*, iv99–iv107. [CrossRef]
20. Hochberg, M.C. Serious joint-related adverse events in randomized controlled trials of anti-nerve growth factor monoclonal antibodies. *Osteoarthr. Cartil.* **2015**, *23*, S18–S21. [CrossRef] [PubMed]
21. Bélanger, P.; West, C.R.; Brown, M.T. Development of pain therapies targeting nerve growth factor signal transduction and the strategies used to resolve safety issues. *J. Toxicol. Sci.* **2018**, *43*, 1–10. [CrossRef]
22. Xu, L.; Nwosu, L.N.; Burston, J.J.; Millns, P.J.; Sagar, D.R.; Mapp, P.I.; Meesawatsom, P.; Li, L.; Bennett, A.J.; Walsh, D.A.; et al. The anti-NGF antibody muMab 911 both prevents and reverses pain behaviour and subchondral osteoclast numbers in a rat model of osteoarthritis pain. *Osteoarthr. Cartil.* **2016**, *24*, 1587–1595. [CrossRef]
23. Miyagi, M.; Ishikawa, T.; Kamoda, H.; Suzuki, M.; Inoue, G.; Sakuma, Y.; Oikawa, Y.; Orita, S.; Uchida, K.; Takahashi, K.; et al. Efficacy of nerve growth factor antibody in a knee osteoarthritis pain model in mice. *BMC Musculoskelet. Disord.* **2017**, *18*, 428. [CrossRef]
24. Aso, K.; Shahtaheri, S.M.; Hill, R.; Wilson, D.; McWilliams, D.F.; Nwosu, L.N.; Chapman, V.; Walsh, D.A. Contribution of nerves within osteochondral channels to osteoarthritis knee pain in humans and rats. *Osteoarthr. Cartil.* **2020**, *28*, 1245–1254. [CrossRef]

25. Ishikawa, G.; Koya, Y.; Tanaka, H.; Nagakura, Y. Long-term analgesic effect of a single dose of anti-NGF antibody on pain during motion without notable suppression of joint edema and lesion in a rat model of osteoarthritis. *Osteoarthr. Cartil.* **2015**, *23*, 925–932. [CrossRef]

26. Sabsovich, I.; Wei, T.; Guo, T.Z.; Zhao, R.; Shi, X.; Li, X.; Yeomans, D.C.; Klyukinov, M.; Kingery, W.S.; Clark, D.J. Effect of anti-NGF antibodies in a rat tibia fracture model of complex regional pain syndrome type I. *Pain* **2008**, *138*, 47–60. [CrossRef]

27. Belzile, E.L.; Deakon, R.T.; Vannabouathong, C.; Bhandari, M.; Lamontagne, M.; McCormack, R. Cost-utility of a single-injection combined corticosteroid-hyaluronic acid formulation vs a 2-injection regimen of sequential corticosteroid and hyaluronic acid injections. *Clin. Med. Insights Arthritis Musculoskelet. Disord.* **2017**, *10*. [CrossRef] [PubMed]

28. Lolignier, S.; Eijkelkamp, N.; Wood, J.N. Mechanical allodynia. *Pflug. Arch.* **2015**, *467*, 133–139. [CrossRef]

29. Orita, S.; Ishikawa, T.; Miyagi, M.; Ochiai, N.; Inoue, G.; Eguchi, Y.; Kamoda, H.; Arai, G.; Toyone, T.; Aoki, Y.; et al. Pain-related sensory innervation in monoiodoacetate-induced osteoarthritis in rat knees that gradually develops neuronal injury in addition to inflammatory pain. *BMC Musculoskelet. Disord.* **2011**, *12*, 134. [CrossRef] [PubMed]

30. Iannone, F.; De Bari, C.; Dell'Accio, F.; Covelli, M.; Patella, V.; Lo Bianco, G.; Lapadula, G. Increased expression of nerve growth factor (NGF) and high affinity NGF receptor (p140 TrkA) in human osteoarthritic chondrocytes. *Rheumatol. Oxf. Engl.* **2002**, *41*, 1413–1418. [CrossRef] [PubMed]

31. Chang, D.S.; Hsu, E.; Hottinger, D.G.; Cohen, S.P. Anti-nerve growth factor in pain management: Current evidence. *J. Pain Res.* **2016**, *9*, 373–383.

32. Woolf, C.J. Central sensitization: Implications for the diagnosis and treatment of pain. *Pain* **2011**, *152*, S2–S15. [CrossRef] [PubMed]

33. Woolf, C.J.; Safieh-Garabedian, B.; Ma, Q.P.; Crilly, P.; Winter, J. Nerve growth factor contributes to the generation of inflammatory sensory hypersensitivity. *Neuroscience* **1994**, *62*, 327–331. [CrossRef]

34. Mapp, P.I.; Sagar, D.R.; Ashraf, S.; Burston, J.J.; Suri, S.; Chapman, V.; Walsh, D.A. Differences in structural and pain phenotypes in the sodium monoiodoacetate and meniscal transection models of osteoarthritis. *Osteoarthr. Cartil.* **2013**, *21*, 1336–1345. [CrossRef]

35. Iwasaki, K.; Sudo, H.; Kasahara, Y.; Yamada, K.; Ohnishi, T.; Tsujimoto, T.; Iwasaki, N. Effects of multiple intra-articular injections of 0.5% bupivacaine on normal and osteoarthritic joints in rats. *Arthroscopy* **2016**, *32*, 2026–2036. [CrossRef]

36. Chaplan, S.R.; Bach, F.W.; Pogrel, J.W.; Chung, J.M.; Yaksh, T.L. Quantitative assessment of tactile allodynia in the rat paw. *J. Neurosci. Methods* **1994**, *53*, 55–63. [CrossRef]

37. Guingamp, C.; Gegout-Pottie, P.; Philippe, L.; Terlain, B.; Netter, P.; Gillet, P. Mono-iodoacetate-induced experimental osteoarthritis: A dose-response study of loss of mobility, morphology, and biochemistry. *Arthritis Rheum.* **1997**, *40*, 1670–1679. [CrossRef]

38. Mankin, H.J.; Dorfman, H.A.; Lippiello, L.; Zarins, A. Biochemical and metabolic abnormalities in articular cartilage from osteoarthritic human hips: II. Correlation of morphology with biochemical and metabolic data. *J. Bone Joint Surg.* **1971**, *53*, 523.e37. [CrossRef]

International Journal of
Molecular Sciences

Article

Knee Osteoarthritis Progression Is Delayed in Silent Information Regulator 2 Ortholog 1 Knock-in Mice

Tetsuya Yamamoto, Nobuaki Miyaji, Kiminari Kataoka, Kyohei Nishida, Kanto Nagai, Noriyuki Kanzaki, Yuichi Hoshino, Ryosuke Kuroda and Takehiko Matsushita *

Department of Orthopaedic Surgery, Kobe University Graduate School of Medicine, Kobe 650-0017, Japan;
ytetsu36@gmail.com (T.Y.); nobuakimiyaji@gmail.com (N.M.); k.s.kimi.h.m.m.cp@gmail.com (K.K.);
kyohei_nishida0610@yahoo.co.jp (K.N.); kantona9@gmail.com (K.N.); kanzaki@med.kobe-u.ac.jp (N.K.);
you.1.hoshino@gmail.com (Y.H.); kurodar@med.kobe-u.ac.jp (R.K.)
* Correspondence: matsushi@med.kobe-u.ac.jp; Tel.: +81-78-382-5985

Abstract: Overexpression of silent information regulator 2 ortholog 1 (SIRT1) is associated with beneficial roles in aging-related diseases; however, the effects of SIRT1 overexpression on osteoarthritis (OA) progression have not yet been studied. The aim of this study was to investigate OA progression in SIRT1-KI mice using a mouse OA model. OA was induced via destabilization of the medial meniscus using 12-week-old SIRT1-KI and wild type (control) mice. OA progression was evaluated histologically based on the Osteoarthritis Research Society International (OARSI) score at 4, 8, 12, and 16 weeks after surgery. The production of SIRT1, type II collagen, MMP-13, ADAMTS-5, cleaved caspase 3, Poly (ADP-ribose) polymerase (PARP) p85, acetylated NF-κB p65, interleukin 1 beta (IL-1β), and IL-6 was examined via immunostaining. The OARSI scores were significantly lower in SIRT1-KI mice than those in control mice at 8, 12, and 16 weeks after surgery. The proportion of SIRT1 and type II collagen-positive-chondrocytes was significantly higher in SIRT1-KI mice than that in control mice. Moreover, the proportion of MMP-13-, ADAMTS-5-, cleaved caspase 3-, PARP p85-, acetylated NF-κB p65-, IL-1β-, and IL-6-positive chondrocytes was significantly lower in SIRT1-KI mice than that in control mice. The mechanically induced OA progression was delayed in SIRT1-KI mice compared to that in control mice. Therefore, overexpression of SIRT1 may represent a mechanism for delaying OA progression.

Keywords: silent information regulator 2 ortholog 1 (SIRT1); osteoarthritis; knock-in mice; knee

Citation: Yamamoto, T.; Miyaji, N.;
Kataoka, K.; Nishida, K.; Nagai, K.;
Kanzaki, N.; Hoshino, Y.; Kuroda, R.;
Matsushita, T. Knee Osteoarthritis
Progression Is Delayed in Silent
Information Regulator 2 Ortholog 1
Knock-in Mice. *Int. J. Mol. Sci.* **2021**,
22, 10685. https://doi.org/10.3390/
ijms221910685

Academic Editor: Masato Sato

Received: 19 July 2021
Accepted: 29 September 2021
Published: 1 October 2021

Publisher's Note: MDPI stays neutral
with regard to jurisdictional claims in
published maps and institutional affiliations.

1. Introduction

Osteoarthritis (OA) affecting the knee joint is one of the most common joint diseases and causes joint pain and disability. Although OA has a multifactorial etiology and eventually affects the entire joint, its central pathological feature involves the progressive loss of articular cartilage [1,2]. Treatment using disease-modifying OA drugs (DMOADs) improves the underlying OA pathophysiology, thereby inhibiting structural damage to prevent or reduce long-term disability and offer potential symptomatic relief [3]. Currently, there are no effective DMOADs available that are suitable for treatment when the cartilage has been lost [2].

Sirtuins are members of the class III histone deacetylase family and regulate diverse cellular activities in aging [4]. Silent information regulator 2 type 1 (SIRT1) is sirtuin homolog and regulates various vital signaling pathways such as DNA repair and apoptosis, myogenic and adipogenic differentiation, mitochondrial biogenesis, and glucose and insulin homeostasis [5]. Studies have shown that regulation of SIRT1 may affect OA progression. In human chondrocytes, SIRT1 inhibits apoptosis and promotes cartilage-specific gene expression [6,7], and SIRT1 inhibition regulates the expression of genes related to OA [8,9]. Additionally, several genetic Sirt1-deficient mouse models exhibit accelerated

OA progression [10–12]. Therefore, SIRT1 may play a role in protecting chondrocytes and preventing OA development.

Resveratrol (3,4,5-trihydroxystlben; RSV) is a natural product isolated from most grape cultivars; it can activate SIRT1 [4]. New molecules were identified that can stimulate sirtuin activities to a greater extent than that obtained using RSV, such as SRT1720, SRT2104, SRT2379, and other molecules [13]. We previously found that the intraperitoneal (i.p.) injection of SRT1720 reduces OA progression in a mouse model [14]. Additionally, both i.p. and intra-articular injections of SRT2104 reduce OA progression in an OA mouse model [15]. The findings of these studies using natural compounds and chemical activators strongly suggest that activation of SIRT1 can inhibit OA progression. However, the direct effect of overexpression of SIRT1 on OA suppression has not yet been investigated.

SIRT1 transgenic mice have been generated in which the SIRT1 cDNA has been knocked into the β-actin locus (SIRT1-KI) [16]. SIRT1-KI mice are characterized as lean mice showing low levels of blood cholesterol, high glucose tolerance, and active metabolism [16]. In previous studies on SIRT1-KI mice, the stiffness of the aorta has been found to be suppressed [17] and the reproductive capacity is prolonged [18]. These studies suggest that SIRT1 overexpression plays beneficial roles in aging-related diseases and SIRT1-KI mice represent useful models to examine the effects of SIRT1 overexpression on aging-related diseases. Although previous studies using chemical SIRT1 activators and natural compounds of SIRT1 activators have suggested beneficial roles of SIRT1 on OA progression, non-direct effects of those activators have been also reported [19]. In addition, effects of constitutive overexpression of SIRT1, which could have potential adverse effects of OA progression, have not yet been examined. Therefore, the aim of this study was to investigate whether OA progression is suppressed in SIRT1-KI mice with OA to examine the effects of SIRT1 overexpression in vivo and to evaluate the possible mechanisms associated with delayed progression.

2. Results

2.1. Features of SIRT1-KI Mice

SIRT1-KI mice were maintained by mating heterozygous SIRT1-KI mice with wild type littermates. SIRT1-KI mice were identified via genotyping. The proportion of SIRT-KI mice was approximately 40% (one to three out of four to six newborn pups) (Figure 1A). During postnatal growth, SIRT1-KI mice showed no significant differences in skeletal and body weight compared to that of the littermate control mice (up to postnatal 12 months), although there was a tendency that the development of ossification was slightly delayed in SRIT1-KI mice compared with wild type mice (Figure 1B–D). Sirt1 mRNA expression was significantly higher in SIRT1-KI mice than in control mice. In articular cartilage samples, Sirt1 was significantly upregulated compared to that in muscle tissues, and there was no significant decrease in SIRT1 expression with growth (Figure 1E). Safranin-O fast staining and hematoxylin-eosin staining of control and SIRT1-KI mouse tissues at 3 weeks of age are shown in Figure 1F.

A

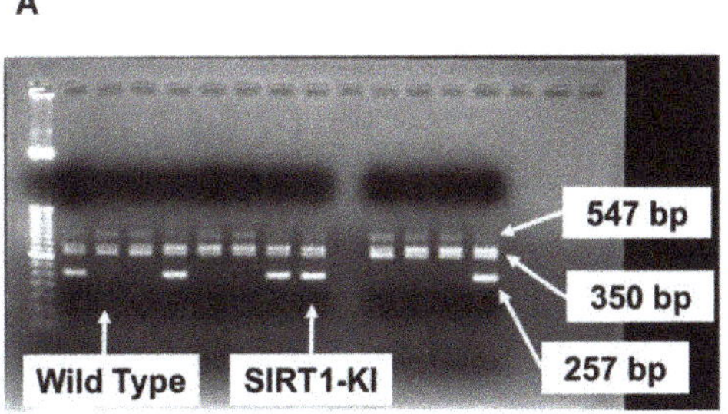

B

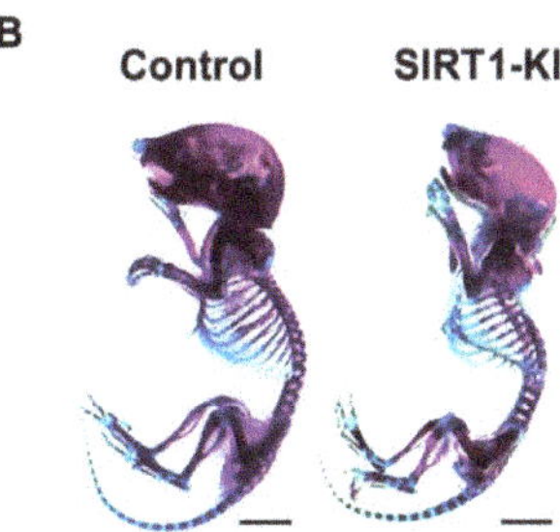

Figure 1. *Cont.*

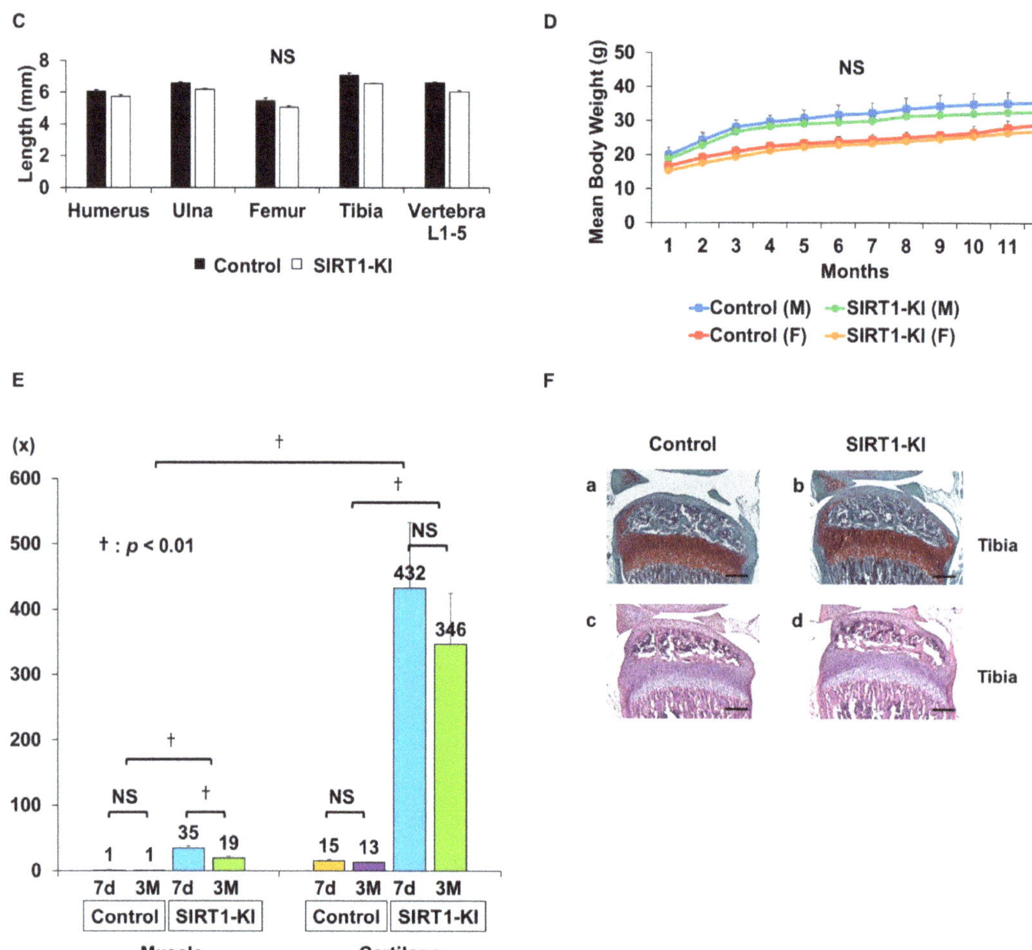

Figure 1. (**A**) Reverse transcription PCR. Genotyping was performed as follows: 257 bp for SIRT1-KI type; and 547 bp for wild type. (**B**) Double-staining with Alizarin red and Alcian blue of the whole skeleton of postnatal day 2 control and SIRT1-KI mice (scale bars = 5 mm). (**C**) Length of long bones and vertebra (first to fifth lumbar spines) of control (n = 5) and SIRT1-KI (n = 5) littermate embryos at postnatal day 2. NS, not significant. (**D**) Mean Body Weight (g) of Mice in the Control, SIRT1-KI at the Indicated Time Points (n = 5 mice/group). NS, not significant. M, male. F, female. (**E**) Sirt1 expression in control mice and SIRT1-KI mice. Relative expression level of Sirt1 mRNA was examined by real-time PCR setting the expression level in the muscle of control mice as 1 (**F**) (**a,b**) Safranin O-fast green staining of the medial knee joint at 3-week-old. Scale bars = 100 μm. (**c,d**) Hematoxylin-eosin staining of the medial knee joint at 3-week-old. Scale bars = 100 μm.

2.2. Reduced Severity of Cartilage Loss in SIRT-KI Mice after Destabilization via Medial Meniscus Surgery

Destabilization of the medial meniscus surgery was performed in the knee joint of 12-week-old mice. The knees of control mice and SIRT1-KI mice were collected at 4, 8, 12, and 16 weeks postoperatively. Histological analysis showed that OA developed gradually in all groups. The OARSI scores of the medial femoral condyle and tibial condyle in the control group were significantly higher than those in the SIRT-KI group at 8, 12, and 16 weeks, but not at 4 weeks (Figure 2).

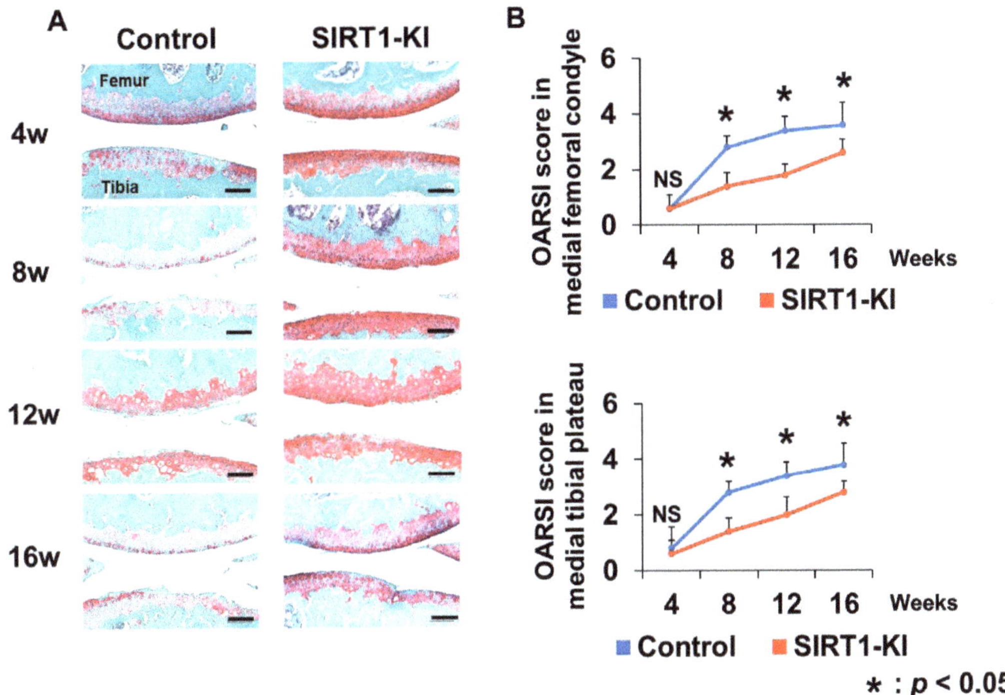

Figure 2. (**A**) Safranin O-fast green staining of the medial knee joint at 4, 8, 12, and 16 weeks postsurgery. Scale bars = 100 μm. (**B**) The Osteoarthritis Research Society International (OARSI) scores of the medial femoral and tibial condyle (n = 5 mice/group for each time point; * $p < 0.05$, NS, not significant.).

2.3. SIRT-KI Mice Showed Lower Levels of Cartilage Matrix-Degrading Enzymes, Apoptosis, and Inflammatory Cytokines than Those in Control Mice after OA Induction via Destabilization of Medial Meniscus

Immunohistochemical analyses revealed that a considerably higher number of SIRT1-positive chondrocytes were detected in the SIRT1-KI group than that in the control group at 8, 12, and 16 weeks after surgery (Figure 3). The proportion of type-II collagen-positive chondrocytes significantly increased in the SIRT-KI group compared to that in the control group at 8 weeks after surgery. The proportions of MMP-13-, ADAMTS-5-, cleaved caspase 3 (apoptotic marker)-, PARP p85 fragment (apoptotic marker)-, NF-κB P65 (inflammatory regulator)-, IL-1β-, and IL-6-positive chondrocytes significantly decreased in the SIRT1-KI group compared to that in the control group at 8 weeks after surgery (Figure 4).

2.4. SIRT1-KI Chondrocytes Showed Higher Expression of Extracellular Matrix Genes than That in the Control Group, whereas Cartilage Degrading Enzyme Genes Were Downregulated

Real-time PCR analysis showed that the mRNA expression of Sirt1, Col2a1, and Acan was significantly higher and Col10a1 expression was lower in chondrocytes of SIRT1-KI mice compared to those in control mouse chondrocytes. Stimulation with IL-1β significantly reduced Sirt1, Col2a1, and Acan mRNA expression in chondrocytes of both SIT1-KI and control mice; however, expression of those genes was significantly higher in chondrocytes of SIT1-KI mice compared to that in control mice.

The stimulation with IL-1β significantly increased Mmp-3, Mmp-13, and Adamts-5 mRNA expression in both chondrocytes of both mouse groups; however, the upregulation was attenuated in chondrocytes of SIRT1-KI mice compared to that in control mice (Figure 5).

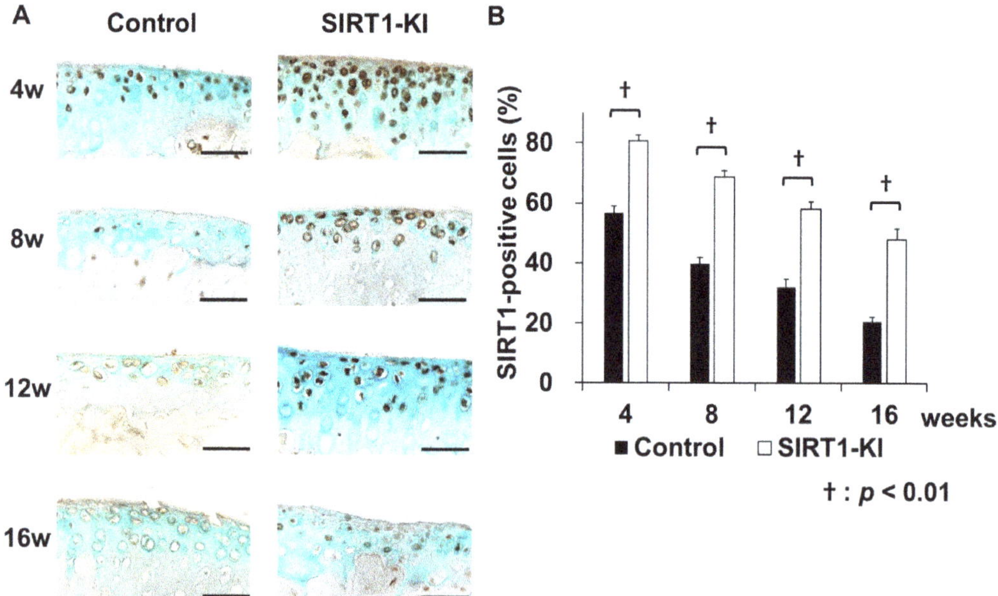

Figure 3. (**A**) Immunohistochemistry of SIRT1 in the medial tibial plateau at 4, 8, 12, and 16 weeks postsurgery. Scale bars = 50 μm. (**B**) The percentage of SIRT1-positive chondrocytes. Three micrographs of the medial tibial plateau were taken under ×40 magnification. The percentage was determined as the positive chondrocytes/total number of chondrocytes at ×100 magnification (n = 5 mice/group for each time point; [†] $p < 0.01$).

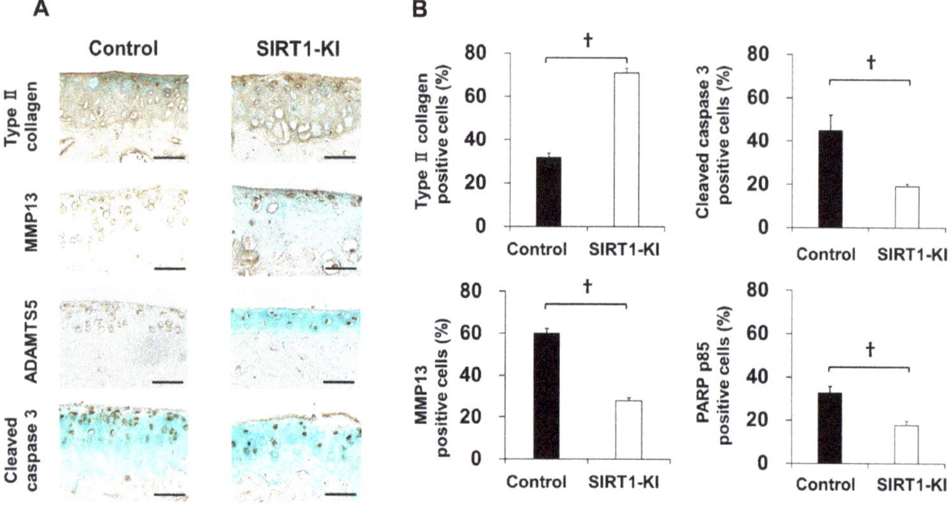

Figure 4. *Cont.*

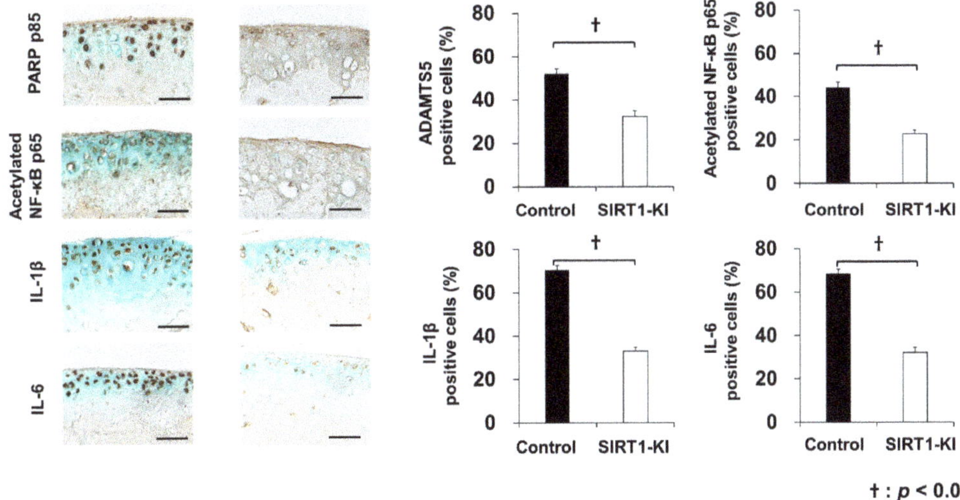

Figure 4. (**A**) Immunohistochemistry of type-II collagen, MMP-13, ADAMTS-5, cleaved caspase 3, PARP p85, acetylated NF-κB P65, IL-1β, and IL-6 in the medial tibial plateau at 8 weeks postsurgery. Scale bars = 50 μm. (**B**) The percentage of type-II collagen-, MMP-13-, ADAMTS-5-, cleaved caspase 3-, PARP p85 fragment-, acetylated NF-κB P65-, IL-1β-, and IL-6-positive cells. Three micrographs of the medial tibial plateau were taken at ×40 magnification. The percentage was determined as the positive cells/total number of cells at ×100 magnification (*n* = 5 mice/group for each time point; † *p* < 0.01).

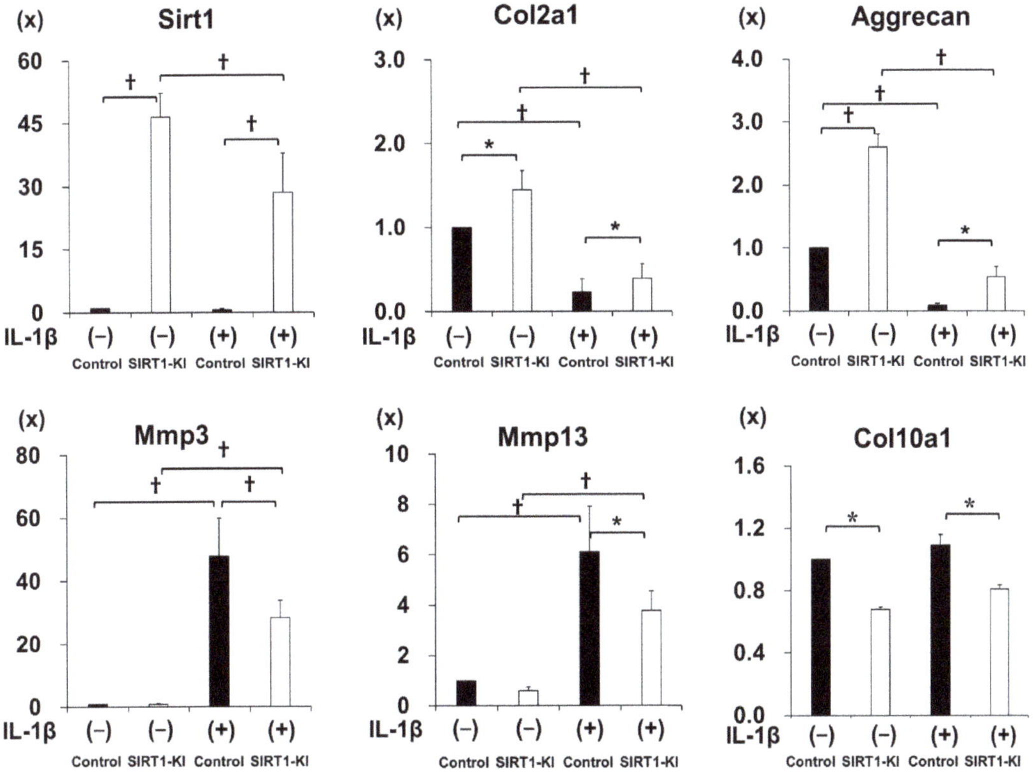

Figure 5. *Cont.*

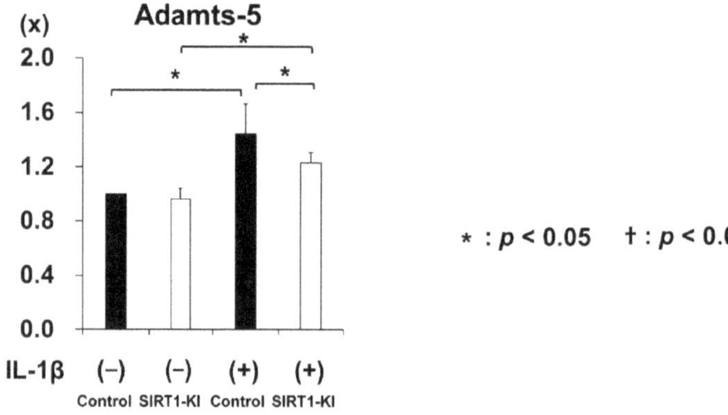

Figure 5. Real-time polymerase chain reaction (PCR) analysis of Sirt1, Col2a1, Col10a1, Mmp-3, Mmp-13, Aggrecan, and Adamts-5 mRNA expression in primary mouse chondrocytes. Primary mouse epiphyseal chondrocytes were cultured with 10 ng/mL IL-1β for 24 h (n = 5 independent experiments; * $p < 0.05$, † $p < 0.01$).

2.5. Microarray Analysis Identified 21 Genes Associated with SIRT1 Overexpression

Microarray analysis was performed to identify the genes associated with and regulated by SIRT1 overexpression. In total, 75 genes were found to be upregulated by more than two-fold whereas 45 genes were downregulated to <50% of their original expression level, without IL-1β stimulation in SIRT1-KI chondrocytes compared to that in control mouse chondrocytes (Figure 6A). Moreover, under stimulation with IL-1β, 119 genes were found to be upregulated by more than two-fold and 104 genes were downregulated to <50% in SIRT1-KI chondrocytes compared to that in control chondrocytes (Figure 6B). The expression of 11 genes was consistently two-fold higher and that of 10 genes was lower in SIRT-KI chondrocytes compared to that in control chondrocytes (Figure 6C) regardless of IL-1β stimulation (Table 1). Among those genes, XIST was found to be consistently downregulated in SIRT1-KI chondrocytes compared with wild type mice (Figure 7).

Table 1. Differentially regulated genes (2-fold) in control mice and SIRT1-KI mice.

No.	Genes	Log Ratio
1	Sirt1	3.484823
2	Acta2	2.169581
3	Aspn	1.389573
4	Postn	1.343505
5	Actg2	1.302517
6	Ccdc3	1.214218
7	Myl9	1.189704
8	Tagln	1.168344
9	Sfrp2	1.061471
10	Pi15	1.019975
11	Cfh	1.001788
12	Gm24405	−1.001405
13	Gm23604	−1.026104
14	Chil1	−1.117696
15	Serpina3c	−1.196805
16	Gm11979	−1.22567
17	Mrpl32	−1.343651
18	Xist	−1.368057
19	Gm24445	−1.382748
20	Gm23335	−1.61242
21	Snora69	−1.69543

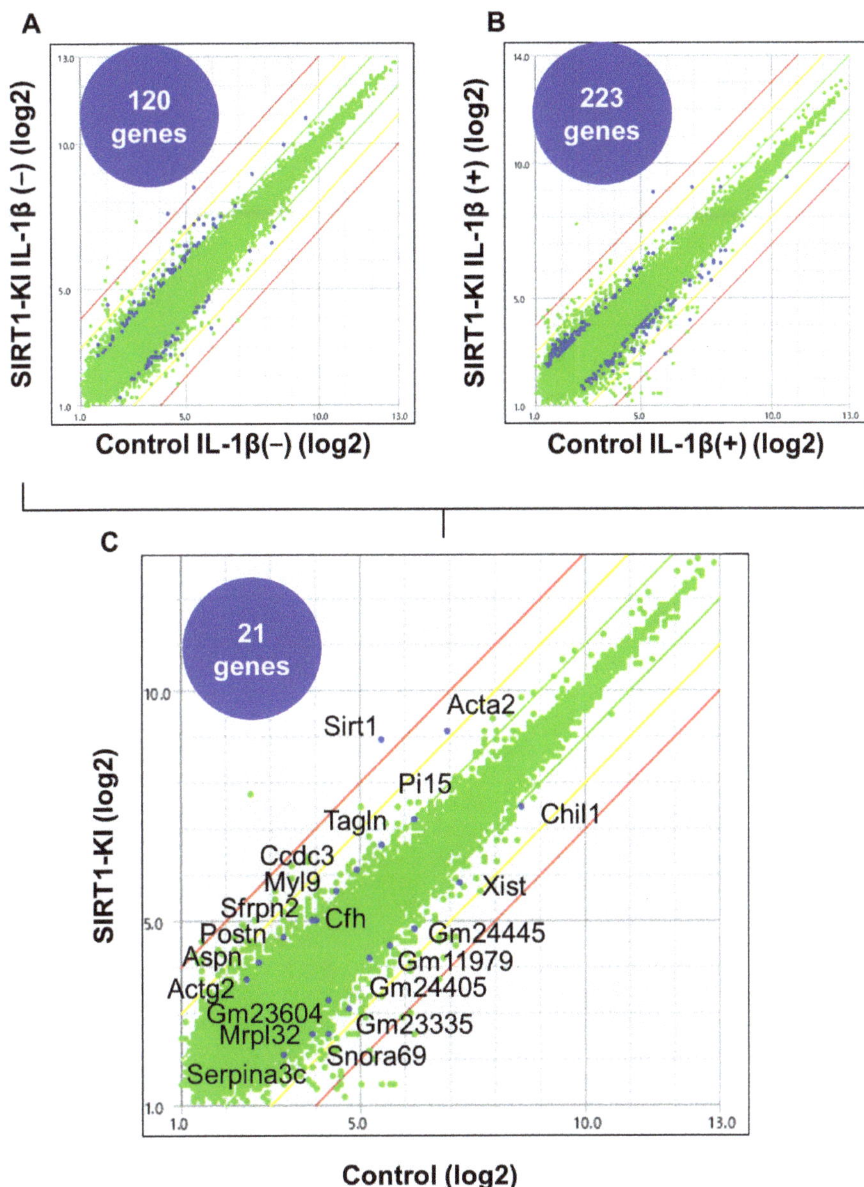

Figure 6. Scatter plots showing the correlation of signal values between two samples from chondrocytes without IL-1β stimulation and chondrocytes with IL-1β stimulation. (**A**) Upon gene expression profiling on SIRT1-KI chondrocytes, 120 genes were found to be upregulated or downregulated, with a two-fold change cut-off in chondrocytes without IL-1β stimulation. (**B**) Upon gene expression profiling on SIRT1-KI chondrocytes, 223 genes were found to be upregulated or downregulated, with a two-fold change cut-off in chondrocytes with IL-1β stimulation. (**C**) There were 21 genes involved in both of the above.

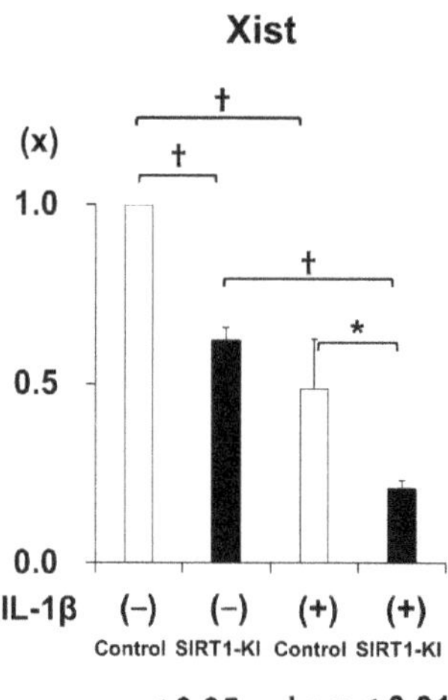

Figure 7. Real-time polymerase chain reaction (PCR) analysis of Xist mRNA expression in primary mouse chondrocytes. Primary mouse epiphyseal chondrocytes were cultured with 10 ng/mL IL-1β for 24 h (n = 5 independent experiments; * $p < 0.05$, † $p < 0.01$).

3. Discussion

The effects of SIRT1 overexpression on OA progression have not yet been investigated in SIRT-KI mice. The main finding of this study was that OA progression was delayed in SIRT1-KI mice compared to that in control mice in an experimental OA model. The delayed OA progression in SIRT1-KI mice was associated with an increased number of type II collagen and SIRT1-positive chondrocytes and decreased number of MMP-13-, ADAMTS-5-, cleaved caspase 3-, PARP p85 fragment-, acetylated NF-κB P65-, IL-1β-, and IL-6-positive chondrocytes compared to that in control mice. Additionally, the expression of extracellular matrix genes was higher in SIRT1-KI chondrocytes than that in control mice regardless of stimulation with IL-1β. Further, the upregulation of MMP-3, MMP-13, and ADMTS5 via stimulation with IL-1β was reduced in SIRT1-KI chondrocytes compared to that in control chondrocytes.

In early studies, the role of SIRT1 in the regulation of apoptosis via deacetylation of P53 has been highlighted as a main functional role of SIRT1 [20]. Thereafter, protective roles of SIRT1 against apoptosis in chondrocytes via various pathways were also demonstrated in previous studies [6,7,21]. Therefore, the reduced number of apoptotic chondrocytes in SIRT1-KI mice during OA development might have partially contributed to the delayed OA progression in SIRT-KI mice.

Regarding regulation of cartilage extracellular matrix gene expression changes by SIRT1, it has been reported that SIRT1 induces Col2a1 and Acan expression via an interaction with SOX9 in human chondrocytes [8,22,23]. Similarly, our previous study showed that the SIRT1 activator, SRT2104, stimulates Col2a1 expression in mouse chondrocytes [15]. The findings of these studies strongly suggest that SIRT1 functions as a positive regulator for extracellular matrix genes. In this study, Col2a1 and Acan expression was increased in SIRT1-KI chondrocytes regardless of IL-1β stimulation. The results of the present study agree with those of previous studies and suggest that the upregulation of extracellular matrix genes via overexpression of SIRT1 may represent a mechanism of delayed OA progression in SIRT1-KI mice. Regarding SIRT1 expression, it was reduced by the treatment with IL-β in SIRT-KI chondrocytes and SIRT1-KI mice during development of OA, although the expression level was still higher than in control chondrocytes and mice. Since SIRT1 was overexpressed under the β-actin promoter, it is possible that IL-1β reduced SIRT1 expression by decreasing promoter activity. Of interest, Yurube et al. reported that β-actin expression was reduced during the development of disc degeneration [24]. Therefore, β-actin expression may be susceptible to cellular stresses and the SIRT1 expression regulated by the β-actin promoter reduced even in SIRT-KI chondrocytes and mice, although detailed mechanism needs to be examined.

The expression of acetylated NF-κB p65 subunit and cartilage-degrading enzymes, including MMP-3, MMP-13, and ADAMTS-5, was decreased in the SIRT1-KI group compared to that in the control group. Additionally, the expression of IL-1β and IL-6 was decreased in the SIRT1-KI group compared to that in the control group. These observations are consistent with our previous findings: the proportion of MMP-13- and acetylated NF-κB p65-positive chondrocytes is decreased in mice treated with SIRT1 activators [14,15], and increased in cartilage-specific Sirt1-knockout mice12 compared to that in control mice. Previous studies have shown that the NF-κB pathway mediates the inflammatory responses of chondrocytes [25–27] and expression of inflammation-related genes, including MMP-3 and MPP-13, IL-6, IL-1, tumor necrosis factor alpha [28]. Yeung et al. have demonstrated that SIRT1 suppresses NF-κB signaling via deacetylation of the NF-κB p65 subunit [29]. Moreover, we found that the overexpression of SIRT1 in chondrocytes decreases the expression of acetylated NF-κB p65 and MMPs induced via IL-1β stimulation in vitro [30]. Therefore, OA development in the SIRT-KI group might have been attenuated via downregulation of cartilage-degrading enzymes in chondrocytes via modulation of the NF-κB pathway. Moreover, the involvement of additional pathways in the regulation of MMP expression by SIRT1 has been suggested, such as LEF1-dependent regulation [31] and p38, JNK, and ERK phosphorylation in chondrocytes [32]. Therefore, it is also possible that the reduction in levels of cartilage-degrading enzymes in SIRT-KI mice is associated with other mechanisms and pathways.

SIRT1 plays protective roles in chondrocytes; however, SIRT1 also has been reported to play important roles in the regulation of inflammatory responses in macrophages. Park et al. reported that SIRT1 activation by resveratrol suppresses the lipopolysaccharide/interferon γ-induced NF-κB activity in macrophages in rheumatoid arthritis, and inflammatory M1 polarization is reduced in SIRT1 transgenic mice. Further, collagen-induced arthritis is attenuated in SIRT1-transgenic mice associated with an increase level of M2 macrophage markers [33]. We have also found that delayed OA progression in mice treated with SRT2014 is associated with reduced M1 and increased M2 macrophage marker levels in the synovium [15]. Although the effects of SIRT1 overexpression were not examined in this study, modulation of macrophage polarization in the synovium may have also contributed to the delayed OA progression in SIRT1-KI mice.

Microarray analysis showed that 11 genes were consistently upregulated by more than two-fold while 10 genes were downregulated to less than 50% of their original expression level in SIRT-KI chondrocytes compared to those in control chondrocytes regardless of IL-1β stimulation. Among those genes, XIST was found to be consistently downregulated in SIRT1-KI chondrocytes. Recently, many studies have demonstrated important roles of lncRNA and miRNAs in the process of OA [34,35]. XIST, a lncRNA regulating X-linked chromosomal inactivation, has been recently suggested to be involved in pathogenesis of OA. The expression of XIST was increased in OA cartilage and knockdown of XIST improved the inflammatory microenvironment in OA via acting on M1 macrophages [36]. Further, XIST knockdown ameliorated IL-1β-induced decreased cell viability and downregulation of COL2A1 and aggrecan [37]. Therefore, XIST might play a role in the delayed OA progression in SIRT1-KI mice. However, further studies are required to elucidate the role of XIST in the SIRT-KI mice.

This study has a few limitations. First, the mean body weight of mice in the control group tended to be higher than that in the SIRT1-KI group, and the difference in body weight might have affected OA progression. However, these differences were not significant; therefore, the effect might have been negligible. Second, only female mice were used in this study and different results may be obtained using male mice. Third, OA progression was examined in a posttraumatic OA model, and the features of OA progression during aging were not studied. Finally, since sham control was not included, we cannot rule out the possibility that the upregulation of cartilage anabolism and downregulation of catabolism were stimulated by the OA induction although those responses were generally opposite to those during OA progression.

4. Materials and Methods

4.1. Mice

Sirt1-overexpressing (Sirt1; C57BL/6-Actbtm3.1 (Sirt1) Npa/J) mice (The Jackson Laboratory, Bar Harbor, ME, USA) were used in this study. Mice were maintained under pathogen-free conditions and were allowed free access to food, water, and activity. All animal experiments were performed according to protocols approved by the Institutional Animal Care and Use Committee at Kobe University Graduate School of Medicine (approval number P150604), Kobe, City.

4.2. Genotyping

SIRT1-KI mice were maintained by mating heterozygous SIRT1-KI mice with wild type littermates. DNA was extracted from mouse tails using DNeasy Blood & Tissue Kit (Qiagen, Valencia, CA, USA) according to the manufacturer's instructions. Polymerase chain reaction (PCR) was performed using three primers: 10,967 primer, 5′-TATGGAATCCTGTGGCATCCATGA-3′; 10,968 primer, 5′-CAAAGCCATGCCAATGTTGTCTCT-3′; 10,969 primer, 5′-GGCACATGCCAGAGTCCAAGTTTA-3′. PCR products were separated on a 2% agarose gel and visualized via ethidium bromide staining. Genotyping was performed as follows: 257 bp for SIRT1-KI type; and 547 bp for wild type.

4.3. Skeletal Preparation and Body Weight Analysis

Postnatal day 2 SIRT1-KI mice and control littermates were skinned, eviscerated, and fixed in 95% ethanol. The tissues were placed in a mixture of acetic acid: 95% ethanol (1:4) for 3 h to lower the pH. Alcian Blue staining was performed and the tissues were placed in potassium hydroxide to remove the soft tissue. After visualizing the cartilage via staining, Alizarin Red staining was performed. The length of long bones and vertebra (first to fifth lumbar spines) of control and SIRT1-KI littermate embryos at postnatal day 2 was measured using a divider. The weight of the mice was measured every month until one year of age.

4.4. Sirt1 Expression in Control and SIRT1-KI Mice

The expression of Sirt1 in muscle (tibialis anterior and quadriceps muscle) and cartilage was examined in 7-day-old and 3-month-old control and SIRT1-KI mice respectively. RNA was isolated using RNeasy Kit (Qiagen), and cDNA was synthesized using High-Capacity cDNA Reverse Transcription Kits (Applied Biosystems, Waltham, MA, USA). Real-time PCR analysis was performed using TaqMan assays (Applied Biosystems) to determine the expression of Sirt1 in duplicate for each sample, and the relative gene-expression levels were normalized to glyceraldehyde 3-phosphate dehydrogenase (Gapdh) expression levels as a reference control, based on the comparative cycle-threshold method.

4.5. OA Model

Based on our previous study [15], female mice were anaesthetized via i.p. injection containing a combination anesthetic (0.3 mg/kg of medetomidine, 4.0 mg/kg of midazolam, and 5.0 mg/kg of butorphanol), and the knee joint was exposed via the medial parapatellar approach. Experimental OA was induced in the knee joint of 12-week-old mice by resecting the medial meniscotibial ligament under a microscope to destabilize the medial meniscus [38]. The joint capsule and skin were closed using 3-0 nylon sutures.

4.6. Cell Culture and Real-Time PCR Analysis

Articular cartilage samples were collected from each knee joint of postnatal day 7 control and SIRT1-KI mice. The epiphyseal chondrocytes were cultured in monolayers in 6-well plates (2 × 105 cells/well) for 48 h. After reaching 80% confluency, the chondrocytes were incubated for 24 h with or without 10 ng/mL interleukin 1 beta (IL-1β) (R&D Systems,

Minneapolis, MN, USA). The concentration of IL-1β was chosen based on the previous study [15]. RNA was isolated using RNeasy Kit (Qiagen), and cDNA was synthesized using High-Capacity cDNA Reverse Transcription Kits (Applied Biosystems). Real-time PCR analysis was performed using TaqMan assays to analyze the expression of Sirt1, Col2a1, Col10a1, Mmp-3, Mmp-13, Acan, Adamts-5, and Xist (Applied Biosystems) in duplicate for each sample, and the relative gene-expression levels were normalized to Gapdh expression levels as a reference control, based on the comparative cycle-threshold method (Table 2).

Table 2. Primers used for real-time PCR.

Gene	Assay ID
Gapdh	Mm99999915_g1
Sirt1	Mm01168521_m1
Col2a1	Mm01309565_m1
Col10a1	Mm00487041_m1
Mmp-3	Mm00440295_m1
Mmp-13	Mm00439491_m1
Acan	Mm00545807_m1
Adamts-5	Mm00478620_m1
Xist	Mm01232884_m1

4.7. Histological Analysis

The control and SIRT1-KI mice were euthanized at 3 weeks of age and at 4, 8, 12, and 16 weeks after surgery (n = 5 mice/group for each time point). The entire knee joints were fixed in 4% paraformaldehyde in 0.1 M phosphate-buffered saline overnight at 4 °C, decalcified for two weeks using 10% ethylenediaminetetraacetic acid, and embedded in paraffin wax. Each specimen was cut into 6 μm slices along the sagittal plane and stained with safranin O-fast green and hematoxylin-eosin. Three slices were selected from each medial femoral condyle and medial tibial plateau, and photographs were taken at a magnification of 40×. The histological OA grade for each field was evaluated using the Osteoarthritis Research Society International (OARSI) cartilage OA histopathology grading system (score 0 to 6) [39]. OA grading was assessed by a single observer (NM) who was blinded to the study groups.

4.8. Immunohistochemistry

Deparaffinized sections were digested using proteinase (Dako Denmark AS, Glostrup, Denmark) for 10 min and treated with 3% hydrogen peroxide (Wako Pure Chemical Industries Ltd., Osaka, Japan) to block endogenous peroxidase activity. After epitope retrieval, the sections were incubated overnight at 4 °C with primary antibodies against the following mouse proteins: SIRT1 (1:50, Millipore, Billerica, MA, USA), type-II collagen (1:100, Abcam, Cambridge, UK), MMP-13 (1:100, Abcam), ADAMTS-5 (1:100, Santa Cruz Biotechnology Inc., Santa Cruz, CA, USA), cleaved caspase-3 (1: 100, Cell Signaling Technology, Tokyo, Japan), poly(ADP-ribose) polymerase (PARP) p85 (1:100, Promega, Madison, WI, USA), acetylated NF-κB p65 (1: 100, Sigma-Aldrich, St. Louis, MO, USA), IL-1β (1:100, Abcam), and IL-6 (1:100, Abcam). The anti-SIRT1 antibody was a mouse monoclonal antibody, and the others were rabbit polyclonal antibodies. Sections were then incubated with peroxidase-labeled anti-rabbit or mouse immunoglobulin (Histofine Simple Stain MAX Po; Nichirei Bioscience, Tokyo, Japan) at room temperature (25–28 °C) for 30 min, and the signals were developed as brown reaction products using the peroxidase substrate, 3,3-diaminobenzidine, with methyl green or hematoxylin counterstaining. As negative controls, a non-immune mouse or rabbit IgG (1:50 dilution) was used instead of the primary antibodies. SIRT1 expression was evaluated at the age of 3 weeks and at 4, 8, 12, and 16 weeks after surgery, and the expression of other proteins was evaluated at 8 weeks after surgery. All images were obtained under a microscope (Biozero; Keyence Corp., Itasca, OH, USA).

4.9. Microarray Analysis

The articular cartilage was collected from each knee joint of postnatal day 7 control wild-type littermates and SIRT1-KI mice. The epiphyseal chondrocytes were cultured in monolayers in 6-well plates (2 × 105 cells/well) for 48 h. After reaching 80% confluency, the chondrocytes were incubated for 24 h with or without 10 ng/mL IL-1β (R&D Systems). RNA was isolated using RNeasy Kit (Qiagen). After performing total RNA extraction, the samples were submitted to Kurabo Industries Ltd. (Okayama, Japan). An Affymetrix GeneChipTM Mouse Gene 2.0 ST Array was used to compare expression data between control and SIRT1-KI mice.

4.10. Statistical Analysis

An unpaired two-tailed Student's *t*-test was performed to compare differences between groups. One-way analysis of variance was performed to compare multiple groups using the Bonferroni method as a post-hoc test; p-values < 0.05 were considered significant. Data were analyzed using BellCurve for Excel (Social Survey Research Information Co., Ltd., Tokyo, Japan).

5. Conclusions

We have demonstrated that OA progression was delayed in the SIRT1-overexpressing SIRT1-KI mice in an experimental OA model. The delayed OA progression was associated with reduction in levels of cartilage-degrading enzymes, apoptotic markers, and acetylated NF-κB p65 in chondrocytes. SIRT1 overexpression may represent a promising strategy for treatment of OA.

Author Contributions: Conceptualization, T.Y., N.M. and T.M.; methodology, T.Y. and T.M.; software, T.Y. and K.N. (Kyohei Nishida); validation, T.Y., K.N. (Kanto Nagai) and N.K.; formal analysis, T.Y.; investigation, T.Y.; resources, T.Y.; data curation, T.Y. and K.K.; writing—original draft preparation, T.Y.; writing—review and editing, T.Y., Y.H. and R.K.; visualization, T.Y.; supervision, T.M. and R.K.; project administration, T.M. and R.K.; funding acquisition, T.M. All authors have read and agreed to the published version of the manuscript.

Funding: This research was funded by Grant-in-Aid for Scientific Research from the Japan Society for the Promotion of Science, grant number 26462294.

Institutional Review Board Statement: The study was conducted according to the guidelines of the Declaration of Helsinki, and approved by the Institutional Animal Care and Use Committee at Kobe University Graduate School of Medicine (P200103 and 28 Jan 2020).

Informed Consent Statement: Not applicable.

Acknowledgments: We would like to thank K. Tanaka, M. Nagata, and M. Yasuda for their technical assistance.

Conflicts of Interest: The authors declare no conflict of interest.

References

1. Felson, D.T.; Neogi, T. Osteoarthritis: Is It a Disease of Cartilage or of Bone? *Arthritis Rheum.* **2004**, *50*, 341–344. [CrossRef]
2. Cai, X.; Yuan, S.; Zeng, Y.; Wang, C.; Yu, N.; Ding, C. New Trends in Pharmacological Treatments for Osteoarthritis. *Front. Pharmacol.* **2021**, *12*, 1–16. [CrossRef]
3. Latourte, A.; Kloppenburg, M.; Richette, P. Emerging pharmaceutical therapies for osteoarthritis. *Nat. Rev. Rheumatol.* **2020**, *16*, 673–688. [CrossRef]
4. Deng, Z.; Li, Y.; Liu, H.; Xiao, S.; Li, L.; Tian, J.; Cheng, C.; Zhang, G.; Zhang, F. The role of sirtuin 1 and its activator, resveratrol in osteoarthritis. *Biosci. Rep.* **2019**, *39*, 1–22. [CrossRef]
5. Hubbard, B.P.; Sinclair, D.A. Small molecule SIRT1 activators for the treatment of aging and age-related diseases. *Trends Pharmacol. Sci.* **2014**, *35*, 146–154. [CrossRef]
6. Gagarina, V.; Gabay, O.; Dvir-Ginzberg, M.; Lee, E.J.; Brady, J.K.; Quon, M.J.; Hall, D.J. SirT1 enhances survival of human osteoarthritic chondrocytes by repressing protein tyrosine phosphatase 1B and activating the insulin-like growth factor receptor pathway. *Arthritis Rheum.* **2010**, *62*, 1383–1392. [CrossRef]
7. Takayama, K.; Ishida, K.; Matsushita, T.; Fujita, N.; Hayashi, S.; Sasaki, K.; Tei, K.; Kubo, S.; Matsumoto, T.; Fujioka, H.; et al. SIRT1 regulation of apoptosis of human chondrocytes. *Arthritis Rheum.* **2009**, *60*, 2731–2740. [CrossRef] [PubMed]

8. Dvir-Ginzberg, M.; Gagarina, V.; Lee, E.J.; Hall, D.J. Regulation of cartilage-specific gene expression in human chondrocytes by SirT1 and nicotinamide phosphoribosyltransferase. *J. Biol. Chem.* **2008**, *283*, 36300–36310. [CrossRef]

9. Fujita, N.; Matsushita, T.; Ishida, K.; Kubo, S.; Matsumoto, T.; Takayama, K.; Kurosaka, M.; Kuroda, R. Potential involvement of SIRT1 in the pathogenesis of osteoarthritis through the modulation of chondrocyte gene expressions. *J. Orthop. Res.* **2011**, *29*, 511–515. [CrossRef] [PubMed]

10. Gabay, O.; Sanchez, C.; Dvir-Ginzberg, M.; Gagarina, V.; Zaal, K.J.; Song, Y.; He, X.H.; McBurney, M.W. Sirtuin 1 enzymatic activity is required for cartilage homeostasis in vivo in a mouse model. *Arthritis Rheum.* **2013**, *65*, 159–166. [CrossRef] [PubMed]

11. Gabay, O.; Oppenheimer, H.; Meir, H.; Zaal, K.; Sanchez, C.; Dvir-Ginzberg, M. Increased apoptotic chondrocytes in articular cartilage from adult heterozygous SirT1 mice. *Ann. Rheum. Dis.* **2012**, *71*, 613–616. [CrossRef]

12. Matsuzaki, T.; Matsushita, T.; Takayama, K.; Matsumoto, T.; Nishida, K.; Kuroda, R.; Kurosaka, M. Disruption of Sirt1 in chondrocytes causes accelerated progression of osteoarthritis under mechanical stress and during ageing in mice. *Ann. Rheum. Dis.* **2014**, *73*, 1397–1404. [CrossRef]

13. Villalba, J.M.; Alcaín, F.J. Sirtuin activators and inhibitors. *BioFactors* **2012**, *38*, 349–359. [CrossRef] [PubMed]

14. Nishida, K.; Matsushita, T.; Takayama, K.; Tanaka, T.; Miyaji, N.; Ibaraki, K.; Araki, D.; Kanzaki, N.; Matsumoto, T.; Kuroda, R. Intraperitoneal injection of the SIRT1 activator SRT1720 attenuates the progression of experimental osteoarthritis in mice. *Bone Jt. Res.* **2018**, *7*, 252–262. [CrossRef] [PubMed]

15. Miyaji, N.; Nishida, K.; Tanaka, T.; Araki, D.; Kanzaki, N.; Hoshino, Y.; Kuroda, R.; Matsushita, T. Inhibition of Knee Osteoarthritis Progression in Mice by Administering SRT2014, an Activator of Silent Information Regulator 2 Ortholog 1. *Cartilage* **2020**, 1–11. [CrossRef]

16. Bordone, L.; Cohen, D.; Robinson, A.; Motta, M.C.; Van Veen, E.; Czopik, A.; Steele, A.D.; Crowe, H.; Marmor, S.; Luo, J.; et al. SIRT1 transgenic mice show phenotypes resembling calorie restriction. *Aging Cell* **2007**, *6*, 759–767. [CrossRef] [PubMed]

17. Machin, D.R.; Auduong, Y.; Gogulamudi, V.R.; Liu, Y.; Islam, M.T.; Lesniewski, L.A.; Donato, A.J. Lifelong SIRT-1 overexpression attenuates large artery stiffening with advancing age. *Aging* **2020**, *12*, 11314–11324. [CrossRef] [PubMed]

18. Long, G.Y.; Yang, J.Y.; Xu, J.J.; Ni, Y.H.; Zhou, X.L.; Ma, J.Y.; Fu, Y.C.; Luo, L.L. SIRT1 knock-in mice preserve ovarian reserve resembling caloric restriction. *Gene* **2019**, *686*, 194–202. [CrossRef]

19. Pacholec, M.; Bleasdale, J.E.; Chrunyk, B.; Cunningham, D.; Flynn, D.; Garofalo, R.S.; Griffith, D.; Griffor, M.; Loulakis, P.; Pabst, B.; et al. SRT1720, SRT2183, SRT1460, and resveratrol are not direct activators of SIRT1. *J. Biol. Chem.* **2010**, *285*, 8340–8351. [CrossRef] [PubMed]

20. Luo, J.; Nikolaev, A.Y.; Imai, S.I.; Chen, D.; Su, F.; Shiloh, A.; Guarente, L.; Gu, W. Negative control of p53 by Sir2α promotes cell survival under stress. *Cell* **2001**, *107*, 137–148. [CrossRef]

21. Oppenheimer, H.; Gabay, O.; Meir, H.; Haze, A.; Kandel, L.; Liebergall, M.; Gagarina, V.; Lee, E.J.; Dvir-Ginzberg, M. 75-Kd Sirtuin 1 Blocks Tumor Necrosis Factor A-Mediated Apoptosis in Human Osteoarthritic Chondrocytes. *Arthritis Rheum.* **2012**, *64*, 718–728. [CrossRef]

22. Bar Oz, M.; Kumar, A.; Elayyan, J.; Reich, E.; Binyamin, M.; Kandel, L.; Liebergall, M.; Steinmeyer, J.; Lefebvre, V.; Dvir-Ginzberg, M. Acetylation reduces SOX9 nuclear entry and ACAN gene transactivation in human chondrocytes. *Aging Cell* **2016**, *15*, 499–508. [CrossRef] [PubMed]

23. Oppenheimer, H.; Kumar, A.; Meir, H.; Schwartz, I.; Zini, A.; Haze, A.; Kandel, L.; Mattan, Y.; Liebergall, M.; Dvir-Ginzberg, M. Set7/9 impacts col2a1 expression through binding and repression of sirT1 histone deacetylation. *J. Bone Miner. Res.* **2014**, *29*, 348–360. [CrossRef]

24. Yurube, T.; Takada, T.; Hirata, H.; Kakutani, K.; Maeno, K.; Zhang, Z.; Yamamoto, J.; Doita, M.; Kurosaka, M.; Nishida, K. Modified house-keeping gene expression in a rat tail compression loading-induced disc degeneration model. *J. Orthop. Res.* **2011**, *29*, 1284–1290. [CrossRef] [PubMed]

25. Roman-Blas, J.A.; Jimenez, S.A. NF-κB as a potential therapeutic target in osteoarthritis and rheumatoid arthritis. *Osteoarthr. Cartil.* **2006**, *14*, 839–848. [CrossRef] [PubMed]

26. Saklatvala, J. Inflammatory Signaling in Cartilage: MAPK and NF-κB Pathways in Chondrocytes and the Use of Inhibitors for Research into Pathogenesis and Therapy of Osteoarthritis. *Curr. Drug Targets* **2007**, *8*, 305–313. [CrossRef] [PubMed]

27. Goldring, M.B.; Marcu, K.B. Cartilage homeostasis in health and rheumatic diseases. *Arthritis Res. Ther.* **2009**, *11*, 224. [CrossRef]

28. Grall, F.; Gu, X.; Tan, L.; Cho, J.Y.; Inan, M.S.; Pettit, A.R.; Thamrongsak, U.; Choy, B.K.; Manning, C.; Akbarali, Y.; et al. Responses to the proinflammatory cytokines interleukin-1 and tumor necrosis factor α in cells derived from rheumatoid synovium and other joint tissues involve nuclear factor κB-mediated induction of the Ets transcription factor ESE-1. *Arthritis Rheum.* **2003**, *48*, 1249–1260. [CrossRef]

29. Yeung, F.; Hoberg, J.E.; Ramsey, C.S.; Keller, M.D.; Jones, D.R.; Frye, R.A.; Mayo, M.W. Modulation of NF-κB-dependent transcription and cell survival by the SIRT1 deacetylase. *EMBO J.* **2004**, *23*, 2369–2380. [CrossRef]

30. Matsushita, T.; Sasaki, H.; Takayama, K.; Ishida, K.; Matsumoto, T.; Kubo, S.; Matsuzaki, T.; Nishida, K.; Kurosaka, M.; Kuroda, R. The overexpression of SIRT1 inhibited osteoarthritic gene expression changes induced by interleukin-1β in human chondrocytes. *J. Orthop. Res.* **2013**, *31*, 531–537. [CrossRef]

31. Elayyan, J.; Lee, E.J.; Gabay, O.; Smith, C.A.; Qiq, O.; Reich, E.; Mobasheri, A.; Henrotin, Y.; Kimber, S.J.; Dvir-Ginzberg, M. LEF1-mediated MMP13 gene expression is repressed by SIRT1 in human chondrocytes. *FASEB J.* **2017**, *31*, 3116–3125. [CrossRef]

32. He, D.S.; Hu, X.J.; Yan, Y.Q.; Liu, H. Underlying mechanism of Sirt1 on apoptosis and extracellular matrix degradation of osteoarthritis chondrocytes. *Mol. Med. Rep.* **2017**, *16*, 845–850. [CrossRef]

33. Park, S.Y.; Lee, S.W.; Lee, S.Y.; Hong, K.W.; Bae, S.S.; Kim, K.; Kim, C.D. SIRT1/Adenosine monophosphate-activated protein Kinase α signaling enhances macrophage polarization to an anti-inflammatory phenotype in rheumatoid arthritis. *Front. Immunol.* **2017**, *8*, 1–11. [CrossRef]

34. Chen, W.K.; Yu, X.H.; Yang, W.; Wang, C.; He, W.S.; Yan, Y.G.; Zhang, J.; Wang, W.J. lncRNAs: Novel players in intervertebral disc degeneration and osteoarthritis. *Cell Prolif.* **2017**, *50*, 1–12. [CrossRef]

35. Díaz-Prado, S.; Cicione, C.; Muiños-López, E.; Hermida-Gómez, T.; Oreiro, N.; Fernández-López, C.; Blanco, F.J. Characterization of microRNA expression profiles in normal and osteoarthritic human chondrocytes. *BMC Musculoskelet. Disord.* **2012**, *13*, 1–14. [CrossRef]
36. Li, L.; Lv, G.; Wang, B.; Kuang, L. XIST/miR-376c-5p/OPN axis modulates the influence of proinflammatory M1 macrophages on osteoarthritis chondrocyte apoptosis. *J. Cell. Physiol.* **2020**, *235*, 281–293. [CrossRef]
37. Liu, Y.; Liu, K.; Tang, C.; Shi, Z.; Jing, K.; Zheng, J. Long non-coding RNA XIST contributes to osteoarthritis progression via miR-149-5p/DNMT3A axis. *Biomed. Pharmacother.* **2020**, *128*, 110349. [CrossRef] [PubMed]
38. Glasson, S.S.; Blanchet, T.J.; Morris, E.A. The surgical destabilization of the medial meniscus (DMM) model of osteoarthritis in the 129/SvEv mouse. *Osteoarthr. Cartil.* **2007**, *15*, 1061–1069. [CrossRef] [PubMed]
39. Pritzker, K.P.H.; Gay, S.; Jimenez, S.A.; Ostergaard, K.; Pelletier, J.P.; Revell, K.; Salter, D.; van den Berg, W.B. Osteoarthritis cartilage histopathology: Grading and staging. *Osteoarthr. Cartil.* **2006**, *14*, 13–29. [CrossRef] [PubMed]

MDPI

St. Alban-Anlage 66

4052 Basel

Switzerland

www.mdpi.com

International Journal of Molecular Sciences Editorial Office

E-mail: ijms@mdpi.com

www.mdpi.com/journal/ijms